A One Way
Mission To
Mars

Colonizing the Red Planet

Edited by

Paul Davies

Beyond Center, Arizona State University

Dirk Schulze-Makuch

School of Earth and Environmental Sciences,
Washington State University

Lana Lan Tao, Executive Managing Director
Journal of Cosmology, Cosmology Science Publishers

To Boldly Go: A One Way Mission To Mars Colonizing the Red Planet
Edited by: Paul Davies, Ph.D., and Dirk Schulze-Makuch, Ph.D.,

Includes bibliographical references

ISBN: 0982955243
         9780982955246

1. Mars, 2. Red Planet, 3. Mission to Mars 4. Astronauts.
5. Psychology, 6. Radiation, 7. Astrobiology, 8. Life on Mars,
9. Colonizing Mars, 10. Medical Health, 11. Robots, 12. Terraforming

Editor-in-Chief, Journal of Cosmology,
Rudy Schild, Ph.D.
Center for Astrophysics, Harvard-Smithsonian, Cambridge

Acknowledgments: With the exception of the introduction, the comments of "volunteers" and the final chapter, all chapters in this book have been peer reviewed and were previously published in the Journal of Cosmology, JournalofCosmology.com, in 2010, 2011.

CONTENTS

Introduction:
Onward to Mars

Paul Davies[1] and Dirk Schulze-Makuch[2]

[1]Beyond Center, Arizona State University [2]School of Earth
and Environmental Sciences, Washington State University

IN A 1967 STAR TREK EPISODE called This Side of Paradise, Captain James T.
Kirk offers a comment on the destiny of mankind. "Maybe we weren't meant for
paradise. Maybe we were meant to fight our way through; struggle, claw our way
up, scratch for every inch of the way. Maybe we can't stroll to the music of the
lute. We must march to the sound of drums."

The story of human history is undeniably one of restless struggle and
exploration. From its African homeland, Homo sapiens spread across the globe
and colonized the frozen wastelands of the arctic, the densest jungles of South
America, the snow-bound high plateau of Asia, the blistering deserts of Australia
and the far-flung pale green dots of the Pacific Islands. Many other species have
expanded their territory, either opportunistically when ecological conditions were
favorable, or by evolving new survival traits that enabled them to adapt to new
environments. Humans, however, possess in addition another quality, one that
bestowed a huge advantage over other species when it came to radiation:
intelligence, specifically intelligence of the tool-making variety. Alone among
Earth's species, humans are able to expand beyond their natural habitat by
adapting the habitat to their needs. From hunting to agriculture, from the use of
fire and clothing to the construction of permanent shelters, from irrigation to
fertilizers, the success of human expansion and colonization owes much to our
ability to plan, build, organize and harness nature in support of our survival and
prosperity. These same human qualities, adapted and amplified, are clearly just as
applicable to the realm beyond our planet as to its surface.

On the threshold of the space age, many romantic science fiction writers and
social commentators saw space exploration as a natural extension of terrestrial
exploration. They drew parallels between mankind's first fumbling footsteps in
space and earlier episodes of human expansion, such as the European
colonization of the Americas or the opening up of Antarctica. Just as those
previous beginnings presaged wholesale migration and transformation, so too was
space – "the final frontier" – portrayed as just another step to greater glory. The

Apollo program of Moon landings was widely regarded as merely the first rung on a stairway that led inexorably to the stars. Stanley Kubrick's film 2001: A Space Odyssey perfectly captured this heady enthusiasm for space-faring humanity. Artists' imaginations ran free with depictions of orbiting space stations and hotels, huge space colonies inside artificial cylinders, bases on asteroids, moons and planets, and the construction of huge starships or space arks for exploration and settlement even farther afield. The physicist and futurist Freeman Dyson even imagined the colonization of comets and icy planetesimals in the frigid outer reaches of the solar system, and toyed with idealistic anarchist models of societies severed by the vast gulfs of outer space from their restrictive terrestrial roots. And to achieve this uplifting vision, all that was required was more of the same that gave us Apollo – bigger rockets, heavier payloads, more money, larger-scale engineering. In this way, humans would extend their millennia-long remorseless geographical and numerical expansion on Earth into the boundless tracts beyond our home planet, driven by the sense of cosmic destiny so well exemplified in the stirring lines of James T. Kirk.

Forty years after Apollo, the dream of a seamless transition from terrestrial to space exploration has faded. Astronauts remain stuck in low-Earth orbit, and although NASA and other space agencies have achieved impressive successes in unmanned planetary exploration, the prospects for even sending humans back to the Moon in the near future look bleak. Why, then, are we publishing a volume – indeed, a manifesto – for human missions to Mars? Is it not beyond the bounds of credulity to suppose that humans will go to Mars in the foreseeable future?

The question separates into two: Why Mars? and Why now? The case for sending humans to Mars is not the result of some recent fad, but has built steadily over many years. Of all the locations beyond Earth Mars offers the most earthlike environment. It has an atmosphere which, although thin, offers a degree of protection from meteorites and huge diurnal temperature changes. It has carbon dioxide and relatively abundant water – essential ingredients for life and a source of oxygen for fuel and breathing – and minerals necessary for manufacturing. The gravitational field is about one-third that of Earth, which would be difficult but not impossible for humans to adapt to. The surface area of Mars (which has no oceans) is comparable to the area of Earth's continents. The length of the Martian day is close to that of Earth. To be sure, Mars is extremely cold, with average temperatures well below freezing, but experience with Antarctica suggests that this is not a show-stopper. Nobody supposes that living on Mars would be easy, but there is no reason to believe it is impossible. For a permanent human settlement, no other body in the Solar System comes close as a favorable destination.

We do not let the case for going to Mars rest with the superficial dictum that Edmond Hillary is supposed to have applied to Mount Everest – "Because it's there!" Rather, we assemble a variety of motives that underpin the push for Mars exploration and settlement. These divide roughly into scientific and humanitarian/political arguments. On the scientific front, Mars is often described as Earth's twin. This is admittedly somewhat of a strained analogy, but about 4 billion years ago Mars was warmer, presumably as a result of massive greenhouse warming produced by a very dense carbon dioxide atmosphere, and this permitted liquid water to flow on the surface. There is evidence for ancient rivers, flood plains, and lakes on Mars, and possibly even oceans in the northern hemisphere. Crucially, Mars has volcanoes; in fact, Olympus Mons is the largest volcano in the Solar System. Geothermal heat can drive the continuous cycling of water – creating so-called hydrothermal systems – and a variety of geological processes that accompany it, familiar from our own planet. Unfortunately this early Martian Eden was doomed. The low gravity and absence of a strong magnetic field to deflect the solar wind resulted in the steady loss of the atmospheric gases, and with that came dramatic cooling, until the surface of the planet turned into the freeze-dried desert we see today. Nevertheless, Mars remains a cornucopia for geologists and planetary scientists, and an important source of clues to how the Solar System formed. By comparing Mars and Earth, scientists can build up a more accurate picture of Earth's geological and climatic history over 4.5 billion years, and better understand the forces that shape our planet.

Perhaps the biggest scientific lure of Mars is the possibility of finding traces of life there. During its early warm wet phase, Mars would certainly have been a congenial environment for biology. Of course, we do not know how life began, but many astrobiologists think that hydrothermal systems offer a very plausible setting. This is where volcanic heat cycles water, delivering a continuous supply of dissolved gases and other substances thought to be necessary to incubate life, as well as providing porous rock surfaces to act as catalysts for pre-biotic chemistry. On Earth today, volcanic vents on the sea floor replicate those conditions, and these systems are replete with heat-loving microorganisms which, from gene sequencing studies, seem to be the deepest branches on the tree of life. By implication, some of these organisms are like living fossils, suggesting (but by no means proving) that life began in a location like this.

Curiously, Mars was in some ways a more favorable environment than Earth for life to get started. Being a smaller planet it cooled quicker, and so was ready for life sooner. It also suffered less from the ferocious bombardment by asteroids and comets that plagued the inner Solar System for 700 million years after its formation. The lengthy bombardment churned up the surfaces of both Mars and

Earth, creating a layer of loosely packed material called regolith, but the lower surface gravity of Mars resulted in a less compacted regolith, and so provided a more accessible refuge for microorganisms to shelter deep underground during this violent early phase. So it is entirely possible that life got going on Mars before it did on Earth.

The same bombardment that imperiled early life also served to eject vast amounts of planetary material into solar orbit. A significant fraction of this rocky ejecta eventually struck other planets, a process that continues to this day. A few dozen meteorites are known to have originated on Mars, and it has been estimated that during the history of the Solar System trillions of tons of Mars rocks have come to Earth. If there was life on Mars about 4 billion years ago, it seems inevitable that some viable Martian microbes would have been conveyed to Earth amid this ejected material. The tantalizing prospect thus arises that perhaps Mars was the cradle of life, which spread to Earth in ejected rocks and established itself here only when the bombardment abated and conditions settled down. If so, then we are all descended from Martians and Mars is our ultimate home.

Naturally many people want to know whether there is life on Mars today. Here the situation is less clear cut. Conditions on the exposed surface of Mars are extremely uncongenial. In addition to the cold and the radiation there is the problem of highly oxidizing soil, which attacks and breaks up organic molecules. In spite of this, it is possible that at least some hardy terrestrial microbes, if shielded beneath a rock or a layer of dust, might survive there today, and of course any indigenous Martian life will have had billions of years to adapt to the harsh conditions. A better environment might be provided by the deep subsurface, or by the polar caps, where microbes might lie entombed in the ice, conceivably revivified by occasional warming periods.

In the 1970s two NASA space craft called Viking went to Mars specifically to look for life in the surface soil. They were equipped with on-board systems that delivered a radioactively tagged nutrient medium to scooped-up soil samples, to see if anything in the soil digested it. The telltale sign of life was whether radioactive carbon dioxide was given off. Both space craft yielded positive results for the so-called Labeled Release experiment (LR). When the soil was heated to 130°C, the cycling of radioactive carbon stopped, a result consistent with Martian microbes being destroyed by the high temperature. The designer of the experiment, Gil Levin, maintains to this day that he discovered life on Mars, but most astrobiologists are much more cautious. Other experiments on Viking were ambiguous or negative, and the mission scientists came to acknowledge that the chemistry of the oxidizing Martian soil is complex and still not well understood, especially in the context of the LR experiment. So the positive LR results remain a mystery, and the issue of life at the surface of Mars is unresolved.

The conditions deep beneath the surface of Mars are likely to be more favorable for life. At depths of between 200 m and 2 km the internal heat would have melted the permafrost, creating briny reservoirs that could host the same sort of deep-living microbes found deep underground on Earth, especially those that use hydrogen and carbon dioxide to make biomass and emit methane. The discovery of trace amounts of methane in the Martian atmosphere has fueled speculation about subsurface microbial ecosystems on Mars. If any Mars life is confined to underground pockets, getting at it wouldn't be easy. Ideally scientists need to inspect a freshly exposed subsurface. In the absence of explosive excavation by a nuclear device – likely to be deemed unacceptable on political and perhaps ethical grounds – the best hope would lie in the investigation of a freshly made impact crater, preferably one that had flooded as a result of the liberation of fluid from a subsurface aquifer. Another possibility is if any residual volcanic hotspots continue to cycle water to the surface from a deep reservoir. Failing this, a Mars base would need to be equipped with heavy equipment to drill boreholes at carefully selected sites. It is hard to imagine a project of this magnitude being pursued by purely robotic missions or during the first wave of human exploration and settlement.

The significance of the discovery of life on Mars would hinge on whether it is found to belong to the same tree as terrestrial life, and therefore share a common origin. If that were the case, it would confirm the "panspermia" hypothesis – that life started on Mars and came to Earth, or vice versa. Whilst it would still be a fantastic scientific discovery, and offer an unprecedented opportunity to study a parallel path of evolution, it would fall short of answering the burning question of our existence, which is whether life is a freak accident confined to the Solar System, or whether it is a truly cosmic phenomenon, widespread in the universe. Astronomers are convinced that earthlike planets are common, perhaps numbering a billion or more in our galaxy alone. Thus there is plenty of real estate on which life might arise. But since we do not know the mechanism by which organic chemistry turned into life, we cannot estimate the odds and say whether biogenesis is an exceedingly improbable process, or one likely to happen sooner or later on most earthlike planets. If Mars life turned out to differ from terrestrial life so much that a common origin was inconceivable, then it would offer strong support for the hypothesis that life starts up readily so that the universe should be teeming with it. That conclusion would in turn transform the prospect that some fraction of inhabited planets might also evolve intelligent life and even beings with the desire to contact us. In short, finding life on Mars is of crucial importance to addressing the question of whether or not we are alone in the universe and the related subject of the destiny of mankind.

The time for a human mission to Mars is now. Following this introductory chapter, Robert Zubrin provides in Chapter 1 convincing arguments for why the colonization of Mars could, and should, begin right away, and how associated mission problems such as radiation, low gravity, and human factors can be overcome. This is followed by a collection of thousand-word essays in Chapter 2 elaborating on the humanitarian and political arguments for going to Mars, including its practical implications and visionary aspects.

In spite of the rhetoric, the primary purpose of a mission to Mars is not to slavishly follow the innate human wanderlust – the urge to expand and conquer new territory and confront new challenges. This would no doubt be a motivating factor for some of the astronauts, and is vividly described in Chapter 3 by Edgar Mitchell and Robert Staretz, but a policy for Mars needs to rest on much more than a simple "new frontier" agenda. Some of this exploration could be done robotically, but robotic exploration has its limits and if we want to fully understand Mars, we need to send humans, a point nicely conveyed in Chapter 4 by former astronaut Harrison Schmitt. Further support of arguments that the time is ripe for a human mission to Mars are provided by Jack Stuster in Chapter 5, who argues that human missions to Mars are challenging, but constitute acceptable risk.

There are two reasons why establishing a permanent human presence on Mars is a good idea. First, it serves as a "lifeboat" in the event of a major catastrophe overwhelming humanity on Earth. Threats to our wellbeing include another asteroid or comet strike, sudden massive volcanism, a global pandemic, runaway climate change, nuclear war or some future technology running out of control. If a self-sustaining human colony existed on Mars, it could keep our species, and the flame of our culture, alive until such time as reverse colonization might be attempted. Estimates for the risk of global catastrophe vary widely, but it would be rash to rule out such a scenario over a time scale of millennia. Set against this is the time needed to build up a Mars colony to the point that it was fully self-sustaining. Clearly it will require a population of at least a few hundreds, and the colony will need to produce all its own food, building equipment, tools, medical supplies etc. It is hard to see how this state could be achieved in less than several centuries. So the time needed to "launch the lifeboat" is comparable to the expectation time for Armageddon (at least in some pessimistic estimates). All the more reason, then, to start soon.

The second reason why going to Mars makes sound political sense is that it would provide a welcome focus and unifying vision for all of humanity, at a time of unprecedented tensions and cultural divisions. As several of our contributors point out, financing a Mars colony, or even a small scale exploration mission, would dwarf the space budgets of any one nation. It could be achieved only by combining the resources of a consortium of space agencies and private sector players from around the world, making it a truly international and pan-cultural project. Because any Mars settlers would be beyond the reach of Earth's authorities, a future Mars colony would need to become politically and socially autonomous. Indeed, it would have to evolve its own institutions and ethical codes adapted to the peculiar and different circumstances of life on another planet.

In Chapter 6 we argue that the quickest way to make the Mars dream come true is to send astronauts one way only. By cutting out the return journey, costs are slashed by a huge factor. Moreover, the riskiest part of going to Mars is take-off, landing, and exposure to space conditions. These would be halved by a one-way mission. Supplies could be sent on ahead, and the first astronauts (perhaps two crews of two each) would be resupplied every two years. Eventually others would join and the colony would grow. There is no doubt that it would be politically easier to raise ongoing funding to maintain and expand an established Mars base than to raise the much larger amount needed for a one-off return mission.

Naturally the astronauts would face serious hazards, both of technological and psychological nature. This theme, the human at the center of a mission being both the greatest asset and the greatest weakness, is explored in detail by Sheryl Bishop in Chapter 7 Even if everything went according to plan, the rigors of life on Mars and the absence of medical help would imply a reduced life expectancy (but perhaps not more than the additional adverse health effects of making the return journey). But this would be no suicide mission; the astronauts could expect to live for several years, during which time they could carry out spectacular scientific work. The problem of musculoskeletal deterioration during the journey to Mars and what countermeasures can be taken are provided by Yixian Qin in Chapter 8, while the risk, albeit deemed small, of possible endemic Martian life to a human crew on Mars is discussed in Chapter 9 by Mihai Netea and co-authors.

Mars has so much to offer in terms of technological development and exploration, from geologic surface anomalies to the possible discovery of extraterrestrial life, as is highlighted by Markus Hotakainen in Chapter 10. We certainly should not look to Mars to solve the population problem on Earth. Nobody is suggesting wholesale migration to Mars, and even if such a thing were feasible, the remorseless arithmetic of exponential growth means that even a whole new planet to occupy would provide only a temporary respite. With unrestrained population growth, pretty soon Mars would be full too. And given that Mars is probably the only other planet in the Solar System on which a large human population could be accommodated, any further major migration would entail interstellar travel, which is not even on any scientific drawing board.

The objective of a human mission to Mars is to eventually establish a human base there. The benefits of extended stays of astronauts on Mars are provided in detail by Douglas Gage in Chapter 11. Details of a possible mission architecture, particularly on when and how to get to Mars, are provided by V. Adimurthy and colleagues in Chapter 12. The location of such a base is critical and Penelope Boston makes a passionate case in Chapter 13 of why a lava cave is the ideal site for such a base as it provides habitat, critical resources to the human crew, and a target location for the search for Martian life. An early Mars scientific settlement, which focuses on maximizing the use of local resources, is envisioned by Bruce Mackenzie and colleagues in Chapter 14. How living on Mars would affect the sex life of astronauts and whether the conditions, particularly the lower gravity, would result in complications for pregnancy and child birth, is difficult to evaluate without any experience on which to base conclusions. Nevertheless, Rhawn Joseph makes an admirable first attempt to wrestle with this delicate issue in Chapter 15. Finally, N.N. Ridder and colleagues discuss in Chapter 16 the possibility of making Mars more earth-like, particularly by utilizing strong fluorine-based greenhouse gases to achieve a significant warming of the Martian atmosphere.

The overall message of this volume is not just that going to Mars is a worthwhile scientific program and a great adventure worthy of Homo sapiens. It is that we can begin the project now. Decades of feasibility studies have shown that Mars missions do not require breakthrough technologies at some dim and distant time in the future. They can be accomplished with existing or readily foreseeable technology (for example, see Chapter 1 by Robert Zubrin). They can also be implemented within realistic budgets if carried out cooperatively, and especially if return journeys are eliminated. Examples how to finance a human mission to Mars are provided by Rhawn Joseph in Chapter 17 and C.A. Carberry and colleagues in Chapter 18. All that is needed is the political will. And there is no lack of volunteers. Since the chapters in this volume appeared in journal form last year, we have been inundated with inquiries by hopeful people from all walks of life eager to obtain a one-way ticket, more than a thousand to date. Some of these more memorable emails are reprinted in Chapter 19, demonstrating the wide breadth of people who are excited about the concept, from children to seniors in retirement, from various cultural backgrounds, including military professionals, computer science students and home makers.

The most straightforward way to accomplish the eventual colonization of Mars is indeed via a series of one-way human missions as we advocate in Chapter 7. In Chapter 20 we attempt a vision of life on Mars after the completion of a series of one-way missions, perhaps in about a century from today. We hope that this volume of fascinating and insightful papers by astronauts, scientists and visionaries will help shift global opinion in favor of what would surely be mankind's most glorious collective project in history.

WHY MARS? THE TIME IS NOW

- 1 -

Human Mars Exploration:
The Time Is Now

Robert Zubrin, Ph.D.
President of the Mars Society

ABSTRACT

This paper explains and details the Mars Direct plan. One unmanned trip up to Mars prepares sources of fuel and water for the main manned mission to come later, the underlying philosophy being travel light and live off the land. After showing that the cost is not prohibitive and well within NASA's budget this article dispenses with the "dragons" that would shake the courage of the ill informed. Those include Radiation, Zero Gravity, Back Contamination, Human factors and Dust Storms. Like the dragons of old, they do not survive the light of factual information. Finally it tackles the most important question: Why Do It? There are several answers, including for the knowledge. There is also the possibility of discovering that life is not unique to Earth. Included is a discussion of why and how human "fossil hunters" would be able to easily outstrip and go far beyond what robotic probes could accomplish. Based on the evidence from the Apollo program the chances are quite high that the challenge of Mars would give a very substantial return for the dollar. And lastly: Mars will not only be explored but colonized!

Key Words: Mars, exploration

1. THE TIME HAS COME

THE TIME HAS COME for America to set itself a bold new goal in space. The recent celebrations of the 40th anniversary of the Apollo Moon landings have reminded us of what we as a nation were once able to accomplish, and by so doing have put the question to us: are we still a nation of pioneers? Do we choose to make the efforts required to continue to be the vanguard of human progress, a people of the future; or will we allow ourselves to be a people of the past, one whose accomplishments are celebrated not in newspapers, but in museums? There can be no progress without a goal. The American space program, begun so brilliantly with Apollo and its associated programs, has spent most of the subsequent four decades without a central goal. We need such an overriding goal to drive our space program forward (Zubrin 1997). At this point of history, that goal can only be the human exploration and settlement of Mars (Mitchell & Staretz, 2010; Schmitt 2010; Schulze-Makuch & Davies 2010).

Some have said that a human mission to Mars is a venture for the far future, a task for "the next generation." Such a point of view has no basis in fact (Zubrin 1997). On the contrary, the United States has in hand, today, all the technologies required for undertaking an aggressive, continuing program of human Mars exploration, with the first piloted mission reaching the Red Planet Mars within a decade. We do not need to build giant spaceships embodying futuristic technologies in order to go to Mars. We can reach the Red Planet with relatively small spacecraft launched directly to Mars by boosters embodying the same technology that carried astronauts to the Moon more than a quarter-century ago. The key to success comes from following a travel light and live off the land strategy that has well-served explorers over the centuries humanity has wandered and searched the globe. A plan that approaches human missions to the Red Planet in this way is known as the "Mars Direct" approach. Here's how it would work.

2. THE MISSION

At an early launch opportunity, for example 2018, a single heavy lift booster with a capability equal to that of the Saturn V used during the Apollo program is launched off Cape Canaveral and uses its upper stage to throw a 40 tonne unmanned payload onto a trajectory to Mars. Arriving at Mars 8 months later, it uses friction between its aeroshield and Mars' atmosphere to brake itself into orbit around Mars, and then lands with the help of a parachute (Zubrin 1997). This payload is the Earth Return Vehicle (ERV), and it flies out to Mars with its two methane/oxygen driven rocket propulsion stages unfueled. It also has with it 6 tonnes of liquid hydrogen cargo, a 100 kilowatt nuclear reactor mounted in the back of a methane/oxygen driven light truck, a small set of compressors and automated chemical processing unit, and a few small scientific rovers.

As soon as landing is accomplished, the truck is telerobotically driven a few hundred meters away from the site, and the reactor is deployed to provide power to the compressors and chemical processing unit. The hydrogen brought from Earth can be quickly reacted with the Martian atmosphere, which is 95% carbon dioxide gas (CO_2), to produce methane and water, and this eliminates the need for long term storage of cryogenic hydrogen on the planet's surface. The methane so produced is liquefied and stored, while the water is electrolyzed to produce oxygen, which is stored, and hydrogen, which is recycled through the methanator. Ultimately these two reactions (methanation and water electrolysis) produce 24 tonnes of methane and 48 tonnes of oxygen. Since this is not enough oxygen to burn the methane at its optimal mixture ratio, an additional 36 tonnes of oxygen is produced via direct dissociation of Martian CO_2. The entire process takes 10 months, at the conclusion of which a total of 108 tonnes of methane/oxygen bipropellant will have been generated. This represents a leverage of 18:1 of Martian propellant produced compared to the hydrogen brought from Earth needed to create it. Ninety-six tonnes of the bipropellant will be used to fuel the ERV, while 12 tonnes are available to support the use of high powered chemically fueled long range ground vehicles. Large additional stockpiles of oxygen can also be produced, both for breathing and for turning into water by combination with hydrogen brought from Earth. Since water is 89% oxygen (by weight), and since the larger part of most foodstuffs is water, this greatly reduces the amount of life support consumables that need to be hauled from Earth.

The propellant production having been successfully completed, in 2020 two more boosters lift off the Cape and throw their 40 tonne payloads towards Mars. One of the payloads is an unmanned fuel-factory/ERV just like the one launched in 2018, the other is a habitation module containing a crew of 4, a mixture of whole food and dehydrated provisions sufficient for 3 years, and a pressurized methane/oxygen driven ground rover. On the way out to Mars, artificial gravity can be provided to the crew by extending a tether between the habitat and the burnt out booster upper stage, and spinning the assembly. Upon arrival, the manned craft drops the tether, aero-brakes, and then lands at the 2018 landing site where a fully fueled ERV and fully characterized and beaconed landing site await it. With the help of such navigational aids, the crew should be able to land right on the spot; but if the landing is off course by tens or even hundreds of kilometers, the crew can still achieve the surface rendezvous by driving over in their rover; if they are off by thousands of kilometers, the second ERV provides a backup. However assuming the landing and rendezvous at site number 1 is achieved as planned, the second ERV will land several hundred kilometers away to start making propellant for the 2020 mission, which in turn will fly out with an additional ERV to open up Mars landing site number 3. Thus every other year 2 heavy lift boosters are launched, one to land a crew, and the other to prepare a site for the next mission, for an average launch rate of just 1 booster per year to pursue a continuing program of Mars exploration. This is only about 15% of the rate that the U.S. currently launches Space Shuttles, and is clearly affordable. In effect, this dogsled approach removes the manned Mars mission from the realm of mega-fantasy and reduces it to practice as a task of comparable difficulty to that faced in launching the Apollo missions to the Moon (Zubrin 1997).

The crew will stay on the surface for 1.5 years, taking advantage of the mobility afforded by the high powered chemically driven ground vehicles to accomplish a great deal of surface exploration. With an 12 tonne surface fuel stockpile, they have the capability for over 24,000 kilometers worth of traverse before they leave, giving them the kind of mobility necessary to conduct a serious search for evidence of past or present life on Mars - an investigation key to revealing whether life is a phenomenon unique to Earth or general throughout the universe. Since no-one has been left in orbit, the entire crew will have available to them the natural gravity and protection against cosmic rays and solar radiation afforded by the Martian environment, and thus there will not be the strong driver for a quick return to Earth that plagues conventional Mars mission plans based upon orbiting mother-ships with small landing parties. At the conclusion of their stay, the crew returns to Earth in a direct flight from the Martian surface in the ERV. As the series of missions progresses, a string of small bases is left behind on the Martian surface, opening up broad stretches of territory to human cognizance.

3. WE CAN AFFORD IT

Such is the basic Mars Direct plan. In 1990, when it was first put forward, it was viewed as too radical for NASA to consider seriously, but over the next several years with the encouragement of then NASA Associate Administrator for Exploration Mike Griffin, the group at Johnson Space Center in charge of designing human Mars missions decided to take a good hard look at it. They produced a detailed study of a Design Reference Mission based on the Mars Direct plan but scaled up about a factor of 2 in expedition size compared to the original concept. They then produced a cost estimate for what a Mars exploration program based upon this expanded Mars Direct would cost. Their result; $50 billion, with the estimate produced by the same costing group that assigned a $400 billion price tag to the traditional cumbersome approach to human Mars exploration embodied in NASA's 1989 "90 Day Report." I believe that with further discipline applied to the mission design, the program cost could be brought down to the $30 to $40 billion range. Spent over ten years, this would imply an annual expenditure on the order of 20% of NASA's budget, or about half a percent of the US military budget. It is a small price to pay for a new world.

4. KILLING THE DRAGONS

Opponents of human Mars exploration frequently cite several issues which they claim make such missions to dangerous to be considered at this time. Like the dragons that used to mark the maps of medieval cartographers, these concerns have served to deter many who otherwise might be willing to enterprise the exploration of the unknown. It is therefore fitting to briefly address them here.

4.1. Radiation: It is alleged by some that the radiation doses involved in a Mars mission present insuperable risks, or are not well understood. This is untrue. Solar flare radiation, consisting of protons with energies of about 1 MeV, can be shielded by 12 cm of water or provisions, and there will be enough of such materials on board the ship to build an adequate pantry storm shelter for use in such an event. The residual cosmic ray dose, about 50 Rem for the 2.5 year mission, represents a statistical cancer risk of about 1%, roughly the same as that which would be induced by an average smoking habit over the same period.

4.2. Zero Gravity: Cosmonauts have experienced marked physiological deterioration after extended exposure to zero gravity. However a Mars mission can be flown employing artificial gravity generated by rotating the spacecraft. The engineering challenges associated with designing either rigid or tethered artificial gravity systems are modest, and make the entire issue of zero-gravity health effects on interplanetary missions moot.

4.3. Back Contamination: Recently some people have raised the issue of possible back-contamination as a reason to shun human (or robotic sample return) missions to Mars. Such fears have no basis in science. The surface of Mars is too cold for liquid water, is exposed to near vacuum, ultra violet, and cosmic radiation, and contains an antiseptic mixture of peroxides that have eliminated any trace of organic material. It is thus as sterile an environment as one could ask for. Furthermore, pathogens are specifically adapted to their hosts. Thus, while there may be life on Mars deep underground, it is quite unlikely that these could be pathogenic to terrestrial plants or animals, as there are no similar macrofauna or macroflora to support a pathogenic life cycle in Martian subsurface groundwater. In any case, the Earth currently receives about 500 kg of Martian meteoritic ejecta per year. The trauma that this material has gone through during its ejection from Mars, interplanetary cruise, and re-entry at Earth is insufficient to have sterilized it, as has been demonstrated experimentally and in space studies on the viability of microorganisms following ejection and reentry (Burchell et al. 2004; Burchella et al. 2001; Horneck et al. 1994, 1995, 2001, Horneck et al. 1993; Mastrapaa et al. 2001; Nicholson et al. 2000). So if there is the Red Death on Mars, we've already got it. Those concerned with public health would do much better to address their attentions to Africa.

4.4. Human Factors: In popular media, it is frequently claimed that the isolation and stress associated with a 2.5 year round-trip Mars mission present insuperable difficulties. Upon consideration, there is little reason to believe that this is true. Compared to the stresses dealt with by previous generations of explorers and mariners, soldiers in combat, prisoners in prisons, refugees in hiding, and millions of other randomly selected people, those that will be faced by the hand-picked crew of Mars 1 seem modest. Certainly psychological factors are important (Bishop 2010; Fielder & Harrison, 2010; Harrison & Fielder 2010; Suedfeld 2010). However, any serious reading of previous history indicates that far from being the weak link in the chain of the piloted Mars mission, the human psyche is likely to be the strongest link in the chain as Apollo astronauts have testified (Mitchell & Staretz 2010; Schmitt 2010).

4.5. Dust Storms: Mars has intermittent local, and occasionally global dust storms with wind speeds up to 100 km/hour. Attempting to land through such an event would be a bad idea, and two Soviet probes committed to such a maelstrom by their uncontrollable flight systems were destroyed during landing in 1971. However, once on the ground, Martian dust storms present little hazard. Mars' atmosphere has only about 1% the density of Earth at sea-level. Thus a wind with a speed of 100 km/hr on Mars only exerts the same dynamic pressure as a 10 km/hr breeze on Earth. The Viking landers endured many such events without damage.

Humans are more than a match for Mars' dragons.

5. WHY DO IT?

But why do it? There are three reasons.

Reason #1: For the Knowledge - During the summer of 1996, NASA scientists revealed a rock ejected from Mars by meteoric impact which showed strong evidence of life on Mars in the distant past (McKay et al., 1996). If this discovery could be confirmed by actual finds of fossils on the Martian surface, it would show that the origin of life is not unique to the Earth, and thus by implication reveal a universe that is filled with life and probably intelligence as well. From the point of view of humanity learning its true place in the universe, this would be the most important scientific enlightenment since Copernicus.

Robotic probes can help out in such a search, but by themselves are completely insufficient (Drake, 2010; Gage 2010; Schmitt 2010). Fossil hunting requires the ability to travel long distances through unimproved terrain, to climb steep slopes, to do heavy work and delicate work, and to exercise very subtle forms of perception and on-the-spot intuition. All of these skills are far beyond the abilities of robotic rovers. Geology and field paleontology requires human explorers, real live rockhounds on the scene (Schmitt 2010). Drilling to reach subsurface hydrothermal environments where extant Martian life may yet thrive will clearly require human explorers as well. Put simply, as far as the question of Martian life is concerned, if we don't go, we won't know.

Reason #2: For the Challenge - Nations, like people, thrive on challenge and decay without it. The space program itself needs challenge. Consider: Between 1961 and 1973, under the impetus of the Moon race, NASA produced a rate of technological innovation several orders of magnitude greater than that it has shown since, for an average budget in real dollars virtually the same as that today ($19 billion in 2010 dollars). Why? Because it had a goal that made its reach exceed its grasp. It is not necessary to develop anything new if you are not doing anything new. Far from being a waste of money, forcing NASA to take on the challenge of Mars is the key to giving the nation a real technological return for its space dollar.

A humans-to-Mars program would also be an challenge to adventure to every child in the country: "Learn your science and you can become part of pioneering a new world." There will be over 100 million kids in our nation's schools over the next ten years. If a Mars program were to inspire just an extra 1% of them to scientific educations, the net result would be 1 million more scientists, engineers, inventors, medical researchers and doctors, making innovations that create new industries, finding new medical cures, strengthening national defense, and increasing national income to an extent that dwarfs the expenditures of the Mars program.

Reason #3: For the Future - Mars is not just a scientific curiosity, it is a world with a surface area equal to all the continents of Earth combined, possessing all the elements that are needed to support not only life, but technological civilization. As hostile as it may seem, the only thing standing between Mars and habitability is the need to develop a certain amount of Red Planet know-how. This can and will be done by those who go there first to explore.

Mars is the New World. Someday millions of people will live there. What language will they speak? What values and traditions will they cherish, to spread from there as humanity continues to move out into the solar system and beyond? When they look back on our time, will any of our other actions compare in value to what we do today to bring their society into being?

Today, we have the opportunity to be the founders, the parents and shapers of a new and dynamic branch of the human family, and by so doing, put our stamp upon the future. It is a privilege not to be disdained lightly.

6. CONCLUSION

In conclusion, the point needs to be made again. We are ready to go to Mars. Despite whatever issues that remain, the fundamental fact is that we are much better prepared today to send humans to Mars than we were to send people to the Moon in 1961, when John F. Kennedy initiated the Apollo program. Exploring Mars requires no miraculous new technologies, no orbiting spaceports, and no gigantic interplanetary space cruisers (Zubrin 1997). We can establish our first outpost on Mars within a decade. We and not some future generation can have the eternal honor of being the first pioneers of this new world for humanity. All that's needed is present-day technology, some 19th century industrial chemistry, some political vision, and a little bit of moxie.

BIOGRAPHICAL INFORMATION:

Dr. Robert Zubrin, an astronautical engineer, is President of the Mars Society and author of "The Case for Mars: The Plan to Settle the Red Planet and Why Must," published by the Simon and Schuster.

REFERENCES

Bishop, S. L. (2010). Moving to Mars: There and back again, Journal of Cosmology, 12, 3711-3722.

Burchella, M. J., Manna, J., Bunch, A. W., Brandob, P. F. B. 2001. Survivability of bacteria in hypervelocity impact, Icarus. 154, 545-547.

Burchell, J. R. Mann, J., Bunch, A. W. (2004). Survival of bacteria and spores under extreme shock pressures, Monthly Notices of the Royal Astronomical Society, 352, 1273-1278.

Drake, B. G. (2010). Human Exploration of Mars: Challenges and Design Reference Architecture 5.0. Journal of Cosmology, 12, 3578-3587.

Fiedler, E. R., & Harrison, A. A. (2010). Psychosocial Adaptation to a Mars Mission, Journal of Cosmology, 12, 3685-3693.

Gage, D. W. (2010). Mars Base First: A Program-level Optimization for Human Mars Exploration. Journal of Cosmology, 12, 3904-3911.

Harrison, A. A., Fiedler, E. R. (2010). Mars, Human Factors and Behavioral Health, Journal of Cosmology, 12, 3685-3693.

Horneck, G. (1993). Responses ofBacillus subtilis spores to space environment: Results from experiments in space Origins of Life and Evolution of Biospheres 23, 37-52.

Horneck, G., Becker, H., Reitz, G. (1994). Long-term survival of bacterial spores in space. Advances in Space Research, Volume 14, 41-45.

Horneck, G., Eschweiler, U., Reitz, G., Wehner, J., Willimek, R., Strauch, G. (1995). Biological responses to space: results of the experiment Exobiological Unit of ERA on EURECA I. Advances in Space Research 16, 105-118.

Horneck, G., et al., (2001). Bacterial spores survive simulated meteorite impact Icarus 149, 285.

McKay, D.S., Everett, K.G., Thomas-Keprta, K.L., Vali, H., Romanek, C.S., Clemett, S.J., Chillier, X.D.F., Maechling, C.R., Zare, R.N. (1996) Search for past life on Mars: possible relic biogenic activity in Martian meteorite ALH84001. Science, 273, 924-930.

Mastrapaa, R.M.E., Glanzbergb, H., Headc, J.N., Melosha, H.J, Nicholsonb, W.L. (2001). Survival of bacteria exposed to extreme acceleration: implications for panspermia, Earth and Planetary Science Letters 189, 30 1-8.

Mitchell, E. D., & Staretz, R. (2010). Our Destiny – A Space Faring Civilization? Journal of Cosmology, 12, 3500-3505.

Nicholson, W. L., Munakata, N., Horneck, G., Melosh, H. J., Setlow, P. (2000). Resistance of Bacillus Endospores to Extreme Terrestrial and Extraterrestrial Environments, Microbiology and Molecular Biology Reviews 64, 548-572.

Schmitt, H. H. (2010). Apollo on Mars: Geologists Must Explore the Red Planet, Journal of Cosmology, 12, 3506-3516.

Schulze-Makuch, D., & Davies, P., (2010). To boldly go: A one-way human mission to Mars, Journal of Cosmology, 12, 3619-3626.

Suedfeld, P. (2010). Mars: anticipating the next great exploration, Journal of Cosmology, 12, 3723-3740.

Zubrin, R. (1997). The Case for Mars: The Plan to Settle the Red Planet and Why Must, Simon and Schuster, NY.

- 2 -
WHY WE MUST GO TO MARS
Contents:

2.1 The Beckoning Red Dot in the Sky
Shirin Haque, Ph.D.
Astronomer, University of the West Indies

RECENTLY, I WAS ON THE ISLAND of Dominica in the Caribbean, roaming the country sides with its mountainous terrain and undeveloped regions - waterfalls, rivers and forests, pristine in their conditions for maybe thousands of years. As I stared from the northern coast I could just make out the faint outline of another island up ahead. My map told me I was staring at the island of Guadeloupe -- but all I could think of was how Christopher Columbus would have felt to see the similar outline I saw and knowing there was land there unexplored yet. "Land ahoy" indeed!!

So too I look up at the pale red dot in the sky and think "planet ahoy!". The reason often perpetuated, for climbing a mountain by those who do it, is because it is there. It is a human instinct in us to explore, discover and often conquer with sometimes disastrous effects to the prized jewel we found.

One of the most thrilling moments of my life with no regrets, was flying over a volcano on the island of Montserrat to record a documentary. The doors of the helicopter were open and the stench of the sulphurous emissions from the volcano stung our noses as the helicopter did its dives. It was scary, but what a rush! Every neuron in my body was firing. It is what it means to feel alive. Going to Mars is an adventure that beckons similarly.

The human spirit of adventure and exploration of the unknown is likely encoded into our genetic makeup to ensure our survival as a species despite the risk and possible death to the soldiers of exploration at the frontier for the sake of the many that follow and the future.

Going to Mars is nothing more than the next logical step in our advancement of discovery and exploration. It must be done. Until we can do it -- we remain restless caged spirits. Sometimes, like in the case of the lunar landings, there was the dynamics of political agendas. Had there not been political agendas, I believe with certainty that humans would have landed on the moon nonetheless. It was the logical step at the time.

The opportunity to make history, to be the early charters risking it all is a small price for the satisfaction of doing it. It is an elixir of life only to experienced. It is a part of us in the deepest sense and what makes us human.

The journey to Mars can be eight or nine months long. In an adult's lifetime, this is quite doable. How many of us for the sake of jobs, research, and studies spend years away from our home base? What of the early sea farers braving the rough oceans of the Atlantic and inclement weather to discover the 'new world'? We have already ventured forth. Modern day science gives a much better idea as to what to expect and preparation for such compared to the explorers of the days of yore who lived on a whim, prayer and luck oftentimes.

Humans' fundamental needs are few and universally defined. Apart from the need for food and shelter, there is a strong need for companionship. It is essential if one is never to return from Mars, that a group of persons be the early settlers, with others to follow. We have seen how the world has shrunken globally with the advent of connectivity how many our own families are now scattered across the globe and we stay in touch with the technology of the day. Such an evolving humanity lends to making the leap to the next level to Mars. To boldly go where no man has ever gone before. "Star Trek" is visionary in its appeal decades later. We don't care to be beamed up, Scotty.

To have survived the journey, to land and step out on to Mars 'terra firma', despite all the preparations will be nothing like we expected. The hope to step on Martian soil, to see the terrain and know you are in a place of dreams is what churns us on. Humans will always travel to places they have studied thoroughly and seen virtual tours and images of because nothing ever beats the real experience.

The first days, the first weeks will be an awakening, and discovering and getting the basics in order for survival. Yes, there will perhaps be a sense of depression with it all, even as anyone in an uncharted territory can feel, but it evaporates quickly as the mind and senses are enraptured and intoxicated with the new discoveries. The skies, the sunsets, planet earth, Phobos and Deimos in the night sky -- how wonderful to be a part of the bigger universe. A chance at a new beginning.

In the truest sense, a journey to Mars is just another sequel to many other such migrations humans have already done -- albeit this is to another planet. But in spirit it is hardly any different when we leave the shores we were born in and grew up in and settle in another country sometimes to never return or Christopher Columbus takes to the seas to discover "new lands" or going to the moon. What we must learn to never ever repeat are the mistakes of the past - slavery, and wiping off the indigenous peoples. As far as we know, there are no Martians to conquer or enslave. But it will be a new land, new challenges, new resources. Never to exploit, if we have learned anything at all.

Buzz Aldrin made the potent remark 'magnificent desolation' with regard to the moon. Clearly the most meaningful thing in his life is likely to be the moon landing. He continues to be an advocate for manned missions to space. Who can know better than one who played that wild card already? And his own words are that "Mars is waiting for your footsteps."

Pitter... patter... God of War, here we come.

2.2 Is A Manned (One-Way) Journey To Mars Our Responsibility?

Johannes J. Leitner, Ph.D.[1,2], and Maria G. Firneis, Ph.D.[2]
[1]University of Vienna, Research Platform on ExoLife,
Tuerkenschanzstrasse 17, A-1180 Vienna, Austria.
[2]University of Vienna, Institute for Astronomy,
Tuerkenschanzstrasse 17, A-1180 Vienna, Austria

LIFE ON EARTH WITH ITS PRODIGIOUS diversity and especially the homo sapiens sapiens as the most intelligent or at least most dominant species on Earth is exposed to permanent threats from inside and outside. Threats from inside as consequences of social conflicts and wars, but also pandemics denote only some of these conceivable scenarios. Impacts from asteroids have caused mass extinctions in the past and still pose the most popular risk for life on Earth. Furthermore gamma-ray bursts, supernovae, solar eruptions, cosmic rays and the stellar evolution of our Sun form additional astronomical hazards for life on our home-world. Certainly the chance for world-wide extinction is very low at present, but not zero. In this context the question is of importance how large is the risk (percentually) to demand a massive, expensive reaction from our side. Human life on Earth, being the most evolved species which we know up to now, according to our moral standards, has to be preserved absolutely. This is our responsibility! Colonizing our Solar System can help to minimize this risk of extinction and a manned journey to Mars should be the first step to initiate the conquest of space.

Why a manned mission to Mars? Can it only be justified by the scenario of a threatened Earth or by the argument of bringing the first human to another planet as gaining his outstanding place in history. While overrating the development of robotics may seem to abolish the necessity of a steering (deciding) human, in reality the trained scientist collecting data on the Moon (Apollo 17) delivered more important data/samples to the Earth, than any robot/untrained colleague before. A space probe has to be configured well in advance with a restricted equipment to clarify specific hypotheses, for which at least part of the solutions have to be known prior to pose the correct questions. A human can decide on the spot in an unexpected situation.

Why a one-way mission to Mars? This has the advantage that a part by part construction of a science-mission habitat could be set-up in a modular way in advance to provide the human investigators with an apparatus-set comparable to terrestrial geological and biological laboratories to perform experiments, which were not anticipated, while a robot could carry out only preconceived investigations. Thus a sample-return mission is obsolete. The greatest advantage is seen in the sociological point. A one-way mission and the necessary supply for humans with food, clothing and techniques (daily utensils) with several replenishment flights would maintain a long-persistent equipment which can outlast the lifespan of a radiation contaminated human. The government thus is not in the position to be the executioner of the astronauts due to the fact that material loss would doom the humans. Nevertheless, a one-way mission implies that the astronauts as well as the first Martian settlers will die on the Red Planet. What to do with their corpses - cremate them, bury them? This problem has not only a sociological implication, but also underlies the question: 'Do the humans have the right to settle on another planet?' We cannot rule out that Mars possesses its indigenous life (several clues as ALH84001, the methane abundance in the atmosphere, the results of the Viking measurements, water(-ice), etc. are known and subject to controversial discussions) making it feasible that life originated in the Martian past. Do we have the right to settle on Mars and to endanger its potential own biological evolution?

Therefore Mars is only a wild-card for any potentially habitable object to be discovered in the future. Do we need a prime directive according to Star Trek - a consensus of non-involvement? From which starting point of life's evolution do we set this directive to be operational due to our own judicial feelings - for a civilization, a planet with plants and animals, for bacterial life, or also for a planet which could host life at present or in the future? Do we need a COSPAR planetary protection policy extended to all celestial bodies?

Yes and No! Yes, we have to ensure that any life-forms on other planets and moons are allowed to carry on their evolution. However we have to ensure our own evolution as well. In case we decide not to settle Mars as a first step into outer space, we are dooming our own civilization which will evidently disappear at the very latest when our Sun turns to the Red Giant stage. This will not happen within the next one hundred years, but it will happen definitely. We believe that the pioneer spirit of our species has not diminished. A one-way mission to Mars and the decision to build a permanent station on Mars will be the first step to ensure our own future.

2.3 Why Send Humans to Mars?
Looking Beyond Science

Pabulo Henrique Rampelotto
Department of Biology, Federal University
of Santa Maria (UFSM), Brazil

IN THE LAST DECADE, the human exploration of Mars has been a topic of intense debate. Much of the focus of this debate lies on scientific reasons for sending, or not sending, humans to Mars. However, the more profound questions regarding why our natural and financial resources should be spent on such endeavor have not been addressed in a significant way. To be successful, the human exploration of Mars needs reasons beyond science to convince the public. People are far more interested in the short-term outcome of exploration than any nebulous long-term benefits. Finding the right balance of science and other factors is critical to convince taxpayers to part with $100 billion or more of their money over the next couple of decades to fund such endeavor. In the following, I briefly explain why the colonization of Mars will bring benefits for humans on Earth, looking beyond scientific reasons.

The engineering challenges necessary to accomplish the human exploration of Mars will stimulate the global industrial machine and the human mind to think innovatively and continue to operate on the edge of technological possibility. Numerous technological spin-offs will be generated during such a project, and it will require the reduction or elimination of boundaries to collaboration among the scientific community. Exploration will also foster the incredible ingenuity necessary to develop technologies required to accomplish something so vast in scope and complexity. The benefits from this endeavor are by nature unknown at this time, but evidence of the benefits from space ventures undertaken thus far point to drastic improvement to daily life and potential benefits to humanity as whole.

One example could come from the development of water recycling technologies designed to sustain a closed-loop life support system of several people for months or even years at a time (necessary if a human mission to Mars is attempted). This technology could then be applied to drought sufferers across the world or remote settlements that exist far from the safety net of mainstream society.

The permanence of humans in a hostile environment like on Mars will require careful use of local resources. This necessity might stimulate the development of novel methods and technologies in energy extraction and usage that could benefit terrestrial exploitation and thus improve the management of and prolong the existence of resources on Earth.

The study of human physiology in the Martian environment will provide unique insights into whole-body physiology, and in areas as bone physiology, neurovestibular and cardiovascular function. These areas are important for understanding various terrestrial disease processes (e.g. osteoporosis, muscle atrophy, cardiac impairment, and balance and co-ordination defects). Moreover, medical studies in the Martian environment associated with researches in space medicine will provide a stimulus for the development of innovative medical technology, much of which will be directly applicable to terrestrial medicine. In fact, several medical products already developed are space spin-offs including surgically implantable heart pacemaker, implantable heart defibrillator, kidney dialysis machines, CAT scans, radiation therapy for the treatment of cancer, among many others. Undoubtedly, all these space spin-offs significantly improved the human's quality of life.

At the economical level, both the public and the private sector might be beneficiated with a manned mission to Mars, especially if they work in synergy. Recent studies indicate a large financial return to companies that have successfully commercialized NASA life sciences spin-off products. Thousands of spin-off products have resulted from the application of space-derived technology in fields as human resource development, environmental monitoring, natural resource management, public health, medicine and public safety, telecommunications, computers and information technology, industrial productivity and manufacturing technology and transportation. Besides, the space industry has already a significant contribution on the economy of some countries and with the advent of the human exploration of Mars, it will increase its impact on the economy of many nations. This will include positive impact on the economy of developing countries since it open new opportunities for investments.

Furthermore, the benefits of close cooperation among countries in space exploration have been made clear on numerous missions. International crews have been aboard the Space Shuttle many times, and the Mir Space Station has hosted space explorers from many nations. After the realization of the International Space Station, human exploratory missions to Mars are widely considered as the next step of peaceful cooperation in space on a global scale. Successful international partnerships to the human exploration of the red planet will benefit each country involved since these cooperation approaches enrich the scientific and technological character of the initiative, allow access to foreign facilities and capabilities, help share the cost and promote national scientific, technological and industrial capabilities. For these reasons, it has the unique potential to be a unifying endeavor that can provide the entire world with the opportunity for mutual achievement and security through shared commitment to a challenging enterprise.

To conclude, the human exploration of the red planet will significantly benefit all the humanity since it has the potential to improve human's quality of life, provide economic returns to companies, stimulate the economy of many nations including developing countries and promote international collaboration.

2.4 From the Pale Blue Dot to the Red Planet: Why Choose to go to Mars?
Harold Geller, Ph.D.

I HAPPEN TO BE OLD ENOUGH to remember the late President John F. Kennedy. In fact, I got to see Kennedy in person, on the campaign trail, while staying at a hotel in Philadelphia with my family, attending my cousin's wedding. I vaguely recall watching Kennedy on television when he made his speech to a joint session of Congress, so famously saying "that this nation should commit itself to achieving the goal, before this decade is out, of landing a man on the Moon and returning him safely to the Earth." Kennedy knew that such a "space project in this period will be more impressive to mankind, or more important for the long-range exploration of space; and none will be so difficult or expensive to accomplish." He spoke about the need "to accelerate the development of the appropriate lunar space craft [and] to develop alternate liquid and solid fuel boosters." He realized also that there was a need to "propose additional funds for explorations which are particularly important for one purpose -- the survival of the man who first makes this daring flight." Finally Kennedy emphasized that "in a very real sense, it will not be one man going to the moon it will be an entire nation." (Kennedy, 1961)

I still recall very vividly that evening in 1969 when Neil Armstrong took his first steps on the Moon. I was watching it on television (black and white, by the way) at my cousin's house in Ventnor City, New Jersey. I was certainly caught in the period, a time when this nation did come together to achieve a daring goal, one that many had considered foolish. After that summer, I was fully convinced that the United States of America would be putting a human on the surface of Mars, before the century was done. I was wrong.

My master's project focused on the data from the Viking mission to Mars. Many scientists from the Viking mission are now gone, and their works are being re-interpreted in their absence. I also worked on a university proposal for a robotic instrument which we had hoped would be chosen to go to Mars and return samples to Earth. Although our proposal was not chosen, it was generally believed at that time that there would be a Mars sample return mission in the 1990's. This did not occur. And this nation did not send humans to Mars in the last century.

It should be noted that President Kennedy did not offer science as the reason to go to the Moon. In fact, Carl Sagan, years later, in his book Pale Blue Dot: A Vision of the Human Future in Space wrote about the story that Kennedy's science advisor, Jerome Wiesner, purportedly had a deal with Kennedy so that there would be no pretending that the mission to the Moon was for science. (Sagan, 1994).

Today, there are many in the science community who vociferously voice their opinion that all "manned space" operations are a waste, and that all science data about the Moon and Mars, can be retrieved via robotic spacecraft. However, as noted by Kennedy, it is not for science that I believe we should one day venture to Mars. It is for human exploration, and a human insatiable curiosity.

On my desk today, there sits a book, written by James T. Bennett, a faculty member at my institution. Bennett mocks all such efforts like the space program. His view is that all of this is a waste of money, and in fact something our founding fathers never dreamed of spending federal tax dollars to support. In fact the spending of federal taxpayer money for things like a space program may be unconstitutional in his opinion. I do not agree.

Bennett uses quotes from Eric Drexler to defend his views, regarding a mission to Mars. He notes that "private enterprise, not taxpayer-fed fantasies of national glory, should drive space exploration." Such was the view of many citizens when Apollo 11 successfully landed on the Moon. Drexler himself refers to such missions as "grand stunts." Bennett concludes that "there is nothing of liberty or freedom in such a scheme; it is little more than a flashy scam, a hornswaggling of the public" (Bennett, 2010).

Even after all this banter about the supposed waste of taxpayer money to go to Mars, while I know not what others may feel, but for me, if given the opportunity I would be willing to go to Mars, even if it meant never returning to Earth. Since Neil Armstrong took those first steps on the Moon, I had dreamed of someone, perhaps even me, stepping on the sands of Valles Marineris, collecting rock samples along the way. I realize that I am too old for even this one-way mission, but there are others who are not. To give one's final days in service to the pursuit of knowledge, to many, is a fine end to a life lived as well as one could. It reminds me of Wolfgang Vishniac, who died on the hills of Antarctica. Sagan wrote of Vishniac in his seminal book Cosmos. Regarding Vishniac's last moments alive, Sagan pondered "perhaps something had caught his eye, a likely habitat for microbes, say, or a patch of green where none should be. We will never know. In the small brown notebook he was carrying that day, the last entry reads, 'Station 202 retrieved. 10 December 1973. 2230 hours. Soil temperature, -10 degrees. Air temperature -16 degrees.' It had been a typical summer temperature for Mars" (Sagan, 1980).

It was Sagan who noted that we stand on "the shores of the cosmic ocean." He continued by saying: "The ocean calls. Some part of our being knows that this is from where we came. We long to return. These aspirations are not, I think, irreverent" (Sagan, 1980). I agree.

REFERENCES

Bennett, James T. (2010). The Doomsday Lobby: Hype and Panic from Sputniks, Martians, and Marauding Meteors. New York: Springer.

Kennedy, John F. (1961). Special Message to the Congress on Urgent National Needs. Boston: John F. Kennedy Presidential Library and Museum [retrieved 24 January 2011] http://www.jfklibrary.org/Research/Ready-Reference/JFK-Speeches/Special-Message-to-the-Congress-on-Urgent-National- Needs-May-25-1961.aspx

Sagan, Carl (1980). Cosmos. New York: Random House.

Sagan, Carl (1994). Pale Blue Dot: A Vision of the Human Future in Space. New York: Random House.

2.5 A One-Way Trip to Mars

J. Richard Gott, III, Ph.D.
Department of Astrophysics, Princeton University

I'VE BEEN ADVOCATING A ONE-WAY colonizing trip to Mars for many years (Gott, 1997, 2001, 2007). Here's what I said about it in my book, Time Travel in Einstein's Universe:

> "The goal of the human spaceflight program should be to increase our survival prospects by colonizing space. ... we should concentrate on establishing the first self-supporting colony in space as soon as possible. ... We might want to follow the Mars Direct program advocated by American space expert Robert Zubrin. But rather than bring astronauts back from Mars, we might choose to leave them there to multiply, living off indigenous materials. We want them on Mars. That's where they benefit human survivability.... Many people might hesitate to sign up for a one-way trip to Mars, but the beauty is that we only have to find 8 adventurous, willing souls" (Gott 2001).

I've been stressing the fact that we should be in a hurry to colonize space, to improve our survival prospects, since my Nature paper in 1993 (Gott 1993). The real space race is whether we get off the planet before the money for the space program runs out. The human spaceflight program is only 50 years old, and may go extinct on a similar timescale. Expensive programs are often abandoned after a while. In the 1400s, China explored as far as Africa before abruptly abandoning its voyages. Right now we have all our eggs in one basket: Earth. The bones of extinct species in our natural history museums give mute testimony that disasters on Earth routinely occur that cause species to go extinct. It is like sailing on the Titanic with no lifeboats. We need some lifeboats. A colony on Mars might as much as double our long-term survival prospects by giving us two chances instead of one.

Colonies are a great bargain: you just send a few astronauts and they have descendants on Mars, sustained by using indigenous materials. It's the colonists who do all the work. If one is worried that funds will be cut off, it is important to establish a self-supporting colony as soon as possible. Some have argued that older astronauts should be sent on a one-way trip to Mars since they ostensibly have less to lose. But I would want to recruit young astronauts who can have children and grandchildren on Mars: people who would rather be the founders of a Martian civilization than return to a ticker-tape parade on Earth. Founding a colony on Mars would change the course of world history. You couldn't even call it "world" history anymore. If colonizing Mars to increase the survival prospects of the human species is our goal, then, since money is short, we should concentrate on that goal. In New Scientist (Gott 1997) I said:

"And if colonization were the goal, you would not have to bring astronauts back from Mars after all; that is where we want them. Instead we could equip them to stay and establish a colony at the outset, a good strategy if one is worried that funding for the space programme may not last. So we should be asking ourselves: what is the cheapest way to establish a permanent, self-sustaining colony on Mars?"

I have argued that it is a goal we could achieve in the next 50 years if we directed our efforts toward that end. We would need to launch into low Earth orbit only about as many tons in the next 50 years as we have done in the last 50 years. But will we be wise enough to do this?

References

Gott, J. R.(1997) A Grim Reckoning," New Scientist, November 15, pp. 36-39.

Gott, J. R. (2001). Time Travel in Einstein's Universe. Boston: Houghton Mifflin, pp. 229, 231.

Gott, J. R. (2007). Longevity of the Human Spaceflight Program" in E. Belbruno, Ed., New Trends in Astrodynamics and Applications III, American Institute of Physics Conference Proceedings, Vol. 886, pp. 113-122. Melville, NY: AIP.

Gott, J. R. (1993). Implications of the Copernican Principle for Our Future Prospects," Nature, 363, 315.

2.6 For Mars Exploration, Rovers are Good, Humans are Better

Steven W. Ruff, Ph.D.

School of Earth and Space Exploration, Arizona State University, USA

MARS IS A REAL PLACE TO ME, at least the few square kilometers of it in Gusev crater where the Mars Exploration Rover Spirit roamed from January of 2004 to March 2010. I had the privilege of naming hundreds of rocks along its route, which perhaps means that I've named more features on Mars than any other person on Earth. Some of those rocks I can even recognize from orbit, thanks to the incredible resolution of the HiRISE camera on the Mars Reconnaissance Orbiter. I've looked at those rocks and the landscapes that contain them using the rover's eyes - its black and white as well as color cameras. The cameras produced stereo 3D images, making the rocks and landscapes even more real to me. I've spent many hours gazing into those scenes and frequently experience the sensation of actually being there.

My job on the Spirit team was to look at rocks not just with cameras, but also with a thermal infrared spectrometer called Mini-TES. I learned that the squiggly lines of emissivity spectra serve to readily distinguish one rock from another, allowing me to recognize common classes. Typically we'd work among a particular class before moving onto another one as the geology changed along Spirit's route. In some cases, earlier rock classes would show up farther down the trail, presenting a spectrum that was like a familiar face to me. In this way, I developed an even more intimate familiarity with this place on Mars.

Spirit outlived even the wildest speculations about its lifespan, making possible the remarkable discoveries about the igneous, aqueous, and aeolian processes that shaped the landscape that it and we roamed. But despite these successes, I became painfully aware of the shortcomings of robotic exploration of Mars. In a word, it is cumbersome. It took years of painstaking effort to explore just those few square kilometers of Gusev crater. Many tens of humans had to participate to guide the rover along a path that was carefully chosen to maximize both safety and science potential. Although Spirit proved to be much more robust and capable than anyone imagined, its speed and mobility were limiting factors. And despite a science payload exquisitely adapted to the tasks it was designed for, surely we failed to recognize and understand important clues to the geologic history we came to investigate. The experience of exploring a planet with a rover is both incredibly exciting and rewarding and incredibly frustrating. It is science by committee modulated by engineering constraints.

Many on the science team echoed the sentiment that a human geologist could have performed the years of exploration done by Spirit in just a few weeks or perhaps days. It's true that Spirit's amazing toolkit is still unavailable to a terrestrial field geologist. But simple tools combined with the eyes, hands, boots, and brain of a human far outstrip the capabilities of a rover, even those of the next generation Mars Science Laboratory. Given the impossibility of real- time interaction between a human and a robotic surrogate across the millions of kilometers separating Earth from Mars, robotic exploration will never replace what is achievable by humans. Here I am focused on the scientific achievements. The ones that arise from humanity expanding into the solar system, by definition, require humans. Robots should never be viewed as a substitute for humans directly experiencing another world.

A one-way mission to Mars is a bold plan that could expedite the gathering of information about an endlessly fascinating place. The exciting possibility of finally learning whether life ever took hold beyond Earth is profound motivation to send human life there. With sufficient resources, skills, and knowledge, human explorers sent to Mars would be adept at exploring for alien life while preserving their own. In the process, the vicarious thrill and satisfaction that Earth- bound humans have experienced even from robotic missions, would be compounded in ways immeasurable. Given the trajectory of human exploration and settlement, it is not a question of whether Mars will become a target but when.

As a geologist, I routinely daydream about roaming the surface of Mars and imagine how I would go about answering questions posed by robotic explorers. A few times a year, those dreams come back to me while I sleep. But in this context, it's not scientific questions that are compelling; it's the surreal wonder of another planet. Likely these ephemeral moments will be the closest I ever come to the real experience. There was a time before the joyous entanglements of a wife and two young daughters when I would have enthusiastically embraced the adventure of a lifetime in a one-way trip to Mars. Perhaps toward the end of my life I will revisit the idea. For now, I will defer to those bold explorers who are eagerly lining up and wondering why we are waiting.

2.7 To Be Human is to Explore
S. Giddings, Ph.D.

WHEN SETTLERS LEFT EUROPE for the New World, many knew that they would not return, and might perish in the undertaking. Yet, they were drawn by an urge to explore, and to find new lives.

Modern society is losing sight of this elemental challenge -- although some still find its calling in the more remote areas of our planet. Climbers on difficult routes in the high mountains commit to their ascents with full knowledge they may not return -- some do not -- for the sake of the human spirit of adventure and exploration, and to experience something bigger than themselves.

Space is the next frontier, and travel to Mars a small hop to a stepping stone in a grander quest. It would be an honor to represent humankind in its exploration, and those with an adventurous spirit will find this calling -- even with no vision of return. The first human visitors to its surface will watch in awe as they see Earth -- barely larger than a star -- rising from the martian horizon. With this distant view of home, they will begin discoveries on how to make a new life, both for themselves, and for our species.

Like with other settlers, their first challenge will be to find a way to survive and sustain life. They will begin to find resources both on the planet, and within themselves. For the first time in history, they will begin to learn whether and how humans can sustain themselves in an environment utterly different from the cradle of our species, Earth. They will begin a journey that may even be the key to humanity's survival.

These pioneers will confront entirely new challenges of physics, engineering, and mobility. With their fragile toehold, they will then begin to explore the richness of a profoundly different geography and geology, possibly even a biology. They will expand the human psyche to a visceral new understanding, by witnessing the cosmos and the smallness of our planet in views never imagined by our ancestors, and never seen by the human eye. They will give the human family a bold yet terribly humbling perspective on our role in a vast cosmos.

2.8 Logbook for Day 283 on Mars Crew 1,

Crew Biologist Cora S. Thiel Reporting

Cora S. Thiel, Ph.D.[1,2], and Vladimir Pletser, Ph.D.[1,3]

[1]EuroGeoMars campaign; Mars Desert Research Station Crew 76-77
[2]Institute of Medical Physics and Biophysics, University of Muenster, Germany
[3]Human Space Flight Directorate; European Space Agency, The Netherlands

5:00 AM

TODAY THE DAY STARTS EARLY, because we want to prepare for an Extra-Vehicular Activity (EVA). We plan to collect soil and rock samples from several locations to continue our search for life on Mars. We are here now for 283 days, more than nine months, and we are sampling twice a week - that means more than 800 samples were analyzed and still no luck. Not the slightest trace of life! Some of our crew members start to be depressed, but I am still convinced that if we keep on going, we will find the right sampling location and we will be successful. Today might be a promising day, because with the ground penetrating radar we found a location last week containing ice in shallow subsurface layers - an ideal condition for life as we know it on Earth.

6:00 PM

The EVA was very difficult and energy-sapping. We were caught in a dust storm and had to hide for more than two hours in a lava tube cave. Luckily, the weather conditions improved and we could continue our way to the chosen experimental sampling site. Sampling was not easy, our motorized drilling machine broke and we had to wait until our Russian crew engineer repaired it. He is a genius! In the end we were able to collect ten samples and returned to the habitat.

Back in the habitat laboratory, I cataloged all samples and started the biological analysis. First of all, I started the MarsBioanalyzer instrument that we use to extract potential biomolecules from soil and rock samples. A first round of experiments is now running using the Polymerase Chain Reaction (PCR), a well established technique, to amplify a single or a few copies of a piece of DNA, a nucleic acid that is used by all living organisms on Earth for long-term storage of information.

10:00 PM

The first half of the experimental analysis is done. I am waiting impatiently for the electrophoresis results. Will we detect life on Mars today?

11:30 PM

Unbelievable, I got a positive signal in the analysis for soil sample number 5! This means I detected DNA, most likely from some kind of extraterrestrial microorganism. What kind of organism is it? Bacteria? I will work through tonight to start the DNA sequencing unit of the MarsBioanalyzer instrument. I have to have certainty about the origin of this DNA. Is it something known? Related to a form of microorganism on Earth? These results could be the breakthrough! It seems that we detected for the first time that we are not alone here. There is life on Mars!

II

ASTRONAUTS ON MARS

Journal of Cosmology, 2010, Vol 12, 3500-3505.
JournalofCosmology.com, October-November, 2010

- 3 -

Our Destiny – A Space Faring Civilization?

Edgar D. Mitchell, Sc.D. [1], Robert Staretz, M.S.

[1]Apollo 14 Lunar module pilot. Sixth person to walk on the Moon.

ABSTRACT

As a species we have always had an incredible curiosity and because of it the thought of exploration and exploitation of new frontiers has always excited our imagination and motivated our efforts. We now stand on the threshold of becoming a space faring civilization. Our very survival certainly for the long term depends upon it and probably for the near term as well. Throughout our history, we have never been able to predict the perils nor the benefits of exploration but in every case humanity has always prevailed over all obstacles and the rewards it has reaped have always far exceeded our expectations. This certainly will be the case with the exploration of Mars and the other planets and the moons of our solar system. Initially these will be purely exploratory missions but eventually exploration will turn to colonization. Ultimately as we continue to develop and our technological capabilities even the stars will be open to our explorations. Will humanity be prepared for the greatest discoveries of the history of our civilization? Will we find other intelligent civilizations far older and incredibly superior than our technological capabilities and collective wisdom? We end with speculation on the values, ethics and consciousness of these civilizations and lessons they may hold for the future of humanity.

Key Words: Mars, Space exploration, Astronauts, Manned Space Exploration Colonization of the Planets, Interstellar Exploration, Humanity's Destiny in Space, Extraterrestrial Civilizations, Evolution of Consciousness.

1. INTRODUCTION

THIS IS AN HISTORIC TIME for humanity and also one of the most challenging times as well. We stand on the threshold of becoming a space faring civilization shedding the bonds that have tied us to Earth since the very beginnings of the planet's history. In the last 40 years, we have looked back at Earth from space, walked on our moon, sent robotic probes to most of the planets, moons and even some of the asteroids of our solar system. We have explored the depths of our galaxy and the visible universe with both Earth and spaced based telescopes and instrumentation. Later this century we will very likely walk on the surface of another planet. Why? Humanity has always had an insatiable appetite to know, for adventure and a remarkable curiosity to explore the unknown. In spite of the sacrifices and challenges required, history has shown over and over the benefits and rewards of exploration have always far exceeded expectations and mostly in ways that were impossible to predict. No doubt such will be the case again in the exploration of space.

There are many other reasons to travel to other worlds and beyond besides the urge to explore the unknown. One is the obvious long term motivation to become an inter-stellar space faring civilization. At some point in the distant future we will have no choice but to leave our home world. Our sun, already a middle aged star, is powered by fusing hydrogen in the nuclear inferno at its core. As the remaining fuel is consumed, the sun will continue to expand in size and with it the intensity of the radiation increasing at the planets. Already the sun's output is 15% greater than it was a few billion years ago and eventually it will destroy all life on the planet. The long term prognosis is that the sun will expand to such a large degree that in due course it will cause our oceans to boil away into the vacuum of space leaving an uninhabitable desert wasteland behind.

More immediate concerns for inter-planetary travel but perhaps less well known by most of humanity are the issues associated with insuring a sustainable future for our civilization. Much of our planet's non renewable resources such as ores and precious metals will not last forever especially with our already large and exponentially growing population. Mining and refining these ores in space for shipment to Earth will be necessary within short order if we are to maintain and broaden our current standard of living on the planet. Establishment of space colonies will also teach us much about sustainability issues and many will have direct applicability to the future of Earth. Until now our planet has had a thriving ecosystem because nature has long ago evolved and fine tuned Earth's biogeochemical processes to maintain its long term stability. That stability is now being threatened by our own doing.

2. SPACESHIP EARTH

The visionary Buckminister Fuller often referred to our planet as "Spaceship Earth". It was his firm belief that we must all work together as a crew of Spaceship Earth if we are to survive let alone continue to thrive upon it, along with all other living creatures that share our beautiful planet. The available evidence suggests that global population growth fueled by our modern technologies of the last 100 years have created an unsustainable trajectory for all life on the planet. Our unprecedented consumption of nonrenewable resources and increasingly strong indications of run-away climate change have been greatly exacerbated by human activity of the last century. Together these factors suggest that we may soon be facing our first mass extinction event due to human activities. All previous extinction events have resulted from natural causes such as large meteor impacts or super-volcanic eruptions. Are we about to experience one due to our own inattention and misperceptions of how nature has maintained Earth's environment over its entire history by our propensity to interrupt her natural processes on a massive scale?

Exploiting resources of the solar system, creating colonies in space, exploration of other planets, establishing colonies on them and eventually travel to other star systems offers us many lessons for a sustainable Earth although initially on a much smaller scale. Of necessity space colonies will have to be mostly self sufficient because of the vast distances from Earth. Aside from the long travel times to reach these remote outposts, the associated costs of shipping supplies and replacements parts will be prohibitively expensive. Our space colonies will be forced to live as close to self sufficiency as possible by utilizing local resources whenever practical. They will also have to make extensive use of recycling, reusing discarded materials and reducing consumption on a scale that has here-to-for been unprecedented. In a very real sense, space colonies will have to emulate consciously what nature has been doing for billions of years on Earth.

Because of vested interests, short sightedness or personal short term gain all at the expense of long term sustainability, our politicians, many of our leaders and most of our citizenry have ignored, misunderstood or misrepresented the magnitude and nature of the issues facing our civilization for far too long. Perhaps even most pressing is our propensity to resolve our differences by violent rather than by peaceful means. Our technologies are now so powerful that not only do they enable us to explore the solar system but they are the very same ones that may lead to our demise if used to promote the goals of one group at the expense of others. Clearly before we can truly call ourselves a space faring civilization we must put aside these petty differences that are often driven by intolerance, greed, the need for power or the need for control that divide us.

Five hundred years ago Christopher Columbus convinced Queen Isabel and King Fernando of Spain to fund a voyage to find an alternate route to India. Imagine the courage it took to make that journey. Even though most scholars at the time knew that the world was spherical, the consensus view of humanity was still that of a flat Earth. Casting aside those concerns of his crew, Columbus was aware that the trip would likely to be fraught with many unknown and unforeseen dangers. There was also the problem of estimating the duration of the trip and determining the provisions required to sustain the expedition on their journey. As the days stretched into weeks and the weeks into months, the mutinous clamor of the crew increased daily. Finally after 2 1/2 months that eventful journey finally sighted land and what followed became our history. Unfortunately for Columbus he never reached India, but instead opened up a whole new world that happened to be in the way.

If we can get beyond the issues described above and some day land on that first foreign planet, surely whole new worlds will open up for us just as they did for those early European explorers. At that point we will finally be ready to assume our destiny as a space faring civilization. We will go not as citizens of the United States, the United Kingdom, China, Russia or any of the other 195 or so countries of this planet but instead as citizens of planet Earth. We will go with a common vision and mission for the betterment of all mankind. No other activity will unite the citizens of the world in a nobler endeavor. After early exploratory missions, our first objective should be a permanent colony on the moon and our second will most likely the establishment of one on Mars. The first manned colonies on both will be far more expensive and far more perilous than Columbus or his crews could have ever imagined. Columbus' journey was financed by the court of Spain. Permanent colonies beyond Earth will be far too expensive for even the richest nations on Earth. They will likely be funded of necessity by a consortium of nations representing all of mankind. Our explorations and colonization will be full of challenges, fraught with dangers, but filled with incredible and unforeseen rewards and benefits to us all and to our progeny for many generations to come.

3. THE HUMAN MISSION TO MARS

Our first manned mission to Mars will not be too different from the first moon exploration, simply short term and exploratory. The long distance and long travel time to Mars of 9 months or more one way (with existing technologies) require special consideration which is the subject of a separate paper. Plans to originate Mars colonization from moon colonies established for deep space exploration have been proposed. It is argued that the moon's lower mass and therefore much lower gravity, 1/6 of Earth's, translates to greatly reduced costs of launching missions into deep space. Many of the raw materials and resources required to sustain the crew on our early interplanetary missions might possibly be mined from the moon greatly reducing the costs. Newly discovered water if available in sufficient quantity might be mined for human use, and its constituents, oxygen for human consumption along with hydrogen for fuel. Carbon, iron and several other elements will also likely be mined for a variety of purposes. Together they will make up a significant portion of the total resources required for our first interplanetary colonists.

Establishing a fully self sufficient colony on the moon as a stepping stone to the planets will not come cheaply and may prove not to be feasible at all. However, the moon will be a great laboratory and learning environment for the kinds of obstacles, living conditions, and hazards that will also have to be faced on Mars or more distant venues. In some cases the hazards on the moon are even more severe than the Martian environment. For example solar radiation, solar wind, micrometeorites, and 500 degree temperature gradients are far more indicative of what our space explorers will experience during the trip to Mars than the extremes that will be encountered on the Martian surface. The knowledge gained and the technologies developed to support permanent bases on the moon will greatly benefit both for our first voyages to Mars as well as the first Martian colonies and even worlds beyond.

4. A FRIENDLY UNIVERSE?

Interplanetary exploration aside, there is no certainty that we will survive the gathering storm on Earth of the man made challenges to our survival. If we do endure, it is likely that we will eventually meet other intelligent technological civilizations in this increasingly apparent life friendly universe that we live in if we haven't already done so. Hopefully these civilizations will have solved once and for all many of the dilemmas currently facing humanity. Clearly any civilization that mastered the technological challenges of interstellar travel will most likely be much older and far more advanced than us in many ways that we cannot even conceive. They will also likely be much wiser in how they utilize their technologies. When we begin a dialogue with them, perhaps our first order of business should be to find out how they managed to get beyond the civilization threatening technological adolescent stage in which we on Earth are now engaged.

What might such a civilization be like? Because they will likely have evolved under very different circumstances and timeframes, they would certainly have taken evolutionary pathways very different from us. Once they reached the technological stage they would have many different alternatives for future evolutionary growth. Some might be based on unpredictable natural events but eventually most will be accomplished by conscious evolution.

It is very easy to fall into a trap by speculating why we have not already encountered signs of extra-terrestrial civilizations using Fermi's 50 year old paradox "Where is everybody?" Anthropic assumptions and arguments must be evaluated with considerable caution and thought. As I (Edgar Mitchell) have often pointed out in my lectures, 140 years ago, my great grandparents migrated from Georgia to Texas in a covered wagon. The automobile, trains, airplanes, indoor plumbing, electric lights, telephone, radio, electronics, etc. had not yet been invented. And, yet, one hundred and forty years later I went to the moon. The rate of technological advancement in such a short period of time has been absolutely astounding. Where might we be in ten thousand, one hundred thousand, or even several million years hence?

If we were somehow able to transport modern technologies back just two thousand years, what would that ancient society make of it? What if we went back ten thousand years and tried to explain to a caveman the operating principles and purposes of the international space station (ISS)? Clearly our ancient ancestors would not have any basis for understanding what we were trying to communicate. To take this to the extreme, how about trying to explain the ISS to an ant colony? Perhaps this analogy is the appropriate one with regard to advanced extra-terrestrial civilizations.

UFO lore aside, one of the common arguments that many use claiming why we have not yet been visited by intelligent civilizations is due to the difficulty in traversing the tremendous distances in interstellar space. Our current understanding of nature tells us that the maximum velocity that any object can reach is the speed of light in a vacuum or 186,000 miles per second. At that rate at trip to the nearest star system Alpha Centauri, would take about 4.2 years. Clearly reaching that speed is currently well beyond the realm of possibilities for humankind. At best we can currently achieve is 1/1000 of light speed so at that speed it would take us 4,200 years to reach this destination. By extrapolation and linear projections using today's technologies, perhaps in several hundred years we will have advanced our technologies to the point where we can reach a reasonable fraction of the speed of light.. At 1/10 the speed of light a trip would only take 42 years or so.

5. THE UNIVERS REMAINS TO BE DISCOVERED

Travel time is not our only problem though. Imagine the supplies needed and the problem of maintaining an appropriate sustainable environment for the crew for a 42 year one way trip. And then there's the issue of keeping the ship operational for such a long period. Think of the energy requirements that would be needed to sustain such a trip just to keep the ships systems working let alone the energy requirements for the ship's propulsion systems. For all these reasons plus several others we have not bothered to mention, many "experts" conclude that we are a very long way away from the practicality of embarking on such a trip. According to them our only hope of interstellar contact would be to stay at home and try to communicate with an alien civilization via some form of communication mechanism such as powerful narrow band radio or laser beams. And, even then the round trip delay of such a message to our closest neighbor would take about 9 years. Certainly this does not seem like an encouraging prospect for two-way interstellar communications. A monologue perhaps.

The speed of light limitation is based on current human knowledge and our current understanding of the laws of nature. But, history has shown over and over how quickly our understanding can change. One hundred years ago the British scientist Lord Kelvin was discouraging physics students from entering the field of physics because, according to him, all the laws of nature had already been discovered and all that remained was to improve the accuracy of nature's constants to 6 decimal points. Of course we now know how naive that opinion was because within just a few years after his pronouncement, relativity, quantum physics, and many new discoveries in astronomy and cosmology burst upon the scene. Without many of these discoveries our modern technology based civilization would not be possible.

Nature reluctantly reveals itself and scientists often have a vested interested in maintaining the status quo. Max Planck, the father of modern quantum theory eloquently pointed this out with his now famous quote "Science advances funeral by funeral". The universe has had at least 13.7 billion years to evolve to its current level of complexity. Surely our science which has only been a formal discipline for the last 400 years is incomplete, in some areas likely incorrect and has a long way to go before it reveals all of nature's secrets to us.

When Columbus made his first voyage across the Atlantic Ocean, the average speed of travel by ship was about 2 miles per hour. Even today, 500 years later, the maximum speed of surface ships traveling in the open ocean has increased only by a factor of 20, enough of an improvement to reduce the ocean transit time from 2 1/2 months to less than a week. But that's based on a linear projection of the past. We have also developed new technologies that have given us exponential improvements in transit time. Today's commercial airplanes cross the ocean in 5 hours or another factor of 24 increase in speed. With rocket propulsion we gain a further reduction of transit time to a mere 30 minutes, a factor of 10 in further reduction of transit time. Could Columbus have ever imagined technologies that could reduce his transit time by a factor of 3600?

6. THE FUTURE AND ADVANCED CIVILIZATIONS

It is highly likely that our not too distant progeny have an understanding of laws of nature that will be far beyond what we in the 21st century understand. Consider the current rate of growth of knowledge which has been following an exponential growth curve; thus doubling the rate of new knowledge acquisition every two years. Think about the fact that our science is on the verge of being able to genetically engineer the human genome. Based at the current rate of progress in molecular and cellular biology and genetics, it is not very difficult to extrapolate that within a mere few hundred or so years we will have the ability to engineer ourselves with capabilities beyond our wildest dreams of today. By then we will probably have conquered all genetic diseases and will have deliberately endowed ourselves with an immune system that can survive any known viral or bacterial infections. Perhaps, barring accidents, we will also be able to live indefinitely. Certainly our planet, our social systems and our cultural systems will not be able to deal with a human population with such extreme longevity.

At the same time surely our space faring technologies will advance to a much more highly developed state. Will we have any choice but to head for the stars? With extreme longevity, will it matter how long it takes at sub light speed to travel to the stars? If our interstellar travelers got bored with such a journey there would always be the possibility that they could used some form of suspended animation or hibernation for most of the journey. Perhaps they will undertake such journeys with a complete self contained colony within their space craft.

Let us now imagine an alien technological civilization thousands or millions of years beyond our own. If we are within several hundred years of having the capabilities described above, what might an advanced civilization be like? Would they have revealed nature's secrets such that they could traverse the void of space by faster than light travel? Could travel to Alpha Centauri be reduced to a matter of a few days similar to today's trip duration to the moon? Could that be accomplished by manipulating gravity or even the speed of light itself, perhaps by warping space or traversing it through wormholes, or, even by some other means we cannot even begin to imagine? Futurist and science fiction author, Arthur C Clarke, is often quoted "With any highly advanced technological civilization, their technologies would seem like magic to us." To think that we have already uncovered all of nature's secrets is certainly sheer folly.

Given the incredible multitude of stars, the vastness of space, and the possibility of several thousand space faring civilizations thousands to millions of years older than us in our galaxy alone, can there be any doubt that sooner or later we will finally meet them? What might be the consciousness, ethics, morality and values of such a civilization? Certainly nature requires no special morality or ethics to tap into and utilize its secrets. Whatever is built into the belief systems of the discoverer will be utilized when these capabilities are finally unveiled and understood. These belief systems can be highly evolved and saintly or war-like and demonic; either way will be fine with nature for it takes no sides. Humans, for example, have uncovered many of the secrets of unleashing the energy stored within atoms. This knowledge can and has been used for peaceful purposes such as the generation of electricity or it has also been used to build nuclear weapons which can destroy us all. Certainly for humans, our morals, values and ethics have not kept pace with our technological prowess.

Perhaps for these reasons, by necessity, extra-terrestrial civilizations have survived and evolved far beyond technological adolescence and have developed the technologies for inter-stellar travel because they have evolved to higher states of consciousness. If so, they will likely have recognized the need to live in harmony with nature and all that that implies. This would likely include highly evolved self-discipline, ethics, and universal spiritual values first less they would have otherwise long ago destroyed themselves by their command of such powerful technologies.

7. THE CONSCIOUS COSMOS

States of consciousness have been studied for centuries by Tibetan Buddhist monks. The most profound state they refer to is known as Nirvikalpa Samadhi. It is a level of the highest spiritual attainment and evolution, the state of deep undifferentiated awareness in which there is only the Self within a transcendent observing entity. There are no thoughts or objects in mind. It is a state beyond time and space. The Self expands and merges into the entire field of mind so that pure awareness is all that exists. After attaining this state one has complete understanding of cosmic wisdom and one feels that he is in complete union with the Creative Source of all that is. At this point all ordinary concerns and everything else become subordinate to this union and lose all meaning. In this state unconditional love is the organizing principle of the universe. We do not mean to imply that all highly advanced extraterrestrial civilizations have reached such a state. Perhaps the oldest and most advanced have. It is likely that if their entire civilization and all individuals comprising it had reached it, they likely would no longer be interested with the Earthly concerns of humans on this little remote planet in the backwaters of the Milky Way Galaxy.

It is more likely that for evolved self reflective beings throughout the cosmos, there is a spectrum of consciousness that reaches from two extremes. On the left end of this spectrum a state of consciousness exists with concern only for promoting the self, accompanied with outright disdain or malevolence towards other living beings. On the opposite end of such a spectrum perhaps is a state of consciousness similar to Nirvikalpa Samadhi. On such a cosmic scale most humans would be placed to left of center busily pursuing their own self interests where their use of technology has far outpaced the values and ethics necessary to use them wisely. Our expectation is that extra-terrestrial intelligences are shifted by varying degrees beyond human civilization more to the right end of the spectrum. Hopefully as humanity takes its place among space faring civilizations we will evolve more to the right as well. If human history teaches us anything, failure to do so is no longer a viable option.

Journal of Cosmology, 2010, Vol 12, 3506-3516.
JournalofCosmology.com, October-November, 2010

- 4 -

Apollo on Mars: Geologists Must Explore the Red Planet

Harrison H. Schmitt, Ph.D.,
College of Engineering, University of Wisconsin-Madison
Former United States Senator, Apollo 17 Astronaut,
12th and last man to set foot on the Moon.

ABSTRACT

Extension of human geological studies to the surface Mars requires the transfer of the principles of geology and the well-honed techniques of field exploration begun by astronauts on the Moon. The principles related to exploration planning and innovation, or to sample collection and documentation, do not change merely by leaving the Earth. Particularly unchanged is the need for human touch, experience, vision, and imagination in fully realizing the scientific and humanistic value inherent in exploration. Natural influences that affect Mars combine those affecting the Earth and Moon with Mars being a body intermediate in size between the two terrestrial planet extremes. In addition, the Martian atmosphere has filtered out all meteors and comets capable of forming craters less than about 30m in diameter. A field geologist's "x-ray" vision still will be required and must take into account the effect of wind or water-transported materials that may obscure or cover underlying rock contacts and structures. Beginning with precursor return missions to the Moon, professional field explorers should be part of every four person crew sent to the Moon so that this paradigm can be ingrained in planning for Mars' exploration as well as providing much enhanced returns from lunar exploration. Crew members and their operational support teams should commit to terrestrial, field based training on real geological problems constrained by realistic operational planning, equipment, communications, and timelines. Mars exploration will not be easy. As with anything worthwhile, risks exist. But the alternative of postponing settlement of the Moon and the exploration of Mars beyond existing generations would leave the future to other explorers.

Key Words: Mars, Moon, Earth, exploration, Mars access, geology training

1. INTRODUCTION

HIGHER THAN THE WALLS OF the Grand Canyon of the Colorado, mountains over 2000 meters high confined the long, narrow valley of Taurus-Littrow. A brilliant sun, brighter than any sun ever experienced on Earth, illuminates the cratered valley floor and steep mountain slopes, starkly contrasted against a blacker than black sky. Exploration of the nearly four billion year-old valley, and the slightly younger volcanic lava rocks and ash partially filling it, culminated the Apollo Program and the first extension of hands-on geological studies from the Earth to the Moon (Schmitt, 1973, 2003). Now, we contemplate the extension of human geological studies to the surface Mars. What will be new and what will be familiar to the first geologist to step into a red Martian sunrise? With the Apollo missions to the Moon, astronauts began the transfer of the principles of geology and the well-honed techniques of field exploration to solid bodies away from Earth.

Historically, geology and exploration have been inexorably linked from the first known geological map, that of Saxony basalts by Johan Charpentier in 1778 (Charpentier, 1778); to William Smith's 1815 geological map of England (Smith, 1815); to Lewis and Clark's exploration of the Louisiana Purchase (Ambrose, 1996); to Eugene Shoemaker's 1958 preliminary photo and telescopic geological map of the area on the Moon that includes the crater Copernicus (Schmitt, et al, 1967); to the Apollo astronauts geological documentation of lunar samples (See Wolfe, et al, 1981); to the most recent tele-robotic observations on and around Mars (Christensen, 2005; NASA, 2010a); with thousands of other examples in between.

The actual placement of geological exploration in time and space does not change the fundamentals of documenting and graphically representing the relative age of natural features; the structure, internal evolution, and alteration of such features; their inferred origins; and their ultimate implications about potential resources useful to sustaining and advancing civilization. (See Hodges & Schmitt, 2011) Nor are the principles related to exploration planning and innovation, or to sample collection and documentation, changed merely by leaving the Earth – if anything, those principles become more important. Particularly unchanged is the need for human touch, experience, vision, and imagination in fully realizing the scientific and humanistic value inherent in exploration. For each new body to be explored in space, however, we must build on our experience in exploring the last place as geologists have done on Earth for centuries. We must continually ask what may be the same and what may be different as we approach a new challenge. In the future exploration of Mars, probably beginning within the first third of the 21st Century, how will Martian geology, human access, exploration approach, and crew characteristics compare with the experience of Apollo? (Schmitt, 1973).

2. GEOLOGY

"GEOLOGY," LITERALLY "EARTH STUDY," as a scientific discipline encompasses all interacting aspects of a planet whether or not they fall into sub-disciplines such as geophysics, geochemistry, geomorphology, geobiology, planetology, etc. This particularly holds true during geological field exploration when observations, mapping, and sampling must integrate all discernable past and present influences that have produced the features being examined. Extremely complex processes affect geological features on Earth due to interactions of the dynamics of the crust, water, atmosphere; impact of objects in and from space; and alterations by the biosphere, including humans.

On the Moon, the influences in the last 3.8 billion years largely have been external, confined to the effects of impacts of objects from space and of energetic particles that constitute the solar wind.

On Mars, we could expect influences that combined those affecting the Earth and Moon as the Red Planet is a body intermediate in size between these two terrestrial planet extremes. Indeed, our growing geological knowledge confirms this as we analyze images and data collected from Mars orbit and from on the ground exploration by tele-robotic landers. Since the first photographs provided by the cameras on Viking landers and orbiters (NASA, 2010b), we have known that geological features formed on Mars resulted from combinations of internal and external processes. For example, remote sensing of Martian landforms earlyon indicated moving water had formed many features and that water-ice lay just below the surface across much of the planet and at the surface as polar caps (Carr, 1981).

Orbital sensors and robotic analyses of Martian minerals subsequently have identified a variety of water-containing clays, sulfate minerals precipitated from briny solutions, ice in the regolith, and subsurface hydrogen that is presumably water-ice (NASA, 2010c). Further, unlike the Moon where un-oxidized iron metal is stable in rocks and surface materials, extensive hematite (Fe_2O_3) deposits have been identified by the Mars Exploration Rover*Opportunity*. This means that the Martian geologist must be prepared to interpret the implications of a much larger spectrum of rock-forming and alteration minerals than we have encountered on the Moon (See Gornitz, 2008).

2.1. Effect of Atmosphere: Mars has a thin atmosphere, now only about one percent the pressure of that around the Earth. The existence of this thin atmosphere around Mars results in major differences in the geological overprint that explorers evaluate and "look" through to identify, analyze, and understand the underlying rock units. On the Moon, the absence of an atmosphere exposes surface materials totally to the extraordinarily hard vacuum of space (~10-12 torr). Meteors and comets down to dust size, traveling at tens of kilometers per second, impact and modify the rocks and previously formed broken rock, glass, and dust at the lunar surface. This dominating impact process has produced a several meter deep covering of fragmental and partially glassy debris called the "lunar regolith" that covers most older volcanic flows and impact-generated formations. Only on steep slopes can the explorer find actual bedrock.

The impact-generated glass and glass coatings on other material in the lunar regolith contain extremely small (nano-phase) particles of iron, making the lunar dust strongly attracted to magnets. This property of the dust will be very important in keeping air, space suits, equipment, and dwellings clean. Of great potential economic importance in space and on Earth (See Schmitt, 2006), solar wind ions of hydrogen, helium-3 and -4, carbon and nitrogen, streaming from the fusion reactions in the sun, accumulate in the lunar regolith due to the absence of atmosphere.

Field exploration on the Moon requires that a geologist have "x-ray" vision, of a sort. To identify contacts between major rock units, the geologist must visualize how the gradual formation and spreading of regolith by impacts has broadened and subdued the contrasts in color and texture that existed when the contact was created. For example, in the valley of Taurus-Littrow explored on Apollo 17 (Schmitt, 1973), the surface expression of an originally sharp contact between dark, fine-grained basalt flows and older, gray fragmental rocks (impact breccias) had been spread over a few hundred meters in 3.8 billon years of lunar history. On the other hand, a similarly sharp contact between a dust avalanche deposit and the dark regolith overlying that same basalt flow had spread only a few tens of meters in the 100 million years since the avalanche occurred. Still, the position of the contacts could be defined if one understood the processes actively modifying them.

Field identification of different rock types within exposed boulders on the lunar surface, however, required understanding of the effects of continuous and ubiquitous micrometeorite bombardment. Such bombardment by extremely high velocity particles creates a high temperature plasma as well as impact melt at the point of impact. This ejected plasma and glass re-deposits on nearby surfaces and produces a thin, brownish, glassy patina or coating over the entire boulder. Like looking through the desert varnish on exposed rocks in the Earth's dry regions, the lunar geologist must quickly scan and interpret what is underneath this patina until fresh rock can be chipped or broken with a hammer. Small impact pits interrupt the lunar patina and contain impact glass of varying colors in the pit, reflecting the variations in the chemical composition of underlying minerals. Where the pit has formed on a white mineral (plagioclase feldspar), a distinctive white spot interrupts the brown patina due to the very fine spider cracks in that mineral with translucent, milky glass at the point of impact. Where a magnesium and iron-rich mineral like olivine has been hit, the glass is green.

In contrast to the Moon, the Martian atmosphere has filtered out all meteors and comets capable of forming craters less than about 30m in diameter. On the other hand, the *Spirit* Exploration Rover discovered iron meteorites on the surface as a consequence of the braking effect of even that thin atmosphere. At latitudes greater than about 40 degrees, water derived from subsurface ice appears to have mixed with rock debris in the ejecta blankets from impact craters to form mudflows that leave lobate edges on these blankets.

Additionally, and unlike the continuous redistribution of impact-generated debris on the Moon, wind-blown dust has dominated migrating material on Mars, probably for most of the last 3.8 billion years. This dust forms by impacts, wind erosion, and reactivation of dust from earlier periods of active water erosion and mineral and glass alteration. Martian dust migrates on a global scale and accumulates non-uniformly as dunes, layers and coatings on older craters and crater ejecta, volcanic flows and deposits, and previous dust accumulations. Some dust dunes will be very soft and may need to be avoided by explorers much like deep, wind-formed snow drifts in the plains and mountain passes of Earth.

The "Martian regolith" thus generally consists of impact ejecta, hydrolyzed debris flows, and flood deposits interstratified with windblown dust deposits. In polarregions, water-ice and carbon dioxide ice and frosts also will be present in this regolith as recently confirmed by the Phoenix lander (LPL, 2010).

As a consequence of a very different geological history, new challenges will face the Martian field geologist. A field geologist's "x-ray" vision will still be required; however, it will be more like that required on Earth where one must take into account the effect of wind or water-transported materials that may obscure or cover underlying rock contacts and structures. Rock surfaces will not have a glassy patina as on the Moon; however, fine, wind-blown dust appears to form a very thin, patina-like coating on many rocks. On the other hand, wind frequently cleans surfaces so that dust coatings do not appear to be a significant problem to rock and mineral identification.

Under-estimation of distances is one effect of the vacuum atmosphere of the Moon that will probably be the same on Mars. Estimating distances is difficult, even on Earth, in clear space in the absence of familiar objects such as houses, trees, bushes, power poles, and the like. From the Neil Armstrong's first observations after landing Apollo 11 on the Moon, we knew that crews would underestimate the distance to lunar features, much like human experiences in the clear air of Earth's deserts and high mountains. This problem can be solved for near and mid-field distances by comparing the known length of one's shadow to what it seems to be and increasing the estimated distance by a factor of about 50 percent. This technique seemed to work quite well for the author.

Although daytime lighting will be a little more diffuse on Mars due to light scattering by atmospheric dust, down-sun back-scatter from dust probably will resemble the intense back-scatter we experienced on the Moon. This is the same phenomenon that seen looking toward one's shadow on snow or when flying over a leafy forest or cropland. In contrast, looking up-sun will be looking into concentrations of shadows; however, back-scattered light will provide significant light into shadows, as can selfdirected reflections off the space suit. These lighting characteristics affect adjustments of camera apertures; but I would hope, that future exploration cameras and video systems would automatically adjust to lighting conditions, unlike the cameras we used during Apollo. Adjusting the f-stop relative to sun-line was necessary for nearly every photograph we took and added to other inefficiencies resulting from the mobility limitations of the space suit and gloves.

2.2. Near-Surface Geological Fabric: In spite of the filtering effect of the Martian atmosphere, impact-related geology dominates the surface and near surface fabric of most exposed Martian formations except those exposed in the walls of rift valleys. In many of those valleys, as well as throughout other regions, layered rocks resembling sedimentary or volcanic strata dominate. Nonethe- less, ejecta, fractures, shock alteration, and remobilized volatiles related to impacts form the primary fabric of the rocks that must be deciphered by the first geologists. Absence of a continuous cover of impact-generated regolith, however, means that many outcrops of underlying Martian bedrock formations will be accessible for normal geological examination and sampling, as images from both Mars Exploration Rovers have documented. Ejecta blankets within a crater diameter of impact craters provide a broken and overturned version of the sequence of layers penetrated by the impact. To a significant level, inaccessible crater walls can be examined and sampled on traverses radial to such craters.

3. ACCESS

FOR THE APOLLO ASTRONAUTS, the Moon was only three and a half days away. Mars, using conventional chemical rockets, is eight to nine months away at best. EVEN using an advanced Helium-3 fusion rocket that allows continuous acceleration and deceleration, Mars will be three or four months away. Fortunately, we have the Moon to provide critical aid in an inherently difficult task. The Moon represents the most efficient and lowest risk path to Mars. It provides the opportunity for systems verification, operational planning, crew training, settlement management, and gathering essential resources, whether hydrogen, oxygen, water, food, or helium-3. The development of helium-3 fusion power for consumption on Earth even can support much of the development costs of heavy lift launch and interplanetary fusion rockets. (Schmitt, 2006).

Once in orbit around Mars, there will be many challenges in entry, descent and landing of large crewed spacecraft (See Manning, 2007). It is currently estimated that to land a crew on Mars will require a mass of forty to sixty metric tonnes at entry into the atmosphere, or more than ten times the landed mass thought feasible today. Rockets can be used for the final phases of landing; however, like any crew's return to Earth from space, one would like to use friction with the Martian atmosphere to help slow down before gliding (like a Space Shuttle) or deploying parachutes (like Apollo's return to Earth) or using rocket engines (like Apollo lunar landings).

The thin Martian atmosphere, however, poses more problems than it yet solves for future explorers. Mars entry will require a shield to protect spacecraft from frictional heating. On the other hand, its atmosphere has too low a density to help much with atmospheric braking prior to deployment of any parachutes.

Relative to parachutes, a forty metric tonne crewed spacecraft would require a parachute the size of the Pasadena Rose Bowl to further slow down before using rockets for an actual landing.

Development and test of such a huge parachute and other entry concepts, such as inflatable ballutes and heat shields, pose many problems.

No matter how it is done, entry and descent will require a detailed knowledge of the atmospheric parameters along the spacecraft's actual trajectory – altitude profiles for wind velocities and directions, pressures, and temperatures. Obtaining such critical information may require a precursor entry vehicle to proceed ahead of the actual landing spacecraft and provide these parameters for immediate use by guidance systems. Many high altitude regions of Mars actually may not be accessible except by post-landing use of rovers or flyers. Further, a direct entry, descent and landing sequence would need to be accomplished in about 90 seconds with nearly continuous high g loads, suggesting that some means of prolonged deceleration and flight through the atmosphere will need to be devised.

The atmosphere, thin as it is, also makes it difficult to use rockets as an alternative to parachutes for descent as the rockets would need to thrust into a hypersonic airflow. Significant development and testing in the equivalent upper Earth atmosphere will be necessary. Also, such rockets and their propellants would add significantly more mass to the spacecraft launched from Earth. A good guess, with our current state of knowledge, would be that Martian landings will be accomplished by a combination of aerodynamic deceleration for initial entry, rocket and aerodynamically controlled skipping, combined rocket and aerodynamic decent, and a controlled rocket landing. Spacecraft concepts and mission operations both will require designs that assume an "abort to land" if any major system's problems occur during descent. Once on the surface, the problem can be addressed and resolved in contrast to aborting a landing after the cost, time, and risk to get to Mars.

Based on our lunar experience as well as the special circumstances at Mars, the landing spacecraft will probably be a two-stage vehicle, that is, a descent stage and a crewed cabin-ascent stage. For entry, this lander will be protected by an aero-shell that would be jettisoned before the decent rocket is fired. One major departure from Apollo 13 lunar landing operations will be the necessary design requirement that all landing aborts be to the surface, rather than back to orbit, with trouble-shooting and repairs of the ascent vehicle done after landing.

During Earth to Mars transit, the upper stage will serve first as an entry, descent and landing simulator for crew proficiency training during the long Earth to Mars transit period and then as an ascent and rendezvous trainer during the stay on the Martian surface. Because of atmospheric friction during ascent, the upper stage will need to be aerodynamically shaped, unlike the Apollo lunar module's Ascent Stage that operated only in vacuum.

With respect to landing humans on Mars, therefore, the bad news is we do not know how...yet. Bright young men and women, on the other hand, will meet these challenges once we decide to go to Mars. Already ideas are developing on how to do accomplish a Mars landing, how to test those ideas, and how to prepare for the missions. Returning to the Moon with these engineering, operational, and training challenges in mind would help lay the foundations for missions to Mars, and, indeed, may be essential for success.

The time required to reach Mars may create some differences in the psychological environment of Martian exploration versus that of the Moon. A minimum of several months to return home versus a few days might affect some individuals in adverse ways. For example, psychologically, I personally felt very at ease while on the Moon. I attribute this to being both highly motivated and highly trained as well as having very great confidence in the support team on Earth. Although a Mars crew will have to be much more self-reliant than a lunar crew due to physical isolation from Earth and communication limitations, nonetheless, motivation, training, team confidence, and survival instincts will be much the same as working on the Moon. Historically, human explorers have been subjected to much longer separations from home than will early Mars crews.

Psychological issues may not be much of a problem, though there are differing opinions on this (Bishop, 2010; Fiedler & Harrison 2010; Harrison & Fiedler 2010, Suedfeld, 2010). Everyone will be extraordinarily busy with normal spacecraft operation and maintenance activities, scientific tasks, physical conditioning, simulation training for future tasks, continuous updating of the plans for exploration, and many other duties. In fact, if the history of long term Earth-orbit space flight to date is any indication, finding personal time to relax may be the main psychological challenge facing the crew.

Access to Mars, therefore, will require addressing a myriad of complex technological and operational issues. If Americans and their partners are serious about such an effort, as they should be, the most important step is to establish clear and focused objectives and milestones to meet those objectives. Throughout history, Americans and their partners have successfully responded to challenges of this nature and magnitude when given clear goals and competent, courageous leadership. Obvious examples are the Transcontinental Railroad (Ambrose, 2000), the Panama Canal (McCullough, 1977), the Manhattan Project (Kelly, 2007), and Apollo (Cortwright, 1975) as well as victory in two World Wars. In this challenge of taking freedom into deep space, time and dithering are not our friends.

4. EXPLORATION APPROACH

MARTIAN EXPLORATION PROBABLY will begin with what is referred to as "surface rendezvous," that is, a three or four person crew landing in the vicinity of a previously landed, un-crewed habitat and back-up ascent vehicle. This approach also permits the use of a guidance beacon on the habitat to guide the crewed vehicle to a nearby, safe location. A tele-operated rover would provide the opportunity to refine exploration planning prior to crew arrival. It then could be driven from the habitat to the lander for the crew's use in transferring themselves and additional equipment to the habitat. With completion of exploration in the area accessible from the habitat, the rover or rovers can be tele-operated to follow-up on the crewed exploration and then to explore the region between one exploration site and the next. The next crew at another location samples could be collected for examination and sorting for return to Earth on Mars.

The primary constraint on exploration efficiency on Mars, like on the Moon, probably will remain those imposed by having to wear a pressurized space suit as protection against the near vacuum. It remains conceivable that space suit technology will evolve so that the suit glove or its equivalent will approach the dexterity of the human hand and that the suit itself will become as mobile as cross-country ski clothing. Conceivably, robotic field assistants may provide a net increase in exploration efficiency.

Until effective future space suit and robotic technology appears, however, many procedural, equipment, and planning enhancements to exploration efficiency will be required. The Apollo 7LB suit used during the exploration of the valley of Taurus- Littrow in December 1972 was a very good suit, and it allowed us to do a remarkable amount of fieldwork in a very hostile environment. Running at ~6km per hour was easy in this 3.7psi suit. Using a cross-country skiing gait, this speed probably could be maintained at a steady pace for several kilometers if need be. With the tools available, astronauts could take samples, document them photographically, and bag them at a reasonable rate. In about 18 hours of exploration, Apollo 17 collected 250 pounds of mostly well-documented rocks and soils.

Much better leg, waist and arm mobility definitely would be desirable; but the A7LB, worked well. What almost didn't work, or at least created significant fatigue and hand trauma, were the suit gloves. Something must be done about the technology of the gloves when we return to the Moon and go on to Mars. The estimated hand efficiency using the glove was reduced due to forearm fatigue to about ten percent of normal within the first 30 minutes of pressurized activity.

Three, 8-9hr pressurized excursions could be performed using the A7LB suit, including preparation time; but it is not clear how many more would be possible with the hand abrasions and nail damage that the glove caused. One advantage of more efficient cardiovascular circulation in one-sixth gravity, however, is that after an eight-hour rest period, there was no residual muscle soreness. In addition, based on the experience of astronauts constructing the International Space Station, we now know that physical training techniques exist for much superior conditioning of arm muscles for continuous hand exertion.

5. CREW CHARACTERISTICS

INTELLIGENT AND DETAILED PLANNING and knowledgeable management, along with spacecraft and equipment designs that reflect the best available experience and technology, comprise the foundation up on which successful Martian exploration will take place. The crew, on the other hand, ultimately will be the guarantor of success (Bishop, 2010; Fiedler & Harrison 2010; Harrison & Fiedler 2010, Suedfeld, 2010). In addition to being fully prepared to contribute specially skills to an integrated team, each member of a Martian crew must be fully and unequivocally comfortable and compatible with a hierarchical command structure and have absolute confidence in their leader. Successes and failures in human history clearly show that such a command structure is required during long-term isolation of small teams of explorers.

The optimum selection of crews for the field exploration of Mars can benefit from the experience of Apollo and from the new experiences that will be inherent in the planned return to the Moon early in the 21st Century. Specifically, initial crews of Martian explorers should be a mix of professional pilots, field geologists and biologists, physicians and engineers, all cross-trained in each other's skill areas. The optimum crew size for early exploration appears to be four – two professional pilots cross-trained as field explorers and systems engineers as was done for Apollo lunar crews; one professional field geologist cross-trained both as a pilot and a field biologist; and one professional field biologist cross-trained as a physician and field geologist. If permitted by mass and other operational constraints, two such crews should be on each early Mars mission so that one crew can act as an orbiting "mission control center" for the surface crew with the roles reversed when the second crew lands at a different site. The extra lander necessary to implement such a strategy also provides important redundancy that helps to insure overall mission success.

The political urgency and test flight nature of early Apollo planning and development left few options for selecting experienced field geologists as regular members of lunar mission crews (Hodges and Schmitt, in press). On a more or less theoretical basis, one could argue that more field geologists could have been selected in the mid-1960s, trained as jet and helicopter pilots as was the author, and assigned to Apollo 15 and 16 in addition to the author's Apollo 17 mission. Pragmatically, however, the vast developmental uncertainties in how Apollo would be accomplished led to a crew population consisting mostly of professional test and military pilots with only one pilottrained field geologist. In addition, the necessary design and operational characteristics of the two Apollo spacecraft required that all members of a three person crew needed to be accomplished, experienced, and confident in the use of the machines and methods necessary for flight. Field geologist "passengers" would not be compatible with these requirements. Unfortunately, the selection process for Apollo "scientist astronauts" was not designed by the National Academy of Sciences or NASA to maximize the number of field geologists sent to pilot training and then fully integrated into the mission crew selection pool.

Beginning with precursor return missions to the Moon, professional field explorers should be part of every four-person crew sent to the Moon so that this paradigm can be ingrained in planning for Mars' exploration as well as providing much enhanced returns from lunar exploration. Also ingrained in all crew members and their operational support teams should be a commitment to as much possible terrestrial, field-based training on real geological problems constrained by realistic operational planning, equipment, communications, and timelines. This commitment governed the exploration training for most Apollo missions to great benefit and should be modernized for future lunar and Martian missions. (Schmitt, et al, in press).

The author highly recommends that all lunar and Martian crew personnel be trained jet and helicopter pilots. No substitute exists for the psychological preparation for solving real problems in complex spacecraft that comes with all members of an astronaut crew being professional pilots whether originally a test pilot or scientist or engineer.

6. CONCLUSIONS

AS THE LUNAR EXPLORATION experience matures during the 21st Century, thought must be given to the major ways that lunar and Martian exploration will differ. For example:

1) The trip to Mars will be measured in months rather than days and during that time crew training, mission planning, and physiological countermeasures must be aggressively pursued. For example, the spacecraft for landing and return to Earth must incorporate the capability to be simulators for proficiency training as well as being functional spacecraft. For Apollo missions, the

last computerized training simulation for landing on the Moon took place less than a week before we actually began powered descent to a landing site. Without an in-flight simulation capability, that gap could be on the order of nine months for Mars trips; clearly too long and thus the need for using the spacecraft for regular training activity. Similarly, regular training for ascent from Mars, Mars orbit rendezvous and departure, and Earth entry on return all would be necessary.

2) The basic approach to landing will be to rendezvous with a previously landed, uncrewed vehicle that has most of what will be needed for the exploration mission, including habitat, rovers, field and analysis equipment, and a backup ascent vehicle.

3) The Earth will not perform the traditional "mission control" functions due to the long delays in communications. The Earth will become "that great data processor in the sky," participating in exploration data analysis and synthesis, weekly planning, systems and consumables monitoring and analysis, maintenance requirements, ascent simulations scenario development and critique, and other activities where live, that is, "real time" interaction with the crew is not required.

    The actual live mission control functions will need to be carried out by an orbiting crew and/or by half the landed crew during alternating non-excursion days. Although we planned the Apollo lunar exploration activities to the degree possible using available photographs, we left significant latitude to the crews to pursue unanticipated targets of opportunity. We pursued that latitude probably to the limit after discovery of the orange volcanic glass in the rim of Shorty Crater with only 30 minutes of time available for observation and sampling. No time was available to discuss this plan with Mission Control, but we knew immediately what needed to be done. Exactly this approach will be required of the crews on Mars, but implemented continuously, with Mission Control on Earth finding out everything tens of minutes after the fact.

4) In light of the expense and historical importance of each Mars' exploration mission, mission philosophy must be totally success oriented. For example, ideally, two landers should be available in

case one cannot be used. Two landers also would allow two separate sites to be explored if both are deemed functional. Further, systems or software anomalies observed during the entry, descent and landing sequence should be resolved by having designed to an "abort to land" rule rather than an abort-to-orbit rule as existed during Apollo landing sequences. Solutions to such problems can be resolved over time and in consultation with the Earth once the crew lands safely. This abort-to-land concept, however, will impact the design and level of redundancy of the landing craft.

5) Crews should be comprised of exceptionally expert and broadly experienced professionals, cross-trained in each other's specialty so that mission success depends not on any one individual but on enhanced mutual capabilities. Professional and training experience in geological field exploration will constitute an essential component of each crew as a whole.

Young people now alive could have the privilege and adventure of settling the Moon, using its resources, studying its contribution to the history of the Earth and solar system, then exploring and settling Mar - if their parents and grandparents provide the opportunity for them to do so. Parents also must make possible that their children have the education and character necessary to do great things.

Mars exploration will not be easy. As with anything worthwhile, risks exist. But the alternative of postponing settlement of the Moon and the exploration of Mars beyond existing generations would leave Moon and the exploration of Mars beyond existing generations would leave the future to other explorers. Further, without the capability to go back to the Moon and on to Mars and the capability to work in deep space leaves all of humankind vulnerable to the impact on Earth of asteroidal and cometary travelers of the solar system. We have little choice but to continue what Americans began on July 20, 1969, and paused after placing the last men on the moon in December 14, 1972.

REFERENCES

Ambrose,, S. E. (1996). Undaunted Courage. Simon and Schuster, NewYork, US.

Ambrose, S. E. (2000). Nothing Like It In The World. Simon & Schuster, New York, US.

Bishop, S. L. (2010). Moving to Mars: There and back again, Journal of Cosmology, 12, 3711-3722.

Carr, M. H. (1981). The Surface of Mars. Yale University Press, New Haven, US.

Charpentier, J. (1778). Mineralogische Geographie der Chursachsischen Lande. Crusiius, Leipzig, Germany.

Cortwright, E. M. (1975). (Ed.) Apollo Expeditions to the Moon. NASA Special Publication 350, U. S. Government Printing Office, Washington DC, US.

Christensen, P. (2005). The many faces of Mars. Scientific American, April.

Fiedler, E. R., & Harrison, A. A. (2010). Psychosocial Adaptation to a Mars Mission, Journal of Cosmology, 12, 3685-3693.

Gornitz, V. (2008). Decoding the mineral history of Mars. Mineral News, pp.12-13.

Harrison, A. A., Fiedler, E. R. (2010). Mars, Human Factors and Behavioral Health, Journal of Cosmology, 12, 3694-3710.

Hodges, K. V., Schmitt, H. H. (2011). A New Paradigm for Advanced Planetary Field Geology Developed Through Analog Exploration Exercises on Earth. In: G. Bryant, editor, Analogs for Planetary Exploration, Special Paper, Geological Society of America, Denver, US, in press.

Kelly, C. C., Editor, (2007). The Manhattan Project. Black Dog & Leventhal, New York, US.

LPL (2010) Uncovering the mysteries of the Martian arctic, http://phoenix.lpl.arizona.edu/mission.php

Manning, R. (2007). Interview by N. Atkinson. http://www.universetoday.com/2007/07/17/the-mars-landin-approach-getting-largepayloads- to-the-surface-of-the-red-planet/.

McCullough, D. (1977). Path Between the Seas. Simon & Schuster, New York, US. NASA (2010a). NASA's Mars exploration program. http://marsprogram.jpl.nasa.gov/science/geology/

NASA (2010b). NASA marks 35th Anniversary of Mars Viking Mission. http://www.nasa.gov/mission_pages/viking/.

NASA (2010c) Found it! Ice on Mars. http://science.nasa.gov/science-news/science-atnasa/ 2002/28may_marsice/

Schmitt, H. H., Trask, N. J., Shoemaker, E. M. (1967). Geological Map of the Copernicus Quadrangle of the Moon. USGS Geological Atlas of the Moon, 1-515, LAC 58, Geological Survey, Washington, D.C, US.

Schmitt, H. H. (1973). Apollo 17 Report on the valley of Taurus-Littrow. Science, 182, 681-690.

Schmitt, H. H. (1973). Exploring Taurus-Littrow. National Geographic Magazine, 144, September 1973, pp. 290-307.

Schmitt, H. H., (2003). Apollo 17 and the Moon. In: Mark, H. (Ed.), Encylopedia of Space and Space Technology. Wiley, New York, US.

Schmitt, H. H. (2006). Return to the Moon. Copernicus-Praxis, New York, US.

Schmitt, H. H., Snoke, A. W., Helper, M. A., Hurtado, J. M., Hodges, K. V., Rice, J. W., Jr. (2011). Motives, Methods, and Essential Preparation for Planetary Field Geology on the Moon and Mars. In: G. Bryant, editor, Analogs for Planetary Exploration, Special Paper, Geological Society of America, Denver, US, in press.

Smith, W. (1815). A delineation of the strata of England and Wales, with part of Scotland. Cary, London, UK.

Suedfeld, P. (2010). Mars: anticipating the next great exploration, Journal of Cosmology, 12, 3723-3740.

Wolfe, E., and 10 colleagues (1981). The Geologic Investigation of the Taurus-Littrow Valley: Apollo 17 Landing Site. Professional Paper 1080, Geological Survey, Washington, DC, US.

III

TO BOLDLY GO: ONWARD TO MARS

Journal of Cosmology, 2010, Vol 12, 3566-3577.
JournalofCosmology.com, October-November, 2010

- 5 -
Acceptable Risk:
The Human Mission to Mars

Jack Stuster, PhD, CPE,
Anacapa Sciences, Inc., Santa Barbara, California

ABSTRACT

This paper examines some of the issues involved in estimating the risks associated with the human exploration of Mars. A brief description of the relevant orbital mechanics is presented in the context of historical Mars mission planning and is followed by a discussion of some of the factors that contribute to expedition risk. The paper concludes with a summary of implications and a recommendation.

Key Words: Mars, Human Exploration, Risk, Exploration of Mars, History

1. ORBITAL MECHANICS

The planets in our solar system are bound to the sun by gravity in elliptical orbits, a discovery that ranks among the most revolutionary in the history of science and instantly rendered the study of calculus useful for all time. However, it is sufficient to this discussion to understand that the Earth and Mars follow orbits that place them in the same relative positions every 26 months and that the absolute distances between the two planets follow a 15-year cycle. At their closest points during the 26-month cycle Mars appears bright red-orange in the night sky; at the closest oppositions during the 15-year cycle, Mars appears twice as large from Earth as it does when it is farthest. Optimum launch opportunities are defined by the lowest possible mass of the spacecraft that must be propelled, because the greater the mass the more fuel is needed, which further increases the mass and cost of the expedition. The mass/energy requirements follow the 26-month cycles within the 15-year cycles; energy requirements within the 26-

month cycle vary by 60 percent and within the 15-year cycle by a factor of ten. The differences are huge and, essentially, define when spacecraft can be launched to intercept Mars.

The physics, math, and engineering necessary to identify the "optimum" scenario for an expedition to Mars were understood by Werner von Braun who included the calculations in an appendix to a science fiction story he wrote to counter boredom while serving the nascent U.S. rocket program in 1947-48; the austere life among the German scientists interned at White Sands, New Mexico, was similar in some ways to what might be expected on a planetary expedition. The manuscript was apparently unremarkable, but von Braun used the technical appendix as the basis of a lecture he gave at the First Symposium on Spaceflight held at the Hayden Planetarium in New York City in 1951. The appendix was published in a special edition of the German journal *Weltraumfahrt* in 1952 and later that year as a book, titled *Das Marsprojekt*. It was translated into English and published in the United States in 1953.

Von Braun's plan involved 70 crewmembers and ten 4,000-ton ships that would be assembled in low-Earth orbit from parts launched by three-stage winged ferry rockets; a staggering 950 launches would be required to lift the components, supplies, and fuel out of Earth's gravity well and to support assembly of the fleet in orbit. Seven of the ten interplanetary ships would resemble tinker toys made of girders and spheres and lack the streamlining necessary for a planetary landing, but three would have bullet-shaped fuselages equipped with wings to glide through the thin atmosphere of Mars. Rocket engines would propel the fleet on a minimum-energy Earth-to-Mars trajectory with crews discarding the empty fuel tanks during the eight-month weightless cruise phase. Rockets would be fired again to slow the fleet for insertion into Mars orbit. After surveying the planet for suitable sites, one of the winged ships would be dispatched to land on skids at the polar ice cap that is seasonally visible through telescopes from Earth. The crew would then make an arduous 4,000-mile mechanized traverse to build a landing strip near the equator for the two remaining winged spacecraft to land on wheels. The wings would be removed and the fuselages elevated to a vertical position in preparation for launch, rendezvous, and return to Earth with the ships remaining in orbit, many months in the future. Von Braun's plan, which is called a conjunction class mission due to the relative positions of the planets at launch, required the least energy (mass and cost) of all launch options, but would subject the crews and equipment to eight-month transits and 16 months on the surface, for a mission duration of nearly three years. Von Braun believed the journey would be possible by the mid-point of the 21st Century.

The Mars Project attracted the attention of author, Cornelius Ryan, who served on the editorial staff of *Collier's*, a weekly magazine with nearly three-million subscribers and a tradition of shaping public opinion and government policy. Ryan commissioned von Braun and other leading space scientists, writers, and astronomical artists of the period to prepare a series of articles based on *The Mars Project*. The series, titled "Man Will Conquer Space Soon!" was published in eight, beautifully-illustrated installments between March 1952 and April 1954. The primary difference between von Braun's original plan and the one described in the magazine series was the addition of a toroidal space station in Earth orbit to facilitate assembly of the interplanetary ships. The donut-shaped structure became the archetypal space station form in popular culture and the articles propelled von Braun to national prominence and fueled the collective imagination of post-war America. Von Braun and Willy Ley promptly produced four books that expanded on the topics covered in the *Collier's* series and included a revised plan for an expedition to Mars that involved 12 crew members, two ships, and "only" 400 launches of fuel, supplies, and components for assembly in orbit. Walt Disney and producer Ward Kimball were among those inspired by von Braun's plans and hired him and others to help develop three episodes for the wildly popular *Disneyland* television program. "Man in Space" was broadcast in March 1955 to an audience of more than 40 million viewers, including President Dwight Eisenhower who called Walt Disney the next day to request a copy of the program that could be shown to key Pentagon officials (Smith, 1978). "Man and the Moon" aired in December 1955 and, like the previous episode, used documentary footage, on-screen appearances by von Braun and others, and narrated animation to provide remarkably accurate predictions of future events; the programs described the likely effects of weightlessness on humans and introduced the public to a new field of study, called Space Medicine. The third episode in the series, "Mars and Beyond," featured ships with solar-powered ion engines suggested by another German scientist, Ernst Stuhlinger, rather than von Braun's chemical rockets; the program was broadcast on 4 December 1957, two months after the Soviet Union shocked the world with the launch of Sputnik, the first artificial satellite.

The technical appendix to von Braun's unsuccessful attempt at creative writing and the subsequent lectures, magazine articles, books, television episodes, and of course Sputnik, led directly to the creation of NASA in 1958. The agency's initial focus was on matching the Soviet's accomplishments and then

FIGURE 1
Exploring Mars by Chesley Bonestell, 1956.

shifted to landing humans on the Moon in response to President Kennedy's famous directive. However, Mars remained the ultimate goal of the German-born scientists, and others who they influenced, despite their immediate concerns. For example, Stuhlinger, who had worked with von Braun during the war and served as director of the space science lab at the Marshall Space Flight Center, proposed a new approach to Mars in 1962 (Portree, 2001). The plan involved five ships, each with a crew of three astronauts, and paid greater attention to reliability and human factors issues than previous mission plans. In particular, the spacecraft were designed to spin to provide acceleration equal to one-tenth of Earth's gravity to mitigate the negative effects of weightlessness on the crew; three of the ships would each carry a 70-ton Mars lander, which provided the triple redundancy that became NASA's standard for ensuring reliability; and all 15 members of the expedition could return to Earth in a single ship if necessary. The most notable consideration in this regard was Stuhlinger's selection of an "opposition class" trajectory, which requires more energy and longer transits to and from Mars than von Braun's conjunction class plan, but a much shorter surface stay and overall mission duration. In other words, an opposition class plan might cost more to implement than a conjunction class mission, but would subject equipment and humans to less risk due to substantially shorter exposures. Stuhlinger argued that it was better to use more fuel, which would incur greater cost, in order to minimize the risk (i.e., increase the probability of a successful outcome). The debate continues to this day.

2. ESTIMATING RISK

Risk is generally understood to be the quantifiable likelihood of loss or less-than-expected results. The concept is something with which we all are familiar. Every decision we make from the most trivial to the most important is attended by some sort of evaluation and consideration of the costs, benefits, likelihood of a successful outcome, and possible negative consequences. Humans are not particularly good at estimating risk, despite our extensive personal experience with the task. Research shows that we have a tendency to underestimate risk in circumstances where we have some control, and to overrate risk when we have little or no control. This is why we tend to be more fearful of flying in a commercial airliner than driving an automobile, even though the odds of dying on the road are 16 times greater than dying in a plane. Riding a motorcycle, which carries a risk of fatality 17 times that of driving a car, more clearly illustrates the influence of perceived control and other subjective factors when making decisions concerning acceptable risk.

Deciding whether or how to send a human expedition to Mars is more complicated than deciding if one should drive, ride, or fly on vacation, but the process is essentially the same: estimate costs, benefits, and probabilities of various outcomes, and then compare the benefits that would result from success to the consequences of failure while simultaneously comparing the probabilities of the outcomes. This approach produces objective recommendations, but requires quantification of all variables. When considering the risks of an expedition to Mars it is possible to calculate the probability of mission-threatening solar activity from historical records and to identify the failure rates of all mechanical components through testing. Even the expected incidences of medical and behavioral problems can be inferred from space analog experience, as illustrated by the following tables, which are derived from a spreadsheet that was configured by the current author to calculate some of the risks and incidence rates that might be expected on Mars expeditions, based on data from Antarctic research stations.

The tables compare conjunction and opposition class expedition plans with the durations of mission segments listed on the row labeled Days. The values in the row labeled Behavioral Problem assume a 3.3% per year incidence rate of serious behavioral problems throughout the durations of the two mission options (i.e., Conjunction Class/Long Stay, 905 days total; and Opposition Class/Short Stay, 661 days total). The row labeled Differential assumes a 3.3% incidence rate during the interplanetary transit phases and a 2% rate while on the surface of Mars, when confinement would probably be less of a factor and other stressors might be offset by the novelty of task performance. A serious behavioral problem was defined as symptoms that normally would require hospitalization and the assumptions were based on incidence rates of behavioral problems reported from Antarctic experience (Matusov, 1968; Gunderson, 1968; Lugg, 1977; Rivolier and Bachelard, 1988).

TABLE 1

Risk Factor/Definition	Days Incidence per 365 days	Conjunction Class				
		Outbound 180	Surface 545	Return 180	Days 905 Long Stay Risk	Expect In Crew of 6
Behavioral Problem	0.033	0.016	0.049	0.016	0.082	0.491
Differential	0.020	0.016	0.030	0.016	0.062	0.374

Opposition Class				
Outbound 313	Surface 40	Return 308	Days 661 Short Stay Risk	Expect In Crew of 6
0.028	0.004	0.028	0.060	0.359
0.028	0.002	0.028	0.058	0.350

More recent incidence rates (Otto, personal communication 2008) suggest that the initial estimates might be low. Substituting 6% for the 3.3% incidence rate during transits causes the expected occurrence of a behavioral problem in a crew of six to increase from .374 to .534 for the long stay option and from .350 to .626 for the short stay option. That is, if the incidence rate of behavioral problems while on the surface were to be one-third to one-half of the rate during transit, the probability of a serious problem occurring becomes greater for the short stay option, due to the substantially longer time that must be spent by the crew confined to the spacecraft than in the long stay option. However, the long stay option generates a higher probability if the incidence rate were to remain constant throughout the mission. A uniform .06 incidence rate would increase the probability of a serious behavioral problem to nearly 90 percent for the Conjunction Class, long stay option.

Similar calculations were made to predict the incidence of physical injury, again based on Antarctic experience, but reduced by half to accommodate likely differences in tasks and protective equipment during a Mars expedition. The table

TABLE 2

Risk Factor/Definition	Days Incidence per 365 days	Conjunction Class				
		Outbound 180	Surface 545	Return 180	Days 905 Long Stay Risk	Expect In Crew of 6
Behavioral Problem	0.060	0.030	0.090	0.030	0.149	0.893
Differential	0.020	0.030	0.030	0.030	0.089	0.534

Opposition Class				
Outbound 313	Surface 40	Return 308	Days 661 Short Stay Risk	Expect In Crew of 6
0.051	0.007	0.051	0.109	0.652
0.051	0.002	0.051	0.104	0.626

shows that a crew of six should expect four injuries that prevent task performance at least temporarily during the long stay option and one during the short stay. Injuries are more likely to occur while working on the surface of the planet than in transit and probably will involve trauma to the hands, based on analog experience.

TABLE 3

Risk Factor/Definition	Days Incidence per 365 days	Conjunction Class				
		Outbound 180	Surface 545	Return 180	Days 905 Long Stay Risk	Expect In Crew of 6
Physical Injury	0.800	0.395	1.195	0.395	1.984	11.901
Adjusted	0.400	0.197	0.597	0.197	0.992	5.951
Differential	0.100	0.049	0.597	0.049	0.696	4.175

Opposition Class				
Outbound 313	Surface 40	Return 308	Days 661 Short Stay Risk	Expect In Crew of 6
0.686	0.088	0.675	1.449	8.693
0.343	0.044	0.338	0.724	4.346
0.086	0.044	0.084	0.214	1.284

It is reasonable to question whether incidence rates for behavioral problems and physical injuries from Antarctic research stations should be used to estimate what might occur during a Mars expedition. Astronauts are subjected to more rigorous selection procedures than Antarctic personnel. On the other hand, Antarctic personnel spend a maximum of 12 months "on the ice," while members of an opposition or conjunction class expedition to Mars, such as those described here, would experience greater isolation and confinement for durations that are nearly two to three times as long.

The risks and likely incidence rates reported here would be recalculated by mission planners to ensure that the most appropriate data are used and that all issues that might influence differential rates are considered. Other risks would, of course, be calculated using relevant data. Although preliminary, the exercise shows that risks can be estimated mathematically. However, the possible *benefits* of an expedition are more difficult to predict, because the subject, by definition, involves the unknown.

3. ESTIMATING BENEFITS

The decision process requires an estimate of the benefits that would derive from a successful expedition outcome. Estimating benefits and defining success in the context of the unknown are fundamentally subjective tasks, the products of which are determined by the explicit and implicit purposes of the expedition. This raises the question, why do nations and individuals explore? A partial list of historical reasons includes, searching for trade routes; surveying to facilitate commercial and/or military navigation; prospecting for new resources; enhancing national or individual prestige; and, contributing to science, which is the most recent addition to the reasons for exploration.

Calculating the likely benefits of exploration is made difficult because some benefits are subjective, intangible, and/or devoid of practical application. For example, Roald Amundsen was the first to reach the South Pole and the first to navigate the Northwest Passage, which contributed to notoriety for Amundsen and national pride for Norway, but neither accomplishment produced tangible benefits. Further, many of the discoveries made by explorers of the past were completely unexpected. The most notable example is Columbus, who was searching for Asia when he landed in the New World.

The polar explorers, whose expeditions usually were funded privately through subscription and the proceeds from book sales and lecture fees, were criticized for risking human lives. The primary criticism of space exploration, which is sponsored by governments, is the belief that there are more pressing social and economic issues that deserve government attention and resources. However, it is important to note that the relatively meager budgets of the world's space agencies contribute substantively to medical, technological, and economic development in addition to achieving explicit scientific objectives. Space exploration also inspires and enriches us subjectively.

It is impossible to predict all likely benefits of an expedition to Mars, but one of the most practical will be an effective countermeasure to bone demineralization, which must be discovered on Earth *before* any humans depart on a long voyage to the Red Planet. Muscles atrophy and bones become porous from disuse in the absence of gravity's normal influences. Two or more hours of strenuous exercise each day will prevent space explorers from being as weak as kittens when they arrive at their destination and specific resistive exercise appears to mitigate bone demineralization. However, exercise alone will not save interplanetary explorers from developing fragile bones. Something else is needed or there will be no expedition to Mars and, for this reason, scores of scientists are working on possible solutions. One or more of the treatments eventually will become available to the general public, which will remove the fear of premature death from a broken bone from the aging process. The cost to North American society resulting from broken hips alone each year is approximately equal to NASA's annual budget.

The potential benefits that will result from developing a countermeasure to bone demineralization are conveniently quantifiable, but most benefits of Mars exploration, other than creating jobs and fostering innovation and the development of technical and scientific skills, will be intangible. In particular, what metric can be used to calculate the value of knowing more about another planet in our solar system, or that life did or did not exist during an earlier epoch on Mars? Identifying all of the likely benefits in order to obtain support or otherwise justify expeditions can be so problematic that some polar explorers gave up in exasperation. For example, Fridtjof Nansen wrote:

> People perhaps still exist who believe that it is of no importance to explore the unknown regions. This, of course, shows ignorance. The history of the human race is a continual struggle from darkness toward light. It is therefore to no purpose to discuss the use of knowledge. Man wants to know, and when he ceases to do so, he is no longer man. (Quoted in Cherry-Garrard, 1930, p. 348.)

Roald Amundsen responded more bluntly to his critics: "Little minds have room only for thoughts of bread and butter."

4. MINIMIZING RISK

Muscle atrophy, bone demineralization, and radiation loading require development of countermeasures to preserve health and performance of crew during long-duration space exploration; the risk of over-exposure to harmful radiation is the least understood of these factors and the one that presents the greatest technical challenges. In contrast, there are four basic design strategies for reducing the risk of component or system failure:

1) Redundancy. Having two spares for every item needed provides protection against an undetected flaw in a primary item and unexpected damage to the replacement. Triple redundancy has been favored by NASA since the Mercury Program and was a primary strategy of previous explorers. Columbus never would have considered departing Spain with fewer than three hulls and probably would not have returned safely had he done so (his flagship sank on Christmas Day 1492). All serious Mars plans also have involved multiple ships for this reason, that is, until recently.

2) Overbuilding. Engineers typically design a structure or system component to withstand a multiple of the maximum stress, load, or pressure that is expected. A 150 percent design rule increases the cost of a retaining wall and the weight of a rocket motor, but the strategy also increases the proba-bility that neither item will fail catastrophically.

3) Graceful degradation. Sudden, catastrophic failure can overwhelm intrinsic precautions and cause a cascade of unexpected negative consequences. Systems should be designed to degrade gradually to allow sufficient time for isolation, replacement, or repair.

4) Maintainability. Systems intended for use in remote and possibly life-threatening situations should be designed in a manner that facilitates repair by human operators and tenders. This strategy includes provisions for accessibility, spare parts, appropriate tools, and procedures/schematics to guide the process.

All four of these strategies should be employed to increase the reliability of mechanical, electrical, and human components of a Mars expedition. However, it must be understood that risk can be mitigated (by countermeasures and risk-mitigation strategies) or reassigned (by insurance, which has limited utility for an explorer when disaster strikes), but it cannot be eliminated without avoiding exposure to the risky conditions.

5. IMPLICATIONS

The primary implication of this discussion concerns the effects of time on human behavior, because time is a factor that can transform almost any issue into a serious problem. As shown by the preceding exercise, attempts to minimize risk by reducing the overall mission duration might actually increase the incidence of behavioral problems if the plan requires longer periods of confinement during transit. Further, the estimates of risk based on Antarctic experience identified tradeoffs between behavioral problems and physical injuries when comparing conjunction and opposition-class expeditions. The optimum solution would be to greatly reduce the time spent in transit by replacing chemical rockets with a faster method of propulsion.

Werner von Braun looked to the Antarctic experiences of his era for guidance when identifying the possible sources of risk for his Mars project, as we have done here. Von Braun concluded,

> I am convinced that we have, or will acquire, the basic knowledge to solve all the physical problems of a flight to mars. But how about the psychological problems? Can a man retain his sanity while cooped up with many other men in a crowded area, perhaps twice the length of your living room, for more than thirty months? ... Little mannerisms—the way a man cracks his knuckles, blows his nose, the way he grins, talks, or gestures— create tension and hatred which could lead to murder. (*Collier's* April 30, 1954 "Can We Get to Mars?" p. 26.)

Such a grim outcome is unlikely based on space analog and previous space exploration experience. However, several expeditions during the heroic and modern periods were jeopardized by the deteriorating mental states of one or more participants. Only these long expeditions of the past come close to the durations currently projected for opposition and conjunction class Mars missions. More experience with planetary-type operations is needed to ensure the reliability of spacecraft and human crew during three years of continuous operation.

Mars has beckoned since the first humans gazed at the night sky and observed its distinctive color and movement. It is our nearest planetary neighbor and has been assumed, since before we reached the Moon, to be our next destination. Mars is a worthy objective for explorers, even if the benefits that might accrue were to be limited to intangible additions to scientific knowledge. But the question must be asked: Are the risks worth the likely benefits at this time? The answer is yes, but there are other options with lower risks and potentially greater benefits.

6. RECOMMENDATION

The Earth and Mars will remain locked in their 15-year cyclical dance until the Sun expands and consumes the inner planets, an event predicted to occur in about five billion years. In other words, Mars does not threaten our existence and will remain available to us for exploration for a very long time. However, there are many known astronomical bodies in our solar system that pose enormous risk to our planet and countless others that have yet to be discovered. The likelihood and potential consequences of an asteroid or comet impacting the Earth were not fully appreciated during the era of initial Mars mission planning, which influenced our expectations, but the many large craters still evident on Earth suggest an on-going threat and recent research has linked such events in the distant past to mass extinctions. At the time of this writing more than 1,000 objects are categorized as Potentially Hazardous Asteroids and astronomers add frequently to the list, with some discoveries made only days before they pass within a lunar distance of Earth.

Six-month expeditions by three-person crews on space ships composed of at least two Orion-like vehicles could explore asteroids in nearby orbits within a few years of deciding to do so, compared to decades to send humans to Mars. The expeditions would gather information about the objects that pose collision threats with the intention of developing methods to protect Earth. The missions also could lead to mining and other exploitation of the asteroids, while at the same time helping to prepare for the human exploration of Mars. Most important, expeditions to asteroids would expose the crews and equipment to risk for durations similar to current tours of duty onboard the International Space Station and the benefits would be practical and possibly Earth-saving information and experience. We should continue making plans to explore Mars and eventually establish a permanent presence; our species is vulnerable and will be less so when no longer limited to one world. We have a subjective attachment to Mars and exploring asteroids should be part of the plan to get there. The destinations are not mutually exclusive, but Mars attracts our attention while the asteroids demand it.

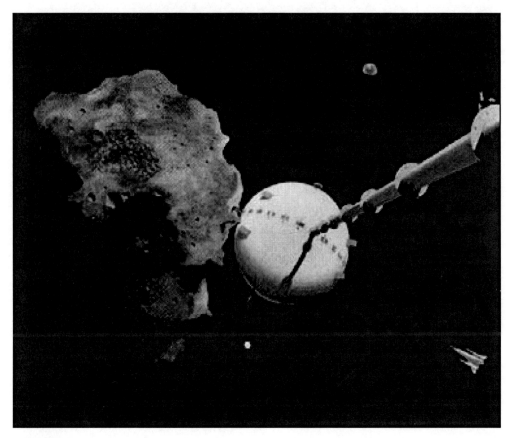

FIGURE 2.
Pittsburgh at L-2 by Chesley Bonestell, 1976.

REFERENCES

Cherry-Garrard, A. (1930). The Worst Journey in the World. New York: Dial Press.

Disney Studios. (1955-1957). Man in Space, Man and the Moon, and Mars and Beyond can be viewed at http://www.youtube.com/watch?v=75vX6O8paGo&fmt=18 and http://kottke.org/08/12/disneys-1955-man-in-space-film.

Gunderson, E.K.E. (1968). Mental health problems in Antarctica. Archives of Environmental Health 17:558-64.

Lugg, D. (1977). Physiological Adaptation and Health of an Expedition in Antarctica with Comment on Behavioral Adaptation. Australian National Antarctic Research Expedition (ANARE) Scientific Report, Series B (4) Number 126. Canberra, Australia: ANARE.

Matusov, A.L. (1968). Morbidity among members of the Tenth Soviet Antarctic Expedition. Soviet Antarctic Expedition 38-256. (Cited in Rivolier and Bachelard, 1988.)

Otto, C. (2008). Personal communication concerning the incidence of behavioral problems at U.S. Antarctic research stations.

Portree, D. S.F. (2001). Humans to Mars: Fifty Years of Mission Planning, 1950–2000, Monographs in Aerospace History Series Number 21. Washington, DC: NASA History Division.

Rivolier, Jean and Claude Bachelard (1988). Studies of analogies between living conditions at an Antarctic scientific base and on a space station. Unpublished manuscript.

Smith, D. R. (1978). They're Following Our Script: Walt Disney's Trip to Tomorrowland. Future, May, p. 55.

von Braun, W. (1953) (Translated by Henry J. White). The Mars Project. Urbana: University of Illinois Press.

Journal of Cosmology, 2010, Vol 12, 3619-3626.
JournalofCosmology.com, October-November, 2010

- 6 -
To Boldly Go: A One-Way Human Mission to Mars

Dirk Schulze-Makuch, Ph.D.[1], and Paul Davies, Ph.D.[2],

[1]School of Earth and Environmental Sciences, Washington State University
[2]Beyond Center, Arizona State University

ABSTRACT

A human mission to Mars is technologically feasible, but hugely expensive requiring enormous financial and political commitments. A creative solution to this dilemma would be a one-way human mission to Mars in place of the manned return mission that remains stuck on the drawing board. Our proposal would cut the costs several fold but at the same time ensure a continuous commitment to the exploration of Mars in particular and space in general. It would also obviate the need for years of rehabilitation for returning astronauts, which would not be an issue if the astronauts were to remain in the low-gravity environment of Mars. We envision that Mars exploration would begin and proceed for a long time on the basis of outbound journeys only. A mission to Mars could use some of the hardware that has been developed for the Moon program. One approach could be to send four astronauts initially, two on each of two space craft, each with a lander and sufficient supplies, to stake a single outpost on Mars. A one-way human mission to Mars would not be a fixed duration project as in the Apollo program, but rather the first step in establishing a permanent human presence on the planet. The astronauts would be re-supplied on a periodic basis from Earth with basic necessities, but otherwise would be expected to become increasingly proficient at harvesting and utilizing resources available on Mars. Eventually the outpost would reach self-sufficiency, and then it could serve as a hub for a greatly expanded colonization program. There are many reasons why a human colony on Mars is a desirable goal, scientifically and politically. The strategy of one-way missions brings this goal within technological and financial feasibility. Nevertheless, attaining it would require not only major international cooperation, but a return to the exploration spirit and risk-taking ethos of the great period of Earth exploration, from Columbus to Amundsen, but which has nowadays been replaced with a culture of safety and political correctness.

Key Words: Mars exploration, Human exploration, Manned Mission, Human Mission, Mars, exploration, lander, bold, risk

1. INTRODUCTION

THE EXPLORATION OF MARS HAS BEEN a priority for the space programs of several nations for decades, yet the prospect of a manned expedition continually recedes in the face of daunting and well-recognized challenges. The long travel time to Mars in zero gravity and high radiation conditions would impose a serious health burden on the astronauts. The costs of developing the launch vehicle and assembling the large amount of equipment needed for the astronauts to survive the journey and their long sojourn on the Martian surface, together with a need to send all the fuel and supplies for a return journey make a manned Mars expedition at least an order of magnitude more expensive than the Apollo program.

In our view, however, many of these human and financial problems would be ameliorated by a one-way mission. It is important to realize that this is not a "suicide mission." The astronauts would go to Mars with the intention of staying for the rest of their lives, as trailblazers of a permanent human Mars colony. They would be resupplied periodically from Earth, and eventually develop some "home grown" industry such as food production and mineral/chemical processing (Zubrin and Baker 1992; Zubrin and Wagner 1997). Their role would be to establish a "base camp" to which more colonists would eventually be sent, and to carry out important scientific and technological projects meanwhile. Of course, the life expectancy of the astronauts would be substantially reduced, but that would also be the case for a return mission. The riskiest part of space exploration is take-off and landing, followed by the exposure to space conditions. Both risk factors would be halved in a one-way mission, and traded for the rigors of life in a cramped and hostile environment away from sophisticated medical equipment. On the financial front, abandoning the need to send the fuel and supplies for the return journey would cut costs dramatically, arguably by about 80 percent. Furthermore, once a Mars base has been established, it would be much easier politically to find the funding for sustaining it over the long term than to mount a hugely expensive return mission.

There are several reasons that motivate the establishment of a permanent Mars colony. We are a vulnerable species living in a part of the galaxy where cosmic events such as major asteroid and comet impacts and supernova explosions pose a significant threat to life on Earth, especially to human life. There are also more immediate threats to our culture, if not our survival as a species. These include global pandemics, nuclear or biological warfare, runaway global warming, sudden ecological collapse and supervolcanoes (Rees 2004). Thus, the colonization of other worlds is a must if the human species is to survive for the long term. The first potential colonization targets would be asteroids, the Moon and Mars. The Moon is the closest object and does provide some shelter (e.g., lava tube caves), but in all other respects falls short compared to the variety of resources available on Mars. The latter is true for asteroids as well. Mars is by far the most promising for sustained colonization and development, because it is similar in many respects to Earth and, crucially, possesses a moderate surface gravity, an atmosphere, abundant water and carbon dioxide, together with a range of essential minerals. Mars is our second closest planetary neighbor (after Venus) and a trip to Mars at the most favorable launch option takes about six months with current chemical rocket technology.

In addition to offering humanity a "lifeboat" in the event of a mega-catastrophe, a Mars colony is attractive for other reasons. Astrobiologists agree that there is a fair probability that Mars hosts, or once hosted, microbial life, perhaps deep beneath the surface (Lederberg and Sagan 1962; Levin 2010; Levin and Straat 1977, 1981; McKay and Stoker 1989; McKay et al. 1996; Baker et al. 2005; Schulze-Makuch et al. 2005, 2008, Darling and Schulze-Makuch 2010; Wierzchos et al. 2010; Mahaney and Dohm 2010). A scientific facility on Mars might therefore be a unique opportunity to study an alien life form and a second evolutionary record, and to develop novel biotechnology therefrom. At the very least, an intensive study of ancient and modern Mars will cast important light on the origin of life on Earth. Mars also conceals a wealth of geological and astronomical data that is almost impossible to access from Earth using robotic probes. A permanent human presence on Mars would open the way to comparative planetology on a scale unimagined by any former generation. In the fullness of time, a Mars base would offer a springboard for human/robotic exploration of the outer solar system and the asteroid belt. Finally, establishing a permanent multicultural and multinational human presence on another world would have major beneficial political and social implications for Earth, and serve as a strong unifying and uplifting theme for all humanity.

2. THE CONCEPT OF A ONE-WAY MISSION TO MARS

A human mission to Mars is undoubtedly technologically feasible, but unlikely to lift off in the very near future, because of the enormous financial and political commitments associated with it. As remarked, however, much of the costs and payload of the mission are associated with bringing the astronauts back to Earth. Furthermore, the returning astronauts would have to go through an intense rehabilitation program after being exposed for at least one year to zero gravity and an extended period to reduced gravity on the surface of Mars. Eliminating the need for returning early colonists would cut the costs several fold while at the same time ensure a continuous commitment to the exploration of Mars and space in general.

The first colonists to Mars wouldn't go in "cold." Robotic probes sent on ahead would establish necessities such as an energy source (such as a small nuclear reactor augmented by solar panels), enough food for two years, the basics for creating home-grown agriculture, one or more rover vehicles and a tool-kit for carrying out essential engineering and maintenance work. In addition, the scientific equipment needed for the colonists to do important research work should be part of the preceding unmanned mission. All this equipment could easily be put into place using current technology before the astronauts set out. The first human contingent would rely heavily on resources that can be produced from Mars such as water, nutrients, and shelter (such as in form of lava tube caves). They also would be continuously resupplied from Earth with necessities that could not be produced from the resources available on Mars. This semi-autonomous phase might last for decades, perhaps even centuries before the size and sophistication of the Mars colony enabled it to be self-sustaining.

The first human contingent would consist of a crew of four, ideally (and if the budget permits) distributed between two two-man space craft to allow for some mission redundancy such as in the Viking mission or for the Mars Exploration Rovers. Also, if any technical malfunction occurs on one space craft, the other craft could come to the rescue. Further, any critical part of equipment after landing would be available in duplicate in case of an emergency.

A one-way human mission to Mars would not be a one-time commitment as was the case with the Apollo program. More than 40 years after the last Apollo mission, no human has set foot on a planetary body beyond Earth. Such a hiatus cannot be afforded if humanity is to commit to a grander vision of space exploration (Davies and Schulze-Makuch 2008; Schulze-Makuch and Irwin 2008). No base on the Moon is needed to launch a one-way human mission to Mars. Given the broad variety of resources available on Mars, the long-term

92

survival of the first colonists is much more feasible than it would be on the Moon.

While the pragmatic advantages of this approach are clear, we anticipate that some ethical considerations may be raised against it. Some in the space agencies or public might feel that the astronauts are being abandoned on Mars, or sacrificed for the sake of the project. However, the situation these first Martian settlers are in, who would of course be volunteers, would really be little different from that of the first white settlers of the North American continent, who left Europe with little expectation of return. Explorers such as Columbus, Frobisher, Scott and Amundsen, while not embarking on their voyages with the intention of staying at their destination, nevertheless took huge personal risks to explore new land, with the knowledge that there was a significant likelihood that they would perish in the attempt. A volunteer signing up for a one-way mission to Mars would do so in the full understanding that he or she would not return to Earth. Nevertheless, informal surveys conducted after lectures and conference presentations on our proposal, have repeatedly shown that many people are willing to volunteer for a one-way mission, both for reasons of scientific curiosity and in a spirit of adventure and human destiny. Others may raise objections based on planetary protection considerations, depending on whether indigenous life exists on Mars or not. However, any Martian biota is almost certainly restricted to microbes that would be adapted to the natural environment of that planet, and would therefore almost certainly not pose a safety concern for the colonists due to their presumably different biochemical make-up (e.g., Houtkooper and Schulze-Makuch 2007). Nevertheless, caution has to be urged since we do not know the biochemistry of the putative Martian biota at this time. Thus, it might be prudent to launch a life detection mission or even a sample return mission prior to a one-way human mission to Mars. On the other hand, if Martian organisms really do pose a hazard to human health, it may be preferable to limit the exposure to the crew of a one-way mission rather than place at risk the entire human population from a botched sample return mission (Rummel et al. 2002).

A much more likely problem is the reverse: that the human habitation would pose a threat to any indigenous Martian micro-organisms, even if all possible precautions would be employed to protect it. Sadly, the battle to protect putative Martian biota from terrestrial organisms has already been compromised by the fact that several unsterilized, or inadequately sterilized, spacecraft have already been sent to Mars. In addition, terrestrial impact ejecta may have conveyed viable Earth microbes to Mars repeatedly over geological time scales (Melosh and Tonks 1993; Davies 1996, 2008; Kirschvink and Weiss 2001). Nor is it clear that terrestrial microbes would be better adapted to life on Mars that they would spread uncontrollably in a way that would completely displace the indigenous organisms. Furthermore, the colonists would likely only affect a small portion of the planet and "nature parks" could be designated with special precautions enforced in respect to human interference. Again, such issues could be addressed by a prior life detection or sample return mission to inform us about any risks to Martian biota and the type of precautions that could be taken to protect it. And while we agree that all reasonable precautions should be taken, we do not think their presence should be an over-riding reason to forever resist sending humans to Mars. Indeed, our presence there would allow us to study indigenous life in detail, further our knowledge about essential characteristics of life, and design methods to actually enhance the prospects of Martian biota (McKay 1982; McKay and Marinova 2001).

3. FIRST STEPS IN THE HUMAN COLONIZATION OF MARS

The success of the project we are proposing would hinge on the quality of preparation. We envisage three stages: careful site selection using existing and future probes to gather relevant data, the establishment of an unmanned base with minimum resources necessary for human habitation, and the dispatch of the first astronauts. We shall not dwell here on the astronautics of the mission, as these have been thoroughly discussed elsewhere (e.g., Zubrin and Wagner 1997).

3.1 Site selection -The final determination of a suitable settlement location would require advance scouting missions that could use geophysical exploration tools like ground penetrating radar to locate subsurface voids from aerial or buoyant platforms. Numerous igneous flow features, including lava tubes (large cave structures formed by rivulets of molten lava) have already been identified on Mars (Boston 2003; Figure 1). Lava tube caves on Mars appear to be much larger than on Earth probably due to the lower gravity on Mars (0.38g compared to 1g on Earth). They are natural caves, and some of them are located at a low elevation in close proximity to the former northern ocean, which means that they could harbor ice deposits inside similar to many ice-containing caves on Earth. Ice caves would go a long way to solving the needs of a settlement for water and oxygen. Mars has a thin but substantial atmosphere mostly consisting of carbon dioxide (95%), but it is approximately $1/100^{th}$ the density of Earth's atmosphere, has no ozone shield and no magnetospheric shielding; thus some natural or artificial shielding to protect from ionizing and ultraviolet radiation will be required. Ice caves would also provide shelter from this radiation. After a candidate cave is located, its interior would need to be robotically explored before selecting it for the colony's first home.

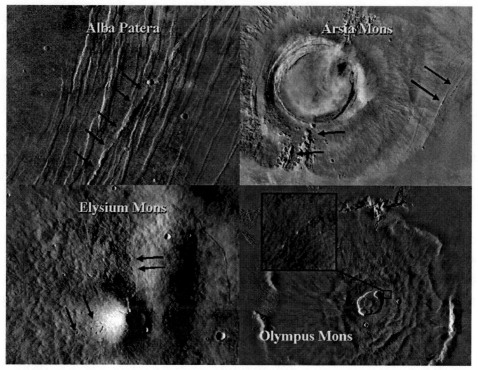

FIGURE 1
Collapsed lava tubes on Mars highlighted by arrows (composite image provided by R.D. "Gus" Frederick, Silverton, Oregon, based on data from NASA). Lava tube caves may serve as a first location to establish a permanent human base on Mars.

3.2. Establishing an unmanned base - After a suitable location is identified, preferentially associated with some natural shelter (e.g., lava tube caves as discussed above) and other nearby resources (water, minerals, nutrients), a base should be established using unmanned probes and robots, including small rover vehicles, to prepare for the arrival of the first human contingent. The base would also be equipped to allow for a more thorough investigation of specific localities of interest. The base would not have to be very sophisticated, but could simply consist of a communication relay and a power generator, perhaps together with a remotely operated telescope (Schulze-Makuch and Irwin 2008). The lander craft should be designed to double as a permanent station, in modular form, to allow later expansion following further one-way missions.

3.3. The first colonists - Crew selection for the initial manned mission would have to take into account several factors. Initially, colonists may be preferred who are beyond their reproductive age, because their life expectancy is likely to be 20 years or less, and secondly, the first settlers will endure some radiation damage to their reproductive organs, both during the trip to Mars and on the Martian surface. One feasible approach for the initial one-way mission would be to send two space crafts with two astronauts each. Ideally, one should be a trained physician, and all should have advanced scientific and technical know-how, and show a strong commitment to scientific research and exploration.

Once the humans arrived at the base, their task would be not unlike that of the early settlers in North America – only the underlying technology and utilized tools would be much more sophisticated. Plants could be grown outside of the caves in an enriched soil underneath a robotically constructed dome, thus providing the inhabitants of the outpost with food and an additional supply of oxygen. Microbes could be used to break down and recycle wastes, thus the human base would constitute its own independent biosphere with some additional resources provided by the Martian environment. Certainly, the first colonists would be exposed to multiple challenges, from physical rigor to psychological strains due to isolation and uncertainties. However, the astronauts will have undergone psychological profiling and training before embarking on the mission, and would remain in constant contact with Earth via normal channels such as email, radio and video links. In the era of modern communications they would in fact feel more connected to home than the early Antarctic explorers (who had no systematic psychological training either). Over time, the human contingent on Mars would slowly increase with follow-up missions. Several cave-centered biospheres would be created, each being in constant communication with other cave-centered biospheres to share experiences on which approaches are working best. At some later time, probably several decades after the first human mission, the colony's population might have expanded to about 150 individuals, which would constitute a viable gene pool to allow the possibility of a successful long-term reproduction program. New arrivees and possibly the use of genetic engineering would further enhance genetic variety and contribute to the health and longevity of the colonists.

While it would undoubtedly take a tremendous effort over many years to establish multiple settlements on Mars, we see no fundamental reason why this plan is not technologically implementable. Some of the heavy lifting hardware has been developed or is in an advanced stage from the recently cancelled Moon program. Work on the permanent unmanned base could be initiated right away, while the human mission and colonization details could be worked out later. We estimate that a reasonable time line for establishing a permanent unmanned base with robots would be 20 years, with the first human contingent arriving shortly thereafter. The main impediment is the narrow vision and the culture of political caution that now pervades the space programs of most nations.

4. CONCLUSIONS

Self-preservation considerations in a dangerous universe and the human exploratory spirit compel us to explore space and colonize other planets. Mars is the planet in our solar system, which is reasonably close and provides an abundance of resources and shelter for such a colonization effort. Nevertheless, the first step for the colonization of Mars will be the most difficult. Here, we propose that the most pragmatic approach to achieve this goal is by establishing a small permanent robotic base followed by a series of one-way missions to Mars. The advantages of a one-way human mission are many-fold including a dramatic reduction of costs, the long-term commitment by the space agency, the public, and the crew, and that no rehabilitation program is needed for crew members when remaining on the low-gravity surface of Mars. The challenges are still monumental, though, foremost because political and financial long-term commitments have to be secured.

REFERENCES

Baker, V.R., Dohm, J.M., Fairén, A.G., Ferré, T.P.A., Ferris, J.C., Miyamoto, H. Schulze-Makuch, D. (2005) Extraterrestrial hydrogeology. Hydrogeology Journal, 13, 51-68.

Boston, P.J. (2003) Extraterrestrial Caves. Encyclopedia of Cave and Karst Science. Fitzroy-Dearborn Publishers, Ltd, London, U.K.

Darling, D., Schulze-Makuch, D. (2010) We Are Not Alone: Why We Have Already Found Extraterrestrial Life. Oneworld, Oxford, UK.

Davies, P. (1996) The transfer of viable micro-organisms between planets. In: Brock, G., Goode, J. (Eds.), Evolution of Hydrothermal Ecosystems on Earth (and Mars?). Proceedings of the CIBA Foundation Symposium No. 20, Wiley, New York, pp. 304-317.

Davies, P. (2008) The Fifth Miracle. Simon & Schuster, New York, USA.

Davies, P., Schulze-Makuch, D. (2008) A one-way human mission to Mars. Astrobiology 8, 310.

Houtkooper, J.M., Schulze-Makuch, D. (2007) A possible biogenic origin for hydrogen peroxide on Mars: the Viking results reinterpreted. Int. J. of Astrobiology, 6, 147-152.

Kirschvink, J. L., Weiss, B.P. (2001) Mars, panspermia, and the origin of life: where did it all begin? Palaeontologia Electronica 4, editorial 2:8p, HYPERLINK "http://palaeo-electronica.org/paleo/2001_2/editor/mars.htm" http://palaeo-electronica.org/paleo/2001_2/editor/mars.htm

Lederberg, J., Sagan, C. (1962) Microenvironments for life on Mars. Proceedings of the National Academy of Sciences (USA), 48, 1473-1475.

Levin, G. V. (2010). Extent life on Mars. Resolving the issues. Journal of Cosmology, 5, 920-929.

Levin, G.V., Straat, P.A. (1977) Recent results from the Viking Labeled Release Experiment on Mars. Journal of Geophysical Research, 82, 4663-4667.

Levin, G.V., Straat, P.A. (1981) A search for a nonbiological explanation of the Viking Labeled Release Life Detection Experiment. Icarus, 45, 494-516.

Mahaney, W.C., Dohm, J. (2010) Life on Mars? Microbes in Mars-like Antarctic environments. Journal of Cosmology, 5, 951-958.

McKay, C.P. (1982) Terraforming Mars. Journal of British Interplanetary Society, 35, 427-433.

McKay, C.P., Stoker, C.R. (1989) The early environment and its evolution on Mars: implication for life. Reviews of Geophysics, 27, 189-214.

McKay, C.P., Marinova, M.M. (2001) The physics, biology, and environmental ethics of making Mars habitable. Astrobiology, 1, 89-101.

McKay, D.S., Everett, K.G., Thomas-Keprta, K.L., Vali, H., Romanek, C.S., Clemett, S.J., Chillier, X.D.F., Maechling, C.R., Zare, R.N. (1996) Search for past life on Mars: possible relic biogenic activity in Martian meteorite ALH84001. Science, 273, 924-930.

Melosh, H.J., Tonks, W.B. (1993) Swapping rocks: ejection and exchange of surface material among the terrestrial planets. Meteoritics, 28, 398-398

Rees, M.J. (2004) Our Final Hour: A Scientist's Warning: How Terror, Error, and Environmental Disaster Threaten Humankind's Future in this Century on Earth and Beyond. Basic Books, New York, USA.

Rummel, J.D., Race, M.S., DeVincenzi, D.L., Schad, P.J., Stabekis, P.D., Viso, M., Acevedo, S.E. (2002) A draft test protocol for detecting possible biohazards in Martian samples returned to Earth. NASA/CP-2002-211842.

Schulze-Makuch, D., Irwin, L.N. (2008) Life in the Universe: Expectations and Constraints (2nd edition). Springer, Berlin, Germany.

Schulze-Makuch, D., Dohm, J.M., Fairén, A.G., Baker, V.R., Fink, W., Strom, R.G. (2005) Venus, Mars, and the ices on Mercury and the Moon: astrobiological implications and proposed mission designs. Astrobiology, 5, 778-795.

Schulze-Makuch, D., Fairén, A.G., Davila, A. F. (2008) The case for life on Mars. Int. J. of Astrobiology, 7, 117-141.

Wierzchos, J., Cámara, B., De Los Rios, A., Davila, A.F., Sánchez Almazo, I.M., Artieda, O., Wierzchos, K., Gómez-Silva, B., McKay, C., Ascaso, C. (2010) Microbial colonization of Ca-sulfate crusts in the hyperarid core of the Atacama Desert: implications for the search of life on Mars. Geobiology, doi: 10.1111/j.1472-4669.2010.00254.x

Zubrin, R.M., Baker, D.A. (1992) Mars direct: humans to the red planet by 1999. Acta Astronautica, 26, 899-912.

Zubrin, R.M., Wagner, R. (1997) The Case for Mars: The Plan to Settle the Red Planet and Why We Must. Touchstone, New York, USA.

IV

PSYCHOLOGY & MEDICAL

Journal of Cosmology, 2010, Vol 12, 3711-3722.
JournalofCosmology.com, October-November, 2010

- 7 -

Moving to Mars: There and Back Again. Stress and the Psychology and Culture of Crew and Astronaut

Sheryl L. Bishop, Ph.D.

University of Texas Medical Branch Galveston, Texas 77555

ABSTRACT

The journey to explore our red neighbor will entail the application of all our terrestrial lessons learned and of some we have yet to discover. A Mars mission represents the extreme in terms of both distance and uncharted environment. The selection, monitoring and support of Mars bound crews will challenge existing technology and knowledge. The human, at the center, represents the greatest strength and the greatest weakness for a Mars mission. Human response to confined and isolated environments has been shown to be characterized by serious stressors and a Mars mission will represent the most extreme of such environments. The impact of such stressors on coping, performance, motivation, behavior, cognitive functioning and psychological well-being must be taken into account. The extraordinary duration of the mission poses special challenges in planning for mission support since very different needs may be driven by particular phases of the mission. Selection, monitoring and support will similarly be significantly affected by anticipating potential differential characteristics and needs across the travel phases to and from Mars and the period on the planet's surface.

Key Words: Mars exploration, Human exploration, Aerospace Psychology, Isolated confined environments, Extreme, unusual environments, Human Factors, Human Performance

1. INTRODUCTION

HUMANS HAVE BEEN DREAMING ABOUT Mars since the ancient Egyptian astronomers mapped its retrograde movement in1534 BCE (Novakovic, 2008).

In 1877, Giovanni Schiaparelli used a 22 cm telescope and created the first map of Mars displaying the famous Martian canals. The canals later proved to be illusionary, but not before their purported existence ignited a period of intense speculation in both scientific (e.g., Percival Lowell) and literary circles (e.g., H.G. Wells) about the red planet and the possibility of life there. We have continued that speculation into modern times even though our technology has debunked the canals, eliminated the prospect of vibrant humanoid civilizations, and revealed our cousin planet to be a cold, harsh and desiccated environment. Our dreams have shrunk to microscopic life forms and plans for colonization tempered by the realities of the existing environment. Yet we would still go.

Across the years of contemplating exploration of Mars, one notable conclusion stands out: The human element is both the greatest strength and the greatest weakness for a Mars mission. From the earliest lunar design projects, there has been a tacit recognition that the success of the group would be the primary limiting factor in long duration missions (La Patra, 1968). There have been any number of scholarly reports from as far back as the beginning of the space era in the early 1960s (most notable the Case for Mars series and the Mars Exploration Studies by the British Interplanetary Society) that have explicitly recognized psychological and sociocultural factors as critical components to mission success. Interpersonal group processes have clearly been shown to impact significantly on behavior and performance in most challenging environments, especially those characterized by isolation and confinement (National Commission on Space, 1986; National Science Board, 1987; Space Science Board, 1987). While considerable improvements and advances have been made in technology, engineering and human factors, understanding the issues dealing with psychosocial human health and well-being are not significantly better in the new millennium than in the earliest days of space travel. While it is true that medical issues have become more specifically defined and identified, we are still grappling with questions that were raised from the beginning: What is the best fit *individual* for a mission to Mars? What is the best *group* for a mission to Mars?

2. THE MARS MISSION CHALLENGE

The inescapable truth is that missions to Mars will not be psychologically comparable to any other undertaking humans have ever attempted. The enormous distance to travel and extraordinary novel environment will separate this chapter of human exploration from all other settings (e.g., historical expeditions to unknown parts of the Earth, wintering-over in Antarctica, long-term submergence in submarines, or even long-duration stays on orbital space stations). Mission scenarios call for crewmembers to endure extraordinary long periods of extreme confinement and isolation (18 months to 3 years) during their voyage to Mars and a stay on the Martian surface that may last from 500 to 1,000 days. Communication and data transmission will be significantly impacted as relative orbital positions between the Earth and Mars require one-way transmission times of 3 to 22 minutes and no real time two-way communication will be possible. Once underway, there will be *no possibility* for any re-supply or short-term rescue. Should something go wrong early in the mission, the crew would be faced with, at a minimum, continuing to Mars for a fly-by of the planet before heading back to Earth, and even then only if orbital alignment makes such a "quick return" possible.

Known psychological risks related to individual performance, behavioral health and crew interactions will be magnified by completely new unpredictable psychological challenges. For instance, given the historical importance of looking back at Earth reflected in numerous reports of astronauts from orbital or lunar flights, there is strong speculation that an "Earth-out-of-view phenomenon" will have a significant psychological impact, referring to the fact that astronauts traveling to Mars will be the first human beings put in a situation where the home planet will be reduced to another dot in space (Kanas & Manzey, 2008; Manzey, 2000). Our closest analogs would have been the early exploratory seafaring missions losing sight of land as they sailed into uncharted open ocean. However, these crews did not have to worry about such fundamental life support elements as breathable atmosphere or explosive decompression. The region may have been unfamiliar, but the physical environment was taken for granted; the only genuine worries being a fear of falling off the edge of the world.

3. THE STRESS OF BOLDLY GOING WHERE NO ONE HAS GONE BEFORE

Despite the fact that most human space missions have successfully met mission goals to date, there is considerable anecdotal and behavioral evidence that a significant number of crewmembers have experienced psychological and interpersonal difficulties arising from the myriad stressors inherent in space missions, especially those involving longer durations (Shayler, 2000).

Confined and isolated environments are characterized by potentially serious stressors: physical demands (e.g., working in suits, reduced gravitational fields, extravehicular activities) as well as inescapable environmental characteristics (e.g., imminent danger, noise, isolation, confinement, loss of normal sensory stimuli, low habitable volume per person, limited opportunities for variety and change and complete dependence on a machine-dominated environment). Anecdotal reports from space and studies conducted in space analogue environments on

Earth (e.g., the Antarctic, submarines, space simulation chambers) have isolated a number of psychosocial issues that may negatively affect crewmembers during multinational space missions (Bishop & Primeau, 2001; Bishop, Santy, & Faulk, 1998; Kanas, 1998; Kanas & Manzey, 2008; Morphew & Maclaren, 1997; Palinkas L., 2003; Sandal, Endresen, Vaernes, & Ursin, 1999). These issues include:

1) tension resulting from external stress and factors related to crew heterogeneity (e.g., differences in personality, gender, and career motivation)

2) variability in the cohesion of the crew

3) improper use of leadership role (e.g., task/instrumental versus emotional/supportive)

4) cultural differences; and

5) language differences.

Post-flight debriefings of space crews have routinely detected strong differences in values among crewmembers arising from their differences in professions, culture, age or nationality, all of which have been cited as reasons for crew tension. Although crews were able to effectively operate and accomplish mission goals on these short duration international space flights (Bluth, 1997; Kozerenko, Sled, & Salnitsky, 1986; Santy, Holland, Looper, & Marcondes-North, 1993), it was recognized that long duration missions would require true accommodation rather than mere tolerance.

It has been argued that the key factor to successful adaptation of the group is the individual's capability to share the general values and aims of the group and to establish empathetic relations with partners. In studies simulating the influence of space flight, crewmembers more favorably perceived those they referred to as "supportive", "loving", "warm", and "socially competent", and who had an orientation toward empathetic interrelations and the social side of life (Gushin, Efimov, Smirnova, Vinokhodova, & Kanas, 1998). Several studies of small isolated groups have shown that in the process of cohesive group formation, crewmembers began to regard each other as very "similar" or "close", sharing common values and beliefs (Gushin, et al., 1997; Gushin, Efimov, Smirnova, Vinokhodova, & Kanas, 1998).

Other studies have demonstrated that those crewmembers who did not perceive others as close and similar to themselves, and who didn't make attempts to understand and share the common group values ran the risk of becoming the "stranger" or the "alien" in the group (Gushin, Efimov, Smirnova, Vinokhodova, & Kanas, 1998; Leon & Koscheyev, 1997; Penwell, 1990). Among those factors mentioned by astronauts and cosmonauts as barriers to identification with others were the language barrier, cultural differences, gender differences, professional variation, and differences in perspective (Herring, 1997; Morphew & Maclaren, 1997). The importance of these factors is supported by studies that have found psychological reactions to space were distinctly different among various cultural backgrounds (Ritsher, 2005).

In contrast, several investigations have actually found evidence of positive, or salutogenic, psychological impacts of isolated, confined environments (Suedfeld & Steel, 2000; Ritscher, Kanas, Ihle, 2005). In early studies of sensory isolation personality characteristics in fact played a significant role in how subjects responded to prolonged and profound sensory and social deprivation. Almost all subjects reported experiencing a variety of illusions and simple to complex hallucinations, with some finding the experience horrifying and others as mystical and spiritually enlightening (Lilly, 1956, 1973, 1977). Those who experience positive growth from their exposure in an extreme environment may be reflective of those who most successfully adapt to the demands of such environments. The identification of the defining characteristics of successful adapters is the holy grail of selecting the best fit individuals for long duration missions.

Yet another unknown is the effects of extraterrestrial environments upon the brain. Bone and muscle loss after even short duration flights are now well documented. However, neurobiological processes dramatically affect personality and intellectual functioning. Early and subsequent studies of prolonged social, sensory, and perceptual isolation in primates and other animals have consistently demonstrated significant effects on learning, memory, perception, and nerve growth versus neural degeneration (Casagrande & Joseph 1980; Joseph & Gallagher 1980; Joseph 1999). From these and other studies it can be concluded that it is of the utmost importance to keep astronauts socially and intellectually engaged and to provide *optimal* amounts of stimulation at all phases of the mission.

The presence of even low levels of chronic stressors, if not met with functional adaptation and/or countermeasures, produce subjective symptoms of stress, persistent performance incompetence, accelerated fatigability, altered mood states, increased rate of infections, and decrements in attention and cognitive functioning (Bishop, Kobrick, Battler, & Binsted, in press; Kanas & Manzey, 2008; Palinkas L., 1991; Sloan & Cooper, 1986; Smith, 1990). Most of the performance effects found so far seem to be associated with more general stress effects related to problems of adaptation to the extreme living and working conditions in a confined and isolated environment that are mediated by individual factors such as personality and culture. In the context of a Mars mission, it has been proposed that the most severe stressors might involve the monotony and boredom resulting from the long periods of low workload, hypo-stimulation, and restricted social contacts due to isolation from family and friends. Yet any mission of significant length will be characterized by periods of hyperarousal stemming from intense work schedules, high activity, excitement, media and public attention, and heightened performance demands which all interact to impact sleep, motivation, attention, physical and mental functioning, e.g., the Earth proximate departure and return phases of the journey, or arrival at Mars and deployment to and return from the surface. The juxtaposition of prolonged periods of low work schedules, minimal activity, repetitive and monotonous station-keeping duties, isolation, confinement, loss of privacy, restricted social contacts, over-familiarity with team-members and environment which undermine motivation, performance and psychosocial functioning, providing countermeasures for this dynamic environment-situation becomes a challenge indeed!

One could argue that the stress of exploration is inherent in any endeavor that is characterized by unknown dangers, isolation and confinement. What makes the stress associated with long duration space missions of special concern? The obvious answer is that there is no exit for participants. Once launched on a trajectory to Moon or Mars, the crew is committed. There is no turning back, no quick rescue, no way out except forward. The implications of the incapability to terminate an expedition are profound. The role of stress and its impact on coping, performance, motivation, behavior, cognitive functioning and psychological well-being must be taken into account in mission planning. All required resources for assessing, monitoring and countering stressors must be resident with the crew given the delays in communication and isolation. Without the ability to return at will, the success of the mission rests with how well planners have selected the best individuals and provided the right resources.

4. WHAT IS "CLOSE ENOUGH"?: THE ROLE OF ANALOGS

Clearly, the need to know what is needed and what works before the launch to Mars demands that substantive research into selection, training, and support be conducted. But how to conduct relevant research when teams in space have been small (1-3 persons) and largely unavailable to the greater scientific community. The National Academy of Sciences in their white paper, "A Risk Reduction Strategy for Human Exploration of Space: A Review of NASA's Bioastronautics

Roadmap" (Longnecker & Molins, 2006), clearly articulated the critical role of analog environments and digital simulation to evaluating and developing mitigation strategies and countermeasures to the risks to the human in space. The value of both natural (e.g., polar expeditions, deep caving teams, saturation diving), man-made (e.g., operational Antarctica bases, submarines) and simulated environments (e.g., bed rest studies, mock EVA simulations, Mars Society habitats, Concordia Antarctic Research facility) to test hardware and instrumentation is incontestable. In addition, the use of these analog environments to test various medical, physiological, behavioral and psychological protocols and the effectiveness of countermeasures has been repeated demonstrated, e.g., the problems associated with adverse effects on thyroid functioning with iodination of drinking water or the superior performance of mixed gendered teams.

The use of space analog environments in which to systematically study individual and group adaptation has had to contend with some significant compromises (Bishop, 2010). For instance, there is no good substitute for microgravity since buoyant environments (i.e., underwater) which can be configured to compensate somewhat for gravity also entail resistance forces to the water itself that are not present in a vacuum. Thus, all analog environments are simulations of greater or lesser fidelity along varying dimensions of interest. An analog for testing equipment or hardware may not have any relevance to assessing how human operators will fare psychologically or as a team. Others may address important components of human adaptation, e.g., confinement, but fail utterly to incorporate true environmental threats. The use of extreme unusual environments (e.g., Antarctica, deep caving, saturation diving) allow an assessment of the impact of true dangerous, unpredictable environments but provide challenges to comparing across specific environments. The spectrum of fidelity to space among terrestrial analogs ranges from laboratory studies where the impact of environmental threat and physical hardship, as well as true isolation and confinement, are limited and, even, sometimes absent, to real teams in real, extreme environments characterized by very little control over extraneous variables. Yet, there is no other alternative. The best approach is to utilize analogs that incorporate capabilities for testing hardware, software and the human operators whose lives depend on successful and effective human-machine and human-human competencies.

5. SELECTING THE "BEST"

So, what kind of person will qualify as the 'best fit'? When selecting best fit *individuals*, it currently appears that we will have to seek those that do not require high levels of sustained stimulation or high levels of interpersonal interaction and support. The burden of support will have to shift away from the "absent" network of family and friends to members of the present crew. The reduced circle of social contacts will challenge crewmembers to be able to balance needs for affiliation with autonomy and privacy needs. Selected individuals will be characterized by effective coping strategies based on self-reliance and autonomy as well as a cooperative group focus. Individuals with high needs for social support will likely be more susceptible to the negative effects of social isolation from friends and family (Sandal, Endresen, Vaernes, & Ursin, 1999) as well as more likely to engage in interpersonal disclosure with team mates that may later lead to discomfort and regret (Suedfeld, 2003). This is an area that needs substantial further attention and research.

The problem of selection is multifold. Individual factors deriving from both nature and nurture clearly contribute to group dynamics. For example, whether lack of privacy is perceived as a stressor, and thus produces detrimental effects on mood and performance, largely depends on both the personality of the individual as well as the cultural background (Raybeck, 1990). Similar effects may be assumed for other stressors as well, like monotony and boredom, time-pressure, or workload. In addition, culture also might be a factor influencing higher-order cognitive processes such as decision-making or the use of schemes in information processing. Studies in analog environments have confirmed that the *kind* of mission has a significant influence upon the type of "best fit" crew needed (Bishop, 2002; Endler, 2004; Palinkas, Gunderson, & Holland, 2000; Sandal, 2000).

Expeditionary missions (e.g., flags and footsteps like the lunar missions) typically engage very task focused individuals with lesser demands for skills in sustained interpersonal interaction whereas missions dedicated to science or establishment of sustainable bases enroll the expertise of individuals with notably different personality profiles and skill sets as well as mission demands for sustained group engagement. The fact that psychological stress has been found in all isolated and confined environments supports a view that these same factors are a risk for long duration lunar and Martian crews as well (Kanas, 2004). Psychological difficulties not only affect individual crewmembers, but they could have potentially disastrous effects on mission performance, e.g., a clinically depressed astronaut might be unable to perform required tasks in an emergency situation. If such problems are not understood and addressed, they will most certainly result in

1) decreased crew morale and compatibility

2) withdrawal or territorial behavior as crewmembers cease to interact with each other

3) scapegoating or singling out individuals for blame for various outcomes as solutions to group conflicts; and

4) the formation of subgroups that compete with each other and destroy crew unity.

6. "WHEN" MAKES A DIFFERENCE IN WHAT IS NEEDED

The task of identifying the best fit and how to train and support them is further complicated by the fact that individual and group functioning will vary dramatically depending on the *phase of the mission*. The differences in mission duration for those environments we have studied to date (11 months for Antarctica; four to six months for MIR or the International Space Station, weeks for expeditionary missions) appear to play a significant role. Although mission task profiles on the surface can be treated as distinct from flight segments, the functioning of the crew cannot be so neatly separated. Both in-flight and planet-side individual and group functioning will be impacted by what comes *before*. Research findings on space and analog missions lasting six weeks or more have demonstrated that psychological and interpersonal issues grow in importance across time and may become some of the most significant factors for a space flight to and from Mars (Kanas & Manzey, 2008). Some studies have found significant psychological and interpersonal difficulties after the halfway point of a mission (Bechtel & Berning, 1991; Gushin, et al., 1997; Kanas & Manzey, 2008; Sandal, Vaernes, & Ursin, 1995). In this perspective, a sense of relief that half of the mission is over is outweighed by the realization that another half is yet to come. Another time model incorporates Rohrer's three sequential phases, which include: 1) initial anxiety; 2) mid-mission depression; and 3) terminal euphoria (Kanas, 1990, 1998; Kanas & Manzey, 2008). Currently, efforts to validate the existence of specific critical phases of adaptation noted in earlier studies of Russian long-duration space missions (Myasnikov & Zamaletdinov, 1998) and in personnel over-wintering on Antarctic research stations has been equivocal (Kanas & Ritsher, 2005; Stuster, Bachelard, & Suedfeld, 2000; Palinkas, 2003; Bergan et al., 1996; Palinkas, Gunderson, & Holland, 2000). When applied to a Mars mission profile, there may be further macro phase cycles bracketing the anticipatory preparation of the trip *to* Mars, the stimulation of exploration while on the surface, and a *return phase* dominated by the long trip home without the excitement of either prior phases.

7. MONITORING AND SUPPORT

By definition, adaptation presumes an accommodation over time. The issue of critical periods in adaptation for Mars is even more important given that such knowledge may enable both crewmembers and Mission Control personnel to prepare for problems and intervene before maladjustments result in operational impact. Given that there will be multiple duration milestones that affect psychological well-being (e.g., half-way points in the trip out and back, halfway points for surface mission, third quarter mission total, planned EVA missions, significant holidays, family events), any viable strategy for a successful mission to Mars will have to encompass countermeasures, monitoring and support that are adaptive to the changing salience of factors particular to mission phases.

The challenges of monitoring and providing psychological support to a Mars mission are inherently different than those for Shuttle, Mir or ISS. In the past, supportive activities have focused on providing increased novelty and stimulation for space crews during times of monotony or asthenia--a depressive state characterized by apathy and listlessness-- (Kanas, et al., 2001) through surprise food deliveries and presents via resupply vehicles, increased contact with Earthside communities and therapists and adjusting schedules to accommodate performance deficits due to maladaptation or stress (Kanas, 1991). For an expeditionary mission, first line capability must reside with the crew since two-way communications with Earth will be delayed by distance for up to 45 minutes. It will be essential for crewmembers to monitor themselves first. All crewmembers will need to be aware of psychosocial issues and trained to recognize a basic level of:

1) individual psychopathology and dysfunctional small group behavior
2) the individual and interpersonal effects of stressors to be expected during the mission
3) crisis intervention techniques and the facilitation of group awareness, cohesion, and team-building; and
4) the appropriate use of tranquilizers and other psychoactive medications, including their usefulness and side effects under conditions of microgravity (Kanas, et al., 2009)

Crewmembers should be monitored for symptoms and signs of developing psychiatric disturbances through a variety of psychological and physiological approaches. Provisions and facilities for restraining and secluding a potentially suicidal, violent, or impulsive crewmember should be made in mission planning.

Internationally approved test batteries for mission monitoring of interrelationships among crewmembers and group performance effectiveness do not yet exist and need to be created.

Countermeasures will need to be both passive (e.g., monitoring through routine cognitive assessment, stress hormonal levels) and active (e.g., scheduled group briefings and personal communication access to support resources). All external contact will, by necessity, be recorded and entail lagged response times. As a consequence, psychological support will primarily depend on the skills of on-board crewmembers and the quality of intelligent tools which are able to detect subtle performance decrements before they lead to overt performance decrements in mission tasks or personal functioning.

In addition to the monitoring of the individual state, the interactions of crewmembers in space will also need to be monitored and conflicts resulting from psychosocial issues will need to be dealt with as they arise. Monitoring of intra- and intergroup relations will involve psychological and physiological assessment at both the individual and group level. Assessment data about the psychosocial climate within the crew will need to be provided to individuals, the team physician and Mission Control evaluators. Analysis of crew behavior and performance should provide immediate feedback to the crew and time-lagged monitoring information to Mission Control support teams. In order to detect such possible adverse effects and to provide appropriate support and countermeasures (e.g. re-scheduling of mission tasks, accommodation of work-rest schedules), tools are needed which are able to discern subtle performance decrements before they lead to overt performance decrements in mission tasks. For monitoring purposes, subjective reports used in combination with more objective and non-invasive methods would be most useful. Specialized computer tests will be needed to investigate cognitive functioning, personality and relationships within the group and to assess individual and group performance (Gushin, Efimov, Smirnova, Vinokhodova, & Kanas, 1998; Kanas, 1991; Kane, Short, Sipes, & Flynn, 2005; Savilov, et al., 1997). The use of computerized tests that may help crewmembers assess their cognitive state and their ability to perform certain behaviors at various times during the mission are currently being assessed. So far, such methods have mostly been utilized in research, and more experience from operational applications are needed. Approaches also need to take the operational constraints of the different mission phases into account. Even with the constraints inherent in interplanetary missions, provision of in-flight support to crewmembers is likely to be a highly important countermeasure to stabilize the emotional state, to ensure optimal well-being of astronauts, and to maintain a close contact between space crew and ground. The methods utilized for this purpose will rely on the availability of effective space-ground communication systems (audio/video transmissions) and high-fidelity and easily available tele-medicine/ tele-psychological store and forward consultation. Attention should be given to enhancing individually tailored leisure time activities that take into account changing interests and needs over the course of the mission (Kanas & Manzey, 2008).

Other important support activities include private psychological consults (even if conducted in store-and-forward formats), providing informal space-ground contact and news from Earth (preferably in the crew member's native language and from homeland news sources), and opportunities to maintain close contact with family and friends on Earth on a regular basis. An important issue that needs careful consideration concerns the variables that affect whether very bad news from home (e.g., death of a close relative) should be transmitted to astronauts in flight when they can't do anything about it. In general, providing support concerning medical and psychological problems for crewmembers' families during the mission can contribute to maintaining the crew member's concentration on the objectives of the mission by relieving them of considerations about possible problems at home and feelings of responsibility. In addition, families should be coached in interacting with their remote family member and be prepared for possible psychological changes in the course of the mission.

Remote monitoring of crewmembers' performance and behavior represents an important basis for early detection of any signs of impaired performance and behavior, and for providing ground-based counseling and advice to the crew (including planning of interventions). It has been shown that performance impairments might be expected primarily during adaptation to the space environment and during specific stress states induced by sleep problems, high workload and/or emotional burden. During the mission, crewmembers may be reluctant to give information regarding emotional stress and adaptation. Willingness to discuss personal matters may be reflective of personality differences as well as expectations of the privacy of information. Experience from multinational airlines suggests that there are cultural differences in attitudes regarding discussion of personal problems and fallibility (Merritt & Helmreich, 1998).

Computer-interactive intervention programs show promise for addressing the autonomous and remote psychological, medical and performance monitoring demands for a Mars mission. Recent suggestions have proposed that a computerized format for identifying problems as well as supplying prevention and intervention information may be more comfortable to crewmembers than disclosing highly personal information to other crewmembers with whom they have to work with in close quarters for an extended period of time. Furthermore, research suggests that computer-based intervention programs delivering cognitive-behavioral, behavior therapy, and self-help instruction are just as or more effective than face-to-face types of intervention for dealing with mild to moderate depression, anxiety disorders, and other types of psychopathology (Bachofen, et al., 1999; Cavanagh & Shapiro, 2004; Proudfoot, Goldberg, Mann, Everitt, Marks, & Gray, 2003).

A number of habitability factors may also enhance coping abilities of crewmembers in dealing with these living conditions. One potential enabling technology involves the use of immersive virtual reality systems and haptic technologies to enhance leisure activities and environmental stimuli. Telepresence and full fidelity audio/video/3-D communication replay capability will provide for more effective psychological support and interaction for crewmembers and to families and friends back on Earth. Both habitability and mission structure factors will play a significant role in generating group fusion (e.g., cohesion, teamwork, interpersonal trust) or group fission (e.g., discord, conflict, miscommunication, reduction in performance and well-being).

8. CONCLUSION

The selection, monitoring and support of the best fit team for a Mars mission begins with the individual but ends with the group. Psychological countermeasures must be implemented before, during and after the mission, and involve crewmembers and their families, as well as relevant ground support personnel. Current knowledge on long-term effects of coping and adaptation is still limited, and assumptions are based on anecdotal data or research in Earth-bound analogs. To determine the generalization of these findings to crews in space, it will be necessary to compare findings from analogs with psychosocial results from actual space missions which has historically been very difficult to obtain. The demarcation of short-term from long-term effects is one important issue for future research, i.e. where adaptation and possibly wellness problems become frequent and/or serious (Dion, 2004). Such knowledge would enable selection, training, and support tailored to meet the specific demands of each mission.

It is important to identify organizational and environmental characteristics that contribute to impaired psychological coping and well-being in multinational crews living and working in space. This should include considering the impact of culture at different levels, e.g. at professional, organizational and national levels. Much more empirical work is needed on defining individual characteristics (e.g., personality, attitude, motivation, needs, skills, coping strategies) and group characteristics that promote optimum coping and adaptation under different kinds of multinational missions. Also the most efficient methods for assessment of such characteristics need to be determined. Such knowledge would be helpful in the development of psychological countermeasures, including selection and crew assignment, psychological training programs and in-flight support systems.

A major challenge in evaluating the efficiency of psychological countermeasures is the identification of valid and reliable performance criteria against which they can be tested. Given the likelihood that Mars crews will include members of both genders, various national and cultural backgrounds and substantially larger crews than ISS, it would be highly desirable that all partners identify and agree to establish a set of common standards and procedures for the selection, training, support, and evaluation of Mars mission crews.

REFERENCES

Bachofen, M., Nakagawa, A., Marks, I., Park, J., Greist, J., Baer, L., et al. (1999). Home self-assessment and self-treatment of obsessive-compulsive disorder using a manual and a computer-conducted telephone interview: replication of a UK-US study. J of Clinical Psychiatry , 60 (8), 545-549.

Bechtel, R., & Berning, A. (1991). The third quarter phenomenon: Do people experience discomfort after stress has passed? In A. Harrison, Y. Clearwater, & C. McKay, From Antarctica to Outer Space: Life in Isolation and Confinement (pp. 260-265). New York: Springer-Verlag.

Bishop, S. (2002). Evaluating teams in extreme environments: Deep caving, polar and desert expeditions. 32nd International Conference on Environmental Systems. San Antonio: Society of Automotive Engineers (SAE).

Bishop, S. L. (2010). From Earth Analogs to Space: Getting There From Here, Chapter 3, In Psychology of Space Exploration: Contemporary Research in Historical Perspective, NASA History Series, D. Vakoch (Ed), in press.

Bishop, S. L., & Primeau, L. (2001). Assessment of Group Dynamics, Psychological and Physiological Parameters during Polar Winter-Over. Proceeding for the Human Systems Conference. Nassau Bay, Texas.

Bishop, S., Kobrick, R., Battler, M., & Binsted, K. (in press). FMARS 2007: Stress and Coping in an Arctic Mars Simulation. Acta Astronautica, doi:10.1016/j.actaastro.2009.11.008.

Bishop, S., Santy, P., & Faulk, D. (1998). Team Dynamics Analysis of the Huautla Cave Diving Expedition: A Case Study. Society of Automotive Engineers (SAE) (1999-01-2009). (S. o. (SAE), Ed.) Retrieved from Society of Aeronautical Engineers: www.sae.org

Bluth, B. (1997). The benefits and delimmas of an international space station. Acta Astronautica, 2, 149-153.

Casagrande, V. A. & Joseph, R. (1980). Morphological effects of monocular deprivation and recovery on the dorsal lateral geniculate nucleus in prosimian primates. Journal of Comparative Neurology, 194, 413-426.

Cavanagh, K., & Shapiro, D. (2004). Computer treatment for common mental health problems. J Clinical Psychology, 60 (3), 239-251.

Dion, K. (2004). Interpersonal and group processes in long-term spaceflight crews: Perspectives from social and organizational psychology. Aviation, Space and Environmental Medicine , 75 (7 Section II), C36-C-43.

Endler, N. (2004). The joint effects of person and situation factors on stress in spaceflight. Aviation, Space and Environmental Medicine, 75 (7), C22-C27.

Gushin, V., Efimov, V., Smirnova, T., Vinokhodova, A., & Kanas, N. (1998). Subject's perceptions of the crew interaction dynamics under prolonged isolation. Aviation Space and Environmental Medicine , 69, 556-561.

Gushin, V., Zaprisa, N., Kolinitchenko, T., Efimov, V., Smirnova, T., Vinokhodova, A., et al. (1997). Content analysis of the crew communication with external communicants under prolonged isolation. Aviation, Space and Environmental Medicine, 68, 1093-1098.

Herring, L. (1997). Astronaut draws attention to psychology, communication. Human Performance in Extreme Environments, 2 (1), 42-47.

Joseph, R. (1999). Environmental influences on neural plasticity, the limbic system, and emotional development and attachment, Child Psychiatry and Human Development. 29, 187-203.

Joseph, R., & Gallagher, R. E. (1980). Gender and early environmental influences on learning, memory, activity, overresponsiveness, and exploration. Developmental Psychobiology, 13, 527-544.

Kanas, N. (1990). Psychological, psychiatric and interpersonal aspects of long-duration space missiong. Journal of Spacecraft and Rockets, 27, 457-463.

Kanas, N. (1991). Psychosocial support for cosmonauts. Aviation, Space and Environmental Medicine, 62, 353-355.

Kanas, N. (1998). Psychosocial issues affecting crews during long-duration international space missions. Acta Astronautica, 42, 339-361.

Kanas, N. (2004). Interactions during space missions. Aviation, Space and Environmental Medicine, 75 (7), C2-C5.

Kanas, N., & Manzey, D. (2008). Space Psychology and Psychiatry. Segundo, CA and Dordrecth, The Netherlands, The Netherlands: Microcosm Press and Springer.

Kanas, N., & Ritsher, J. (2005). Psychosocial issues during a Mars mission. AIAA 1st Space Exploration Conference. Orlando, FL.

Kanas, N., Salnitsky, V., Gushin, V., Weiss, D., Grund, E., Flynn, C., et al. (2001). Asthenia-does it exist in space? Psychosomatic Medicine , 63, 874-880.

Kanas, N., Sandal, S., Boyd, J., Gushin, V., Manzey, D., North, R., et al. (2009). Psychology and culture during long-duration space missions. Acta Astronautica , 64, 659-677.

Kane, R., Short, P., Sipes, W., & Flynn, C. (2005). Development and validation of the spaceflight cognitive assessment tool for Windows (WINSCAT). Aviation, Space and Environmental Medicine, 76 (1 Suppl), B183-B191.

Kozerenko, O., Sled, A., & Salnitsky, V. (1986). Selected problems of psychological support of prolonged space flights. Proceedings of the 38th Congress of the IAF (pp. 386-398). Washington, D.D. : AIAA.

La Patra, J. (1968). Moon lab: Preliminary design of a manned lunar laboratory. Stanford/NASA Ames workshop study.

Leon, G., & Koscheyev, V. (1997). Cross-cultural polar expedition teams as an analog to long duration space missions. Proceedings of the American Aerospace Mediccal Association. Chicago.

Lilly, J. C. (1956). Mental effects of reduction of ordinary levels of physical stimuli on intact, healthy persons. Lilly, John C. Psychiatric Research Reports, 5, 1-9.

Lilly, J. C. (1973). The Center of the Cyclone. Bantam Books. v Lilly, J. C. (1977). The Deep Self: The Tank Method of Physical Isolation, New York, Simon and Schuster.

Longnecker, D., & Molins, R. (2006). Bioastronautics Roadmap: a risk reduction strategy for human exploration of space. David Longnecker and Ricardo Molins, Editors, Committee on Review of NASA's Bioastronautics Roadmap, National Research Council.

Manzey, D. (2000). Monitoring of mental performance during spaceflight. Aviation, Space and Environmental Medicine (71), A69-75.

Merritt, A., & Helmreich, R. (1998). Culture at work in aviation and medicine: National, organizational and professional influences. Brookfield, VT: Ashgate.

Morphew, M., & Maclaren, S. (1997). Blaha suggests need for future research on the effects of isolation and confinement. Journal of Human Performance in Extreme Environments, 2 (1), 52-53.

Myasnikov, V., & Zamaletdinov, I. (1998). Psychological states and group interactions of crew members in flight. Human Performance in Extreme Environments, 3, 44-56. National Commission on Space. (1986). Pioneering the Space Frontier. New York: Bantam Books.

National Science Board. (1987). The Role of the National Science Foundation in Polar Regions. Washington, D.C.: National Academy of Sciences.

Novakovic, B. (2008). Senenmut: An Ancient Egyptian Astronomer. Publications of the Astronomical Observatory of Belgrade, October (85), 19–23.

Palinkas, L. (1991). Effects of physical and social environments on the health and well-being of Antarctic winter-over personnel. Environment and Behavior, 23, 782-799.

Palinkas, L. (2003). The psychology of isolated and confined environments: Understanding human behavior in Antarctica. American Psychologist, 58 (5), 353-363.

Palinkas, L., Gunderson, E., & Holland, A. (2000). Predictors of behavior and performance in extreme environments: The Antarctic space analogue program. Aviation, Space and Environmental Medicine, 71, 619-625.

Penwell, L. (1990). Problems of intergroup behavior in human spaceflight operations. Spacecraft, 27 (5), 464-469.

Proudfoot, J., Goldberg, D., Mann, A., Everitt, B., Marks, I., & Gray, J. (2003). Computerized, interactive, multimedia cognitive-behavioural program for anxiety and depression in general practice. Psychological Medicine , 33 (2), 217-227.

Raybeck, D. (1990). Proxemics and Privacy in Confined Environments. In A. Harrison, Y. Clearwater, & C. McKay, From Antarctica to Outer Space: Life in Isolation and Confinement (pp. 324-325). New York: Springer Verlag.

Ritsher, J. (2005). Cultural factors and the International Space Station. Aviation, Space and Environmental Medicine, 76, B135-B144.

Ritsher, J.B., Kanas, N., & Ihle, E.C. (2005). Psychological adaptation and salutogenesis in space: Lessons from a series of studies. Paper presented at the 15th International Academy of Astronautics, Humans in Space Symposium, Gratz, Austria.

Sandal, G. (2000). Coping in Antarctica: Is it possible to generalize results across settings? Aviation, Space and Environmental Medicine, 71 (9 Suppl), A37-A43.

Sandal, G., Endresen, I., Vaernes, R., & Ursin, H. (1999). Personality and coping strategies during submarine missions. Military Psychology, 11, 381-404.

Sandal, G., Vaernes, R., & Ursin, H. (1995). Interpersonal relations during simulated space missions. Aviation, Space and Environmental Medicine, 66, 617-624.

Santy, P., Holland, A., Looper, L., & Marcondes-North, R. (1993). Multicultural factors in the space environment: Results of an international shuttle crew debrief. Aviation, Space and Environmental Medicine, 4 (N1), 28-31.

Savilov, A., Bayevsky, R., Bystrytskaya, A., Gushin, V., Manovtsev, G., Nichiporuk, I., et al. (1997). On-board equipment based study of the dynamic of psycho-physiological and biomedical responses of the operators during 135-day isolation in teh Mir orbital station mock-up. Aerospace and Ecological Medicine, 5, 28-38.

Shayler, D. (2000). Disasters and accidents in manned spaceflight. Chichester: Springer-Praxis.

Sloan, S., & Cooper, C. L. (1986). Stress coping strategies in commercial airline pilots. J. of Occupational Med., 28 (1), 49-52.

Smith, J. (1990). Long periods in space flight may take physiological, psychological toll among crew. JAMA, 163 (3), 347-351.

Space Science Board. (1987). A Strategy for Space Biology and Medical Science. Washington, D.C.: National Academy Press.

Stuster, J., Bachelard, C., & Suedfeld, P. (2000). The relative importance of behavioral issues during long duration ICE missions. Aviation, Space and Environmental Medicine, 71 (9), A17-A25.

Suedfeld, P. (2003). Canadian Space Psychology: The future may be almost here. Canadian Psychology, 44 (2), 85-92.

Suedfeld, P., & Steel, G. (2000). The environmental psychology of capsule habitats. Annual Review of Psychology, 51, 227-253.

Journal of Cosmology, 2010, Vol 12, 3778-3780.
JournalofCosmology.com, October-November, 2010

8

Challenges to the Musculoskeleton During a Journey to Mars: Assessment and Counter Measures

Yi-Xian Qin, Ph.D.

Department of Biomedical Engineering
State University of New York at Stony Brook, Stony Brook, NY

ABSTRACT

Musculoskeletal deterioration due to microgravity is a serious concern which can greatly impact the success of a mission to Mars. Accumulated data from short and median duration space missions (0.5-6 months) have indicated that microgravity environment significantly alters the musculoskeletal system with evidence of systematic bone loss and muscle atrophy, e.g., bone loss at an average rate of 1.5% per month. It is predicted that extended durations of space missions, such as a mission to Mars and possibly back to Earth, over a total period of 18 to 36 months of space flight, will significantly affect and increase risk of deterioration the bone and connective tissues, and which may be compounded by many unknowns. At present it is difficult to make reasonable predictions of bone loss in a prolonged space mission. This is due to the lack of on board assessments, weak understanding of the mechanism and countermeasurements, as well as limitations in the ability to monitor the effects of treatments. Musculoskeletal complications are also major health problems on Earth, i.e., osteoporosis, and delayed healing of fractures. The data from astronauts who have flown prolonged periods of time in space and the results from studies on Earth may help to maximize our understanding of the risks and challenges for the musculoskeletal tissues during a mission to Mars. For example, understanding age-related bone loss pattern may help us to predict the rate of bone loss in extended long-term space mission. Development of low mass, compact, noninvasive diagnostic and treatment technologies, i.e., using ultrasound, also offers great potential in monitoring longitudinal risks of bone loss, and may help to prevent and treat potential bone fractures. This paper will address critical questions in the Bioastronautics Roadmap related to musculoskeletal complications, predictions of bone loss in the extended space mission and potential assessments and countermeasurements during and following extended long space missions.

Key Words: microgravity, spaceflight, bone, muscle, countermeasures, Mars, bone loss, muscle atrophy, risk of fracture, ultrasound bone densitometry, osteopenia, aging, long term space mission

1. INTRODUCTION

MUSCULOSKELETAL DETERIORATION and its associated complications, i.e., disuse osteopenia and muscle atrophy, are significant threats for astronauts, which increase the risks of fractures during long-term space missions including staying in space station, lunar mission and during the trip to Mars. It has been over 40 years since the beginning of space exploration. Accumulated data have demonstrated that space flight, particularly in lone-term missions, has detrimental effects on bone and muscle. Results from short-term space missions (2-12 weeks) indicated that space flight with microgravity alters calcium metabolism and bone mineral density (BMD) in several hundreds of men and women who have flown to space. As human space exploration now plan and prepare to extend the mission beyond the orbital, for example, manned missions to Mars through extended manned vehicle with 18 to 36 months of duration, one can imagine that the risks and the challenges to the musculoskeletal system will be tremendous. However, there is so little progress has been made in understanding the significance of the problem. There is almost a complete lack of on-board measurements for assessing longitudinal bone loss and muscle atrophy, as well as associated evaluations of countermeasurement outcomes. Relatively short-term flight data and ground-based human research may help to estimate the risks for the long-term space mission.

Development of new technologies will also lead to a better understanding of the barriers to long-term space exploration and to assist in the development of countermeasures to assure safe and productive missions. As defined by the National Research Council report entitled: A Strategy for Research in Space Biology and Medicine in the New Century (Osborn 1998) and the Vision for Space Exploration of the Human Research Program (HRP), there are several areas of fundamental scientific investigations that must be addressed to meet these goals. As an important recommendation, which echo those made by previous reports published by the National Research Council's Space Studies

Board (Dutton 1992; Smith 1991), the principal physiologic hurdle to man's extended presence in space is the osteopenia, musculoskeletal deterioration and fracture that parallel to reduced gravity, and by the NASA's Bioastronautics Roadmap and Human Research Program risk plans.

To review musculoskeletal changes as an effect in the short-term space mission, Earthbased disuse osteoporosis results may help to predict the potential risks in the musculoskeletal system for the mission to Mars. Development of related technologies that has low mass, compact, highly reliable features will greatly impact the diagnosis and even therapeutics for bone loss and muscle atrophy during the long-duration space missions and even ground operations.

2. MICROGRAVITY INDUCED BONE LOSS

Elucidation of microgravity-induced skeletal disorders will lead to a better understanding of barriers to long-term space exploration and assist in the development of countermeasures to assure safe and productive missions. Osteopenia is a disease characterized by the long-term loss of bone tissue, particularly in the weight-supporting skeleton (Riggs & Melton, III 1995; Riggs et al. 2006; Riggs 2009). On average, the magnitude and rate of the loss is staggering; astronauts lose bone mineral in the lower appendicular skeleton at a rate approaching 1.5%-2% per month (Lang 2006; LeBlanc et al. 2000b; LeBlanc et al. 2007). While osteopenia can affect the whole body, complications often occur predominantly at specific sites of the skeleton with great load bearing demands. The greatest BMD loss has been observed in the skeleton of the lower body, i.e., the pelvic bones ($-11.99 \pm 1.22\%$) and the femoral neck ($-8.17 \pm 1.24\%$), while there was no apparent decay found in the skull region (LeBlanc et al. 1996; LeBlanc et al. 2000b; LeBlanc et al. 2007) (see Figure 1).

Moreover, it is apparent that full recovery of bone mass may never occur (Goode 1999; Shackleford et al. 1999; LeBlanc et al. 2000b), potentiating skeletal complications in the astronaut's later life. Similar results were found in the bed rest studies (LeBlanc et al. 2000b; LeBlanc et al. 2007). In a -6 degrees head-down tilt of 7-day bed rest model for microgravity, it was observed that there was a decreased bone formation rate in the iliac crest. Thus, assuming in a 2.5~3-year return-trip to Mars, approximately half of an astronaut's bone density may vanish, severely jeopardizing their health and well-being. The progressive adaptation of the human biological system for short and long-term space flight still remain largely unknown, i.e., current exercise countermeasure protocols can not sufficiently prevent bone loss (Grigoriev et al. 1998). One of the reasons is

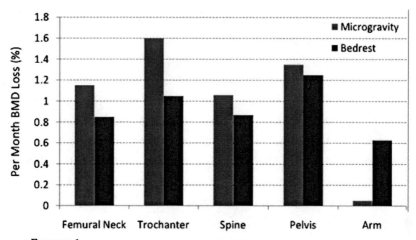

FIGURE 1

Short-term (< 6 months) space mission and bedrest induced the loss of bone mineral density, averaged per month loss. Data adapted from reference LeBlanc et al. (LeBlanc et al. 2000b; LeBlanc et al. 2007).

that it is extremely difficult in monitoring continuous adaptive decay of bone loss during a space flight. In order to understand these effects, we need a better description of human adaptation in space, and then with this information to create prevention and countermeasure strategies through new analysis and technology.

3. OSTEOPOROSIS ON EARTH

Osteopenia and osteoporosis are reductions in bone mass or density that lead to deteriorated and fragile bones. Osteoporosis diminishes both the structure and strength of bone, each considered to be critical in defining the ability of the bone to resist fracture. It is the leading cause of bone fractures in postmenopausal women and in the elderly population for both men and women. About 13 to 18 percent of women aged 50 years and older and 3 to 6 percent of men aged 50 years and older have osteoporosis in the US alone (Melton, III et al. 2010; National Osteoporosis Foundation 2010).

Approximately 24 million people suffer from osteoporosis in the United States with an estimated direct annual cost of over $18 billion to national health programs. Aging is one of the primary factors to induce osteoporotic bone loss in both women and men (Riggs et al. 2000; Riggs et al. 2002), as well as disuse osteopenia (Bloomfield 2010). Using areal BMD assessed by dual energy x-ray absorptiometry (DXA), longitudinal studies with multiple populations over a 25-year period have revealed that loss of BMD is sensitive to aging in both women and men (see Figure 2). Both women and men lose BMD starting at around age

40, with women losing BMD more rapidly than men starting in their early 50s (Clarke & Khosla 2010). From age 50, women lose trabecular BMD rapidly in their vertebrae, pelvis, and ultradistal wrist, while men do not show an accelerated BMD during this time period. About 10 years after 50, women lose approximately 1.3%/year of BMD in cortical bone and 3.4%/year in trabecular bone. During the same period, men lose about 1.0%/year and 1.2%/year of BMD in cortical and trabecular bones, respectively (Figure 2). The rate of age-related bone loss is slower in the following decades after the 60s in both women and men. For example, from 60-70, women and men lose approximately 0.7%/year and 1.0%/per year in cortical and trabecular BMD, respectively. In an averaged age-related BMD trend, combined women's and men's cortical and trabecular bones' data, a nonlinear BMD loss pattern was demonstrated (see Figure 3).

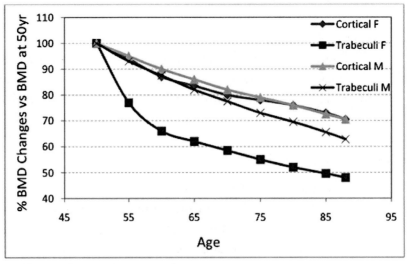

FIGURE 2

Age-related BMD loss in both cortical and trabecular bones in women and men. Longitudinal data were assessed using DXA. Data are adapted from Khosla A, Riggs BL: Pathophysiology of age-related bone loss and osteoporosis. Endocrinal Metab Clin N Am. 34(4): 1017, 2005 (Khosla & Riggs 2005).

4. MICROGRAVITY INDUCED BONE LOSS AND AGE-RELATED BMD REDUCTION

From the pattern of age-related bone loss (Fig. 2), it is suggested that bone loss occurs more rapid in women than men, particularly in the trabecular bone regions, e.g., women lose approximately 33% more trabecular bone mineral than men at age of 70. However, it is observed that there are relatively similar rates of

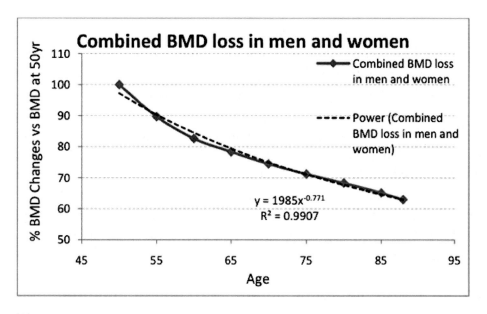

(A)

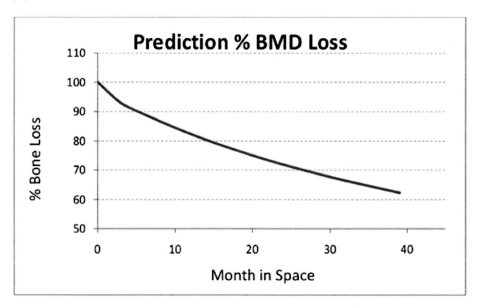

(B)

FIGURE 3

(A) *Age-related bone loss after 50 years old in combined men and women in both cortical and trabecular bones. The dashed line is a simulation using a Power equation (y = 1985x-0.771), which is strongly correlated to the combined osteoporotic bone loss in BMD (R² = 0.9907).*

(B) *Based on the similarity between age-related annular % bone loss rate and monthly % bone loss in microgravity, a pattern of bone loss prediction has shown that 12 months space mission may induce approximately averaged 18% bone loss in the skeleton following a power prediction relation, %bone loss = 1985*(50+m) -0.771.*

age-related cortical bone loss between men and women (Fig. 2). Based on the skeletal loss data in short-term space missions, significant BMD loss was observed among the astronauts population. Although the observation of bone loss occurred in both male and female astronauts, the physical condition of this group is among the healthiest in the population, e.g., controlled nutrition, active physical exercise, and closely medical monitor. If a combined agerelated BMD loss (i.e., averaging both cortical and trabecular bone loss in both men and women) is used, a phenomenological comparison may be applied between the aging population and astronauts in long term space missions, in which the age-related annual BMD loss rate may be similar to the monthly bone loss in space (Fig. 1, 3). This estimation may use age-related bone loss to predict the loss with spaceflight, which presumes that the astronauts (both male and female) will lose about the same amount of bone mass in one month as (combined female and male) aging Americans lose in one year. Based on this prediction, a power relation is proposed as %bone loss = 1985*(50+m) -0.771. It is estimated that long term space missions will significantly reduce bone density because of microgravity, in which it is predicted that a 30-month space mission would result in total of approximately 32% bone loss.

5. MUSCLE ATROPHY

Lower extremity muscle volume was also altered by disuse. Exposure to a 6-month space mission resulted in a decrease in muscle volume of 10% in the quadriceps, 19% in the gastrocnemius and soleus (LeBlanc et al. 2000a; LeBlanc et al. 2007; Ohshima 2010; Fitts et al. 2010; Trappe et al. 2009). Computed tomography measurements of the muscle cross sectional area (CSA) indicated a decrease of 10% in the gastrocnemius and 10-15% in the quadriceps after short-term missions (Narici et al. 2003). Similar results were concluded after spinal cord injury (SCI), where patients suffered significant 21%, 28% and 39% reductions in CSA at the quadriceps femoris, soleus and gastrocnemius muscles, respectively (Gorgey & Dudley 2007; Shah et al. 2006). In addition to the effects on whole muscle volume, muscle fiber characteristics were also altered due to inactivity (Roy et al. 2008; Zhong et al. 2005). There are two primary muscle fibers: slow (type I) fibers which play an important role in maintaining body posture while fast (type II) fibers that are responsive during physical activities. Under disuse conditions, all fiber types were decreased in size, by 16% for type I and by 23-36% for type II (Roy et al. 2008; Stewart et al. 2004; Zhong et al. 2005). The atrophied soleus muscles also underwent a shift from type I (-8% in fiber numbers) to type II fibers (Roy et al. 2008; Stewart et al. 2004).

Clinical muscle stimulations have been examined extensively in SCI patients to strengthen skeletal muscle and alleviate muscle atrophy with promising outcomes (Griffin et al. 2009; Lim & Tow 2007; Shields & Dudley-Javoroski 2007). A few physical training studies further investigated this electrical stimulation technique to determine its effect s on osteopenia. These studies showed mixed results with respect to bone density data (Dudley-Javoroski & Shields 2008; Griffin et al. 2009; Yang et al. 2009). Using dual energy X-ray absorptiometry, BeDell et al. found no change in BMD of the lumbar spine and femoral neck regions after functional electrical stimulation-induced cycling exercise, while Mohr et al. showed a 10% increase in BMD of the proximal tibia following 12 months of similar training (BeDell et al. 1996; Belanger et al. 2000). In a 24- week study of SCI patients in whom 25 Hz electrical stimulation was applied to the quadriceps muscles daily, Belanger and colleagues reported a 28% recovery of BMD in the distal femur and proximal tibia, along with increased muscle strength (Belanger et al. 2000). A number of reported animal studies also indicated that muscle stimulation can not only enhance muscle mass, but also bone mineral density (Swift et al. 2010). Both animal and human studies seem to strongly support that functional disuse can result in significant bone loss and muscle atrophy.

6. NONINVASIVE OSTEOPOROSIS DIAGNOSIS

At present, osteoporotic bone loss is commonly assessed by measures of bone mineral density that reflects *in vivo* conditions of bone mass. Noninvasive measurements of BMD would be of value in predicting the risks of fracture, in assessing the severity of the disease, and in following responses to treatment. Several methods are available for the ground-based clinical measurement of bone mass, with the most commonly used methods being DXA and computed tomography (CT), and emerging technology like micro-CT. These modalities are capable to assess both bone density and structure properties of bone, e.g., trabecular orientation.

DXA is currently the gold standard technique being used because of its relative precision (~2%) and its whole body and/or multi-site imaging abilities (spine, hip, wrist and total skeleton). However, because the source and the detector cross the whole bone (including layers of cortex and trabeculae), current techniques are apparently insensitive for quantifying trabecular bone mass separately. That is, a certain percentage of bone mass must be lost before significant radiation attenuation occurs. In addition, DXA and CT, have limitations on X-ray ionizing radiation (70-140KVp, 10 mrem). There is a lack of precision due to surrounding soft tissue variations, and issues of non-repeatability due to patients' position. Another limitation of the X-ray based approaches may be a lack of the ability to assess bone's integrity information on mechanical property which may be directly related to bone's potentials to resist fracture. Due to these issues, the quality of bones (i.e., structure and strength properties), whether in normal or osteopenic, remains unknown (i.e. it is extremely difficult to monitor the strength and the conductivity in vivo). As a result, an improved diagnostic tool is needed to evaluate both the quantity and the quality of bone, which will help in the early detection and therefore the possible prevention and treatment for this disease.

7. BONE QUALITY AND FRACTURE

If not only bone mineral density but also bone quality, e.g. stiffness and/or modulus, can be monitored or determined instantly during the space mission, then one can better understand the skeleton adaptation on a daily basis. In the case of osteopenia and/or osteoporosis, fractures can occur without a singular traumatic event. Such a database would provide detailed and progressive information of the skeletal modeling/remodeling response during the condition of microgravity, which would potentially direct treatment regimens to prevent or recover bone loss.

Unfortunately, a skeleton at risk of fracture cannot simply be determined by the amount of bone mass (e.g., BMD) that exists; to a larger degree, the quality of the bone is just as important. While a formal definition of bone quality is somewhat elusive, at the very least it incorporates architectural, physical and biological factors that are critical to bone strength, such as bone morphology (i.e., trabecular connectivity, cross sectional geometry), properties of the tissue's material (e.g., stiffness, strength), and its chemical composition and architecture (e.g., calcium concentration, collagen orientation, porosity, permeability). The ability to directly assess both bone density and quality (i.e., strength) would have great impact in predicting the risk of fracture.

8. QUANTITATIVE ULTRASOUND AND BONE QUALITY ASSESSMENT

Recently, new methods in QUS have emerged with the potential to estimate cancellous bone modulus more directly. The primary advantage of ultrasonic techniques (UT) is that it is capable of measuring not only bone quantity, but also bone quality, i.e., estimation of the mechanical property of bone. Over the past 15 years, a number of research approaches have been developed to quantitate bone mass and structural stiffness using UT (Langton & Njeh 2008; Zheng et al. 2009). The advantages of QUS in comparison to X-ray-based technologies may be that it's relatively simple and an inexpensive format that is non-invasive and free of ionizing radiation.

Research efforts have been made on bone quality assessments using image based QUS (Laugier et al. 2000; Qin et al. 2001; Qin et al. 2002). Preliminary results have shown that ultrasound imaging is capable to detect bone quality non-invasively through a developing scanning confocal acoustic navigation (SCAN) system (Qin et al. 2002; Qin et al. 2004; Qin et al. 2008; Xia Y. et al. 2005; Xia et al. 2007) (see Figure 4).

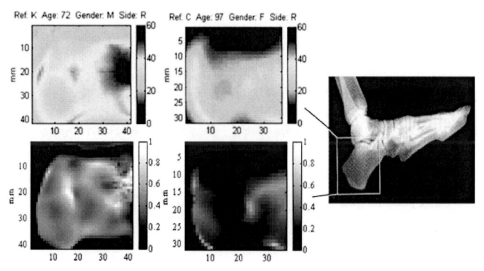

FIGURE 4

Calcaneus transmitted ultrasound images measured by the scanning confocal acoustic navigation technology. Up-left: broadband ultrasound attenuation image for a 72 years old male with normal bone mineral density. Up-right: broadband ultrasound attenuation image for a 97 years old female with severe osteoporotic bone loss. Bottoms: combined ultrasound attenuation and wave velocity images for the corresponding two representative cases, respectively. Right: regular X-ray image for the foot.

If QUS bone densitometry can be developed to provide "true" bone quality parameter-based diagnostic tools, i.e., directly related to bone's structural and strength properties, and to target to multiple and critical skeletal sites such as hip and distal femur, QUS would have a greater impact on the diagnosis of bone diseases (e.g., osteoporosis) than currently available bone densitometry. The SCAN is intended to provide true images reflecting the bone's structural and strength properties at a particular skeletal site of a peripheral limb and potentially in deep tissues like the great trachanter (Nicholson et al. 2001). The technology may further provide both density and strength assessments in the region of interests for the risks of fracture (Qin et al. 2002; Xia et al. 2005).

Most of the available systems measure the calcaneus using plane waves that utilize either water or gel coupling, e.g., Sahara (Hologic Inc., MA), QUS-2 (Metra Biosystems Inc., CA), Paris (Norland Inc., WI), and UBA 575 (Walker Sonix Inc., USA). At the beginning of the year, a bone scanning densitometry device for calcaneus ultrasound measurement was also made available using array plane ultrasound waves (GE-Lunar, Inc., USA). Using several available clinical devices, studies in vivo have shown the ability of QUS to discriminate patients with osteoporotic fractures from agematched controls (Njeh et al. 1997). It has been demonstrated that QUS predicts risks of future fracture generally as well as DXA (Bauer et al. 1997; Njeh et al. 1997). Table 1 has illustrated current quantitative ultrasound devices for calcaneus measurements.

TABLE 1

Summary of current QUS devices for human calcaneus

Device	Performance	Resolution	Predicted Parameter
Sahara (Hologic)	Index	Non-image	Z score
QUS-2 (Metra Biosystem)	Index	Non-image	Z score
UBA 575 (Walker Sonix)	Index	Non-image	Z score
Achilles (GE-Lunar)	Index+image	image, 5mm	Z score
UBIS 5000 (DMS)	Index+image	Image, 2mm	Z score
DTU-one (Osteometer)	Index+image	Image, 2mm	Z score
New SCAN	Image+index	Image, 1mm	Stiffness, BMD, Z

9. QUANTITATIVE ULTRASOUND PARAMETERS IN BONE MEASUREMENT

Ultrasound may be applied to bone using a number of fundamental physical mechanisms: ultrasonic wave propagation velocity (UV) or speed of sound (SOS), energy attenuation (ATT), broadband ultrasound attenuation (BUA), and critical angle ultrasound parameters. Large perspective studies have confirmed that QUS measurements of BUA and UV can identify those individuals at risks of osteoporotic fracture as reliably as BMD (Langton & Njeh 2008; Zheng et al. 2009; Frost et al. 2001). It has been shown that both BUA and UV are decreased in individuals with risk factors for osteoporosis, i.e., primary hyperparathyroidism (Gomez et al. 1998), kidney disease (Wittich et al. 1998), and glucocorticoid use (Blanckaert et al. 1997). The proportion of women classified into each diagnostic category was similar for BMD and QUS. The strength of trabecular bone is an important parameter for bone quality. *In vitro*studies have correlated the UV with stiffness in trabecular bone samples (Xia et al. 2007). This indicates that ultrasound has the potential to be advantageous over the Xray based absorptiometry in assessing the quality of bone in addition to the quantity of bone. By determining the wave velocity through bone, the elastic modulus of bone specimens can be estimated.

10. CLINICAL APPLICATION & THERAPEUTIC EFFECTS OF ULTRASOUND

Low-intensity pulsed ultrasound (LIPUS) is a form of mechanical energy that is transmitted through and into biological tissues as an acoustic pressure wave and has been widely used in medicine as a non-invasive therapeutic tool that has shown an accelerated rate of healing of fresh fractures (Favaro-Pipi et al. 2010; Qin et al. 2006). Substantial basic science data demonstrated that ultrasound has a strong positive influence on each of the three key stages of the healing process (inflammation, repair, and remodeling) because it enhances angiogenic, chondrogenic, and osteogenic activities. The mechanism of ultrasound therapy for fracture healing may be related to the differential energy absorption of ultrasound that gives rise to acoustic streaming. Its resultant fluid flow as a mechanotransduction signal (Qin et al. 2003; Cowin 2002; Cowin & Cardoso 2010) and induced gradients of mechanical strain are recognized as strong determinants of bone-modeling. It has been confirmed by clinical evidence that ultrasound has a role in the treatments of delayed unions and nonunions. The efficacy of LIPUS stimulation in acceleration of the normal fracture repair process was even observed when performed with a diagnostic sonographic device (Heybeli et al. 2002). However, the effectiveness of localized LIPUS with optimized intensity has been not fully investigated. Therapeutic ultrasound and some operative ultrasound use intensities as high as 1- 30 W/cm², which may cause considerable heating in living tissues. The use of ultrasound as a surgical instrument involves even higher levels of intensity (5 to 300 W/cm²), and sharp bursts of energy are used to fragment calculi, to initiate the healing of nonunions, to ablate diseased tissues such as cataracts, and even to remove methylmethacrylate cement during revision of prosthetic joints (Wells 2001). The intensity level used for imaging, which is five orders of magnitude below that used for surgery, is regarded as non-thermal and nondestructive. Current therapeutic ultrasound uses plane waves and exposes the energy to a broad range of tissues. The radical changes of density inherent in a healing callus may result in a significant lose in the amounts of energy in the pathway of the ultrasound. A localized and targeted/guided exposure of LIPUS may overcome these limitations and dramatically increase the efficiency of the treatment. A proposed combined diagnostic and therapeutic QUS with focal scan model can overcome these.

Thus, a non-invasive assessment of trabecular bone strength and density is extremely important in predicting the risks of fracture in space and ground operations. QUS has emerged with the potential to directly detect trabecular bone strength. To overcome the current hurdles such as soft tissue and cortical shell interference, improving the "quality" of QUS and the application of the technology for future clinical usages, the development of image based ultrasound technology will concentrate on several main areas: (1) increasing the resolution, sensitivity, and accuracy in diagnosing longitudinal bone loss to predict the risk of fracture and evaluate treatment effects; (2) predicting local trabecular bulk stiffness and the microstructure of bone, and generating a physical relationship between ultrasound parameters and bone quality; and (3) providing targeted treatment for stress fracture and traumatic fracture using ultrasound. An integrated diagnostic and therapeutic ultrasound system may ultimately provide noninvasive, non-radiation, portable and low weight modality for longitudinal bone loss in extended space mission.

11. SUMMARY

Based on the results of musculoskeletal loss in a 6-month stay on the international space station, it is clear that musculoskeletal tissues will face accelerated decay under a microgravity environment. A trip to Mars and possibly back to Earth, can take between 18-30 months in space, and will have significant impacts on bone and muscle. It is important that the prediction of such a loss may be implemented with age-related human osteoporotic bone mineral density loss, which in turn may provide understanding of the rate and the pattern of bone loss in an extended microgravity environment. The development of suitable technologies, like low weight and noninvasive ultrasound, may impact both diagnosis and countermeasure treatment in future long duration space missions.

Acknowledgements: This work is kindly supported by the National Space Biomedical Research Institute through NASA contract NCC 9-58, the National Institute of Health (R01 AR52379 and R01AR49286), and the US Army Medical Research and Materiel Command. The author wishes to thank Maria Pritz and Minyi Hu for their excellent technical assistance on this manuscript.

REFERENCES

Bauer DC, Gluer CC, Cauley JA, Vogt TM, Ensrud KE, Genant HK & Black DM (1997). Broadband ultrasound attenuation predicts fractures strongly and independently of densitometry in older women. A prospective study. Study of Osteoporotic Fractures Research Group. Arch.Intern.Med. 157:629-634.

BeDell KK, Scremin AM, Perell KL & Kunkel CF (1996). Effects of functional electrical stimulation-induced lower extremity cycling on bone density of spinal cord-injured patients. Am.J Phys.Med.Rehabil. 75:29-34.

Belanger M, Stein RB, Wheeler GD, Gordon T & Leduc B (2000). Electrical stimulation: can it increase muscle strength and reverse osteopenia in spinal cord injured individuals? Arch.Phys.Med.Rehabil. 81:1090-1098.

Blanckaert F, Cortet B, Coquerelle P, Flipo RM, Duquesnoy B, Marchandise X & Delcambre B (1997). Contribution of calcaneal ultrasonic assessment to the evaluation of postmenopausal and glucocorticoid-induced osteoporosis. Rev.Rhum.Engl.Ed 64:305-313.

Bloomfield SA (2010). Disuse osteopenia. Curr.Osteoporos.Rep. 8:91-97.

Clarke BL & Khosla S (2010). Physiology of bone loss. Radiol.Clin.North Am. 48:483-495.

Cowin SC (2002). Mechanosensation and fluid transport in living bone. J.Musculoskelet.Neuronal.Interact. 2:256-260.

Cowin SC & Cardoso L (2010). Fabric dependence of wave propagation in anisotropic porous media. Biomech.Model.Mechanobiol. 64:305-313.

Dudley-Javoroski S & Shields RK (2008). Dose estimation and surveillance of mechanical loading interventions for bone loss after spinal cord injury. Phys.Ther. 88:387-396.

Dutton JA 1992 Setting Priorities For Space Research:Opportunities and Imperatives. Washington D.C.: Space Studies Board National Resarch Council National Academy Press.

Favaro-Pipi E, Bossini P, de Oliveira P, Ribeiro JU, Tim C, Parizotto NA, Alves JM, Ribeiro DA, Selistre de Araujo HS & Muniz Renno AC (2010). Low-Intensity Pulsed Ultrasound Produced an Increase of Osteogenic Genes

Expression During the Process of Bone Healing in Rats. Ultrasound Med.Biol.64:305-313.

Fitts RH, Trappe SW, Costill DL, Gallagher PM, Creer AC, Colloton PA, Peters JR, Romatowski JG, Bain JL & Riley DA (2010). Prolonged space flight-induced alterations in the structure and function of human skeletal muscle fibres. J.Physiol 588:3567-3592.

Frost ML, Blake GM & Fogelman I (2001). Quantitative ultrasound and bone mineral density are equally strongly associated with risk factors for osteoporosis. J.Bone Miner.Res. 16:406-416.

Gomez AC, Schott AM, Hans D, Niepomniszcze H, Mautalen CA & Meunier PJ (1998). Hyperthyroidism influences ultrasound bone measurement on the Os calcis. Osteoporos.Int. 8:455-459.

Goode A (1999). Musculoskeletal change during spaceflight: a new view of an old problem. Br.J.Sports Med. 33:154.

Gorgey AS & Dudley GA (2007). Skeletal muscle atrophy and increased intramuscular fat after incomplete spinal cord injury. Spinal Cord. 45:304-309.

Griffin L, Decker MJ, Hwang JY, Wang B, Kitchen K, Ding Z & Ivy JL (2009). Functional electrical stimulation cycling improves body composition, metabolic and neural factors in persons with spinal cord injury. J Electromyogr.Kinesiol. 19:614-622.

Grigoriev, A. I., Oganov, V. S., Bakulin, A. V., Poliakov, V. V., Voronin, L. I., Morgun, V. V., Shnaider, V. S., Murashko, L. V., Novikov, V. E., LeBlanc, A., and Shackelford, L. (1998). Clinical and Psychological Evaluation of Bone Changes Among Astronauts after Long Term Space Flights (Russian). Aviakosmicheskaia I Ekologicheskaia Meditsina 32[1], 21-25.

Heybeli N, Yesildag A, Oyar O, Gulsoy UK, Tekinsoy MA & Mumcu EF (2002). Diagnostic ultrasound treatment increases the bone fracture-healing rate in an internally fixed rat femoral osteotomy model. J.Ultrasound Med. 21:1357-1363.

Khosla S & Riggs BL (2005). Pathophysiology of age-related bone loss and osteoporosis. Endocrinol.Metab Clin.North Am. 34:1015-30, xi.

Lang TF (2006). What do we know about fracture risk in long-duration spaceflight? J Musculoskelet.Neuronal.Interact. 6:319-321.

Langton CM & Njeh CF (2008). The measurement of broadband ultrasonic attenuation in cancellous bone--a review of the science and technology. IEEE Trans.Ultrason.Ferroelectr. Freq.Control 55:1546-1554.

Laugier P, Novikov V, Elmann-Larsen B & Berger G (2000). Quantitative ultrasound imaging of the calcaneus: precision and variations during a 120-Day bed rest. Calcif.Tissue Int. 66:16-21.

LeBlanc A, Lin C, Shackelford L, Sinitsyn V, Evans H, Belichenko O, Schenkman B, Kozlovskaya I, Oganov V, Bakulin A, Hedrick T & Feeback D (2000a). Muscle volume, MRI relaxation times (T2), and body composition after spaceflight. J Appl.Physiol 89:2158-2164.

LeBlanc, A., Schneider, V., and Shackelford, L. 1996. Bone Mineral and Lean Tissue Loss after Long Duration Spaceflight. Trans.Amer.Soc.Bone Min.Res. 11S, 567.

LeBlanc A, Schneider V, Shackelford L, West S, Oganov V, Bakulin A & Voronin L (2000b). Bone mineral and lean tissue loss after long duration space flight. J.Musculoskelet.Neuronal.Interact. 1:157-160.

LeBlanc AD, Spector ER, Evans HJ & Sibonga JD (2007). Skeletal responses to space flight and the bed rest analog: a review. J.Musculoskelet.Neuronal.Interact. 7:33-47.

Lim PA & Tow AM (2007). Recovery and regeneration after spinal cord injury: a review and summary of recent literature. Ann.Acad.Med.Singapore 36:49-57.

Melton LJ, III, Christen D, Riggs BL, Achenbach SJ, Muller R, van Lenthe GH, Amin S, Atkinson EJ & Khosla S (2010). Assessing forearm fracture risk in postmenopausal women. Osteoporos.Int. 21:1161-1169.

Narici M, Kayser B, Barattini P & Cerretelli P (2003). Effects of 17-day spaceflight on electrically evoked torque and cross-sectional area of the human triceps surae. Eur.J Appl.Physiol 90:275- 282.

National Osteoporosis Foundation (2010). Get the Facts on Osteoporosis. http://www.nof.org. Nicholson PH, Muller R, Cheng XG, Ruegsegger P, Van der PG, Dequeker J & Boonen S (2001). Quantitative ultrasound and trabecular architecture in the human calcaneus. J.Bone Miner.Res. 16:1886-1892.

Njeh CF, Boivin CM & Langton CM (1997). The role of ultrasound in the assessment of osteoporosis: a review. Osteoporos.Int. 7:7-22.

Ohshima H (2010). [Musculoskeletal rehabilitation and bone. Musculoskeletal response to human space flight and physical countermeasures]. Clin.Calcium 20:537-542.

Osborn M 1998 A Strategy for Research in Space Biology and Medicine in the New Century. Washington D.C: Space Studies Board , National Research Council, National Academy Press.

Qin L, Lu H, Fok P, Cheung W, Zheng Y, Lee K & Leung K (2006). Low-intensity pulsed ultrasound accelerates osteogenesis at bone-tendon healing junction. Ultrasound Med.Biol. 32:1905-1911.

Qin Y-X, Lin W & Rubin C (2001). Interdependent relationship between Trabecular Bone Quality and Ultrasound Attenuation and Velocity Using a Scanning Confocol Acoustic Diagnostic System. J Bone Min Res 16:S470-70.

Qin Y-X, Xia Y, Lin W, Chadha A, Gruber B & Rubin C (2002). Assessment of bone quantity and quality in human cadaver calcaneus using scanning confocal ultrasound and DEXA measurements. J Bone Min Res 17:S422.

Qin YX, Kaplan T, Saldanha A & Rubin C (2003). Fluid pressure gradients, arising from oscillations in intramedullary pressure, is correlated with the formation of bone and inhibition of intracortical porosity. J Biomech. 36:1427-1437.

Qin YX, Xia Y, Lin W, Cheng J, Muir J & Rubin C (2008). Longitudinal assessment of human bone quality using scanning confocal quantitative ultrasound. J.Acoust Soc.Am. 123:3638.

Qin YX, Xia Y, Lin W, Rubin C & Gruber B (2004). Assessment of trabecular bone quality in human calcaneus using scanning confocal ultrasound and dual x-ray absorptiometry (DEXA) measurements. J.Acoust.Soc.Am. 116:2492.

Riggs BL (2009). Bone turnover in age-related osteoporosis: early contributions of Pierre Delmas. Osteoporos.Int. 20:1289-1290.

Riggs BL, Khosla S & Melton LJ, III (2000). Primary osteoporosis in men: role of sex steroid deficiency. Mayo Clin.Proc. 75 Suppl:S46-S50.

Riggs BL, Khosla S & Melton LJ, III (2002). Sex steroids and the construction and conservation of the adult skeleton. Endocr.Rev. 23:279-302.

Riggs BL & Melton LJ, III (1995). The worldwide problem of osteoporosis: insights afforded by epidemiology. Bone 17:505S-511S.

Riggs BL, Melton LJ, III, Robb RA, Camp JJ, Atkinson EJ, Oberg AL, Rouleau PA, McCollough CH, Khosla S & Bouxsein ML (2006). Population-based analysis of the relationship of whole bone strength indices and fall-related loads to age- and sex-specific patterns of hip and wrist fractures. J.Bone Miner.Res 21:315-323.

Roy RR, Pierotti DJ, Garfinkel A, Zhong H, Baldwin KM & Edgerton VR (2008). Persistence of motor unit and muscle fiber types in the presence of inactivity. J Exp.Biol. 211:1041-1049.

Shackleford L, LeBlanc A, Feiveson A, Oganov V & . (1999). Bone loss in space: Shuttle/Mir experience and bed rest countermeasure program. 1st Biennial Space Biomed Inv.Workshop 1:86-87.

Shah PK, Stevens JE, Gregory CM, Pathare NC, Jayaraman A, Bickel SC, Bowden M, Behrman AL, Walter GA, Dudley GA & Vandenborne K (2006). Lower-extremity muscle crosssectional area after incomplete spinal cord injury. Arch.Phys.Med.Rehabil. 87:772-778.

Shields RK & Dudley-Javoroski S (2007). Musculoskeletal adaptations in chronic spinal cord injury: effects of long-term soleus electrical stimulation training. Neurorehabil.Neural Repair 21:169-179.

Smith LD (1991). Assesment of Programs in Space Biology and Medicine. Washington, D.C.: Space Studies Board, National Research Council, National Academy Press.

Stewart BG, Tarnopolsky MA, Hicks AL, McCartney N, Mahoney DJ, Staron RS & Phillips SM (2004). Treadmill training-induced adaptations in muscle phenotype in persons with incomplete spinal cord injury. Muscle Nerve 30:61-68.

Swift JM, Nilsson MI, Hogan HA, Sumner LR & Bloomfield SA (2010). Simulated resistance training during hindlimb unloading abolishes disuse bone loss and maintains muscle strength. J.Bone Miner.Res 25:564-574.

Trappe S, Costill D, Gallagher P, Creer A, Peters JR, Evans H, Riley DA & Fitts RH (2009). Exercise in space: human skeletal muscle after 6 months aboard the International Space Station. J.Appl.Physiol 106:1159-1168.

Wells PN (2001). Physics and engineering: milestones in medicine. Med.Eng Phys. 23:147-153. Wittich A, Vega E, Casco C, Marini A, Forlano C, Segovia F, Nadal M & Mautalen C (1998). Ultrasound velocity of the tibia in patients on haemodialysis. J Clin Densitometry 1:157-163.

Xia Y., Lin W. & Qin Y.X. (2005). The influence of cortical end-plate on broadband ultrasound attenuation measurements at the human calcaneus using scanning confocal ultrasound. The Journal of the Acoustical Society of America 118:1801-1807.

Xia Y, Lin W & Qin YX (2007). Bone surface topology mapping and its role in trabecular bone quality assessment using scanning confocal ultrasound. Osteoporos.Int. 18:905-913.

Yang YS, Koontz AM, Triolo RJ, Cooper RA & Boninger ML (2009). Biomechanical analysis of functional electrical stimulation on trunk musculature during wheelchair propulsion. Neurorehabil.Neural Repair 23:717-725.

Zheng R, Le LH, Sacchi MD & Lou E (2009). Broadband ultrasound attenuation measurement of long bone using peak frequency of the echoes. IEEE Trans.Ultrason.Ferroelectr.Freq.Control 56:396-399.

Zhong H, Roy RR, Siengthai B & Edgerton VR (2005). Effects of inactivity on fiber size and myonuclear number in rat soleus muscle. J Appl.Physiol 99:1494-1499.

Journal of Cosmology, 2010, Vol 12, 3846-3854.
JournalofCosmology.com, October-November, 2010

- 9 -
Infection Risk of
a Human Mission to Mars

Mihai G. Netea, Ph.D.[1,2],
Frank L. van de Veerdonk, Ph.D.[1,2],
Marc Strous, Ph.D.[2,3],
and Jos W.M. van der Meer, Ph.D.[1,2],

[1]Department of Medicine and [2]Nijmegen Institute for Infection,
Inflammation and Immunity (N4i), Radboud University
Nijmegen Medical Center, Nijmegen, The Netherlands.
[3]Max Planck Institute for Marine Microbiology, Bremen, Germany

ABSTRACT

Liquid water has almost certainly been a feature on Mars in its earlier history, and the presence of extinct or present life on Mars cannot be excluded. However, based on our current understanding of host-pathogen relationships and evolutionary processes, we may conclude that the chance of a human mission to Mars to encounter pathogenic microorganisms is small, albeit not zero. A set of safety measures to prevent, diagnose and eventually treat infections with Martian microorganisms should be considered, and such measures could even further diminish any potential biohazards. Overall, the scientific, technological and economical benefits of a mission to Mars will heavily outweigh the low probability of an encounter with a pathogenic microbe, and therefore this should not be an impediment for pursuing human exploration of Mars.

Key Words: Mars, Interplanetary Travel, Disease, Infection, Plague, Planetary Protection

1. INTRODUCTION

A HUMAN MISSION TO MARS IS ONE of the most important steps in the exploration of the Solar System in the decades to come (Mitchell and Staretz 2010; Zubrin 2010). While the benefits of a human mission to Mars are obvious from a scientific to technological view, one has to insure that such a mission is undertaken with the appropriate safety measures (Straume et al., 2010; Stuster 2010). If the history of space exploration can be used as a lesson, the highest risk for such a mission will be catastrophic vehicle failure. However, visiting an alien planet will also come with a certain degree of biological risk, in terms of infection or contamination of either the astronauts, the technical crew on the ground, or even Earth ecosystems upon return of the mission. In this article we will review the current knowledge concerning the biohazard of potential Martian microbes.

2. THE PROBABILITY OF LIFE AND PATHOGENIC MICROORGANISMS ON MARS

The probability of life on Mars is one of the most exciting questions to the scientific community (Houtkooper and Schulze-Makuch, 2010; Levin 2010; Leuko et al. 2010; Sephton 2010), but no firm answers have been given during the decades of space exploration. Many believed that life on Earth began in water (Russell and Kanik, 2010), and it is also currently thought that liquid water is one of the prerequisites of life on any planet (Brack, 1999). Mars mapping by Mariner 9 and by Viking 1 and 2 revealed channels resembling riverbeds, and information collected by the Mars Global Surveyor (MGS) strengthened the case for early surface water on Mars (Malin and Edgett, 2000a). Another investigation based on photographic images provided by MGS shows relatively young geological features compatible with the presence of liquid water possible as recent as a few million years ago (Malin and Edgett, 2000b). Even more striking, data from Mars Exploration Rovers discovered round pebbles scattered on the surface of *Meridiani Planum*, suggesting that this region has once been submerged (Squyres, et al., 2004).

Current water on Mars appears to be concentrated in the Northern as well as Southern Polar Ice Cap (Titus, et al., 2003), and as subsurface ice in other areas of the planet (Picardi, et al., 2005). In addition to water, one would consider the presence of organic substances as an additional argument for possible biological processes (Levin 2010). In several Martian meteorites organic matter of extraterrestrial origin has been identified (Jull, et al., 2000; Wright, et al., 1989). However, the *in-situ* experiments searching for Martian life carried out by Viking 1 and 2 in 1976 could not provide unequivocal signs of life or metabolic activity

(Klein, et al., 1976), though some investigators believe otherwise (Levin, 2010). Specifially, Levin (2010), argues that since the Viking Mission Labeled Release experiment, which exploits the sensitivity of ^{14}C respirometry, provided positive responses at both the Viking 1 and 2 sites on Mars, that this evidence favors the presence of life. Even so, the consensus favored chemical or physical agents in the Martian surface material, not life.

Studies on the Martian meteorite ALH84001 that reported the discovery of carbonate granules resembling microfossils, have been hailed as the first probable direct evidence of life outside Earth (McKay, et al., 1996). Within the same meteorite, magnetite crystals with properties compatible with biogenic terrestrial magnetite have also been found (Thomas-Keprta, et al., 2002). However, these reports remain controversial, as non-biological processes have also been proposed to explain the features found in ALH84001 meteorite (Scott, 1999).

The detection of methane and formaldehyde in Mars's atmosphere could be another indication that microbes exist on Mars (Formisano, et al., 2004; Yung et al., 2010). Although the presence of methane could be a mere sign of geographical processes (e.g. serpentinization), it is well known that most of the methane in Earth's atmosphere is produced by microbes. Furthermore, methane concentrations have been found to vary in space and time, suggesting that it is produced or released in a dynamic process and this could support the idea of present life on Mars (Sephton 2010).

All in all, the direct arguments for microbial life on Mars remain equivocal. While proof of liquid water, plate tectonics and a dense atmosphere on early Mars is convincing, the current habitats on Mars are currently extremely hostile. The absence of a magnetic field results in intense ionizing radiation on the Martian surface. The associated damage to DNA or RNA based life forms would presumably wipe out any forms of life we know down to 7.5 m in the sediment (in the absence of active DNA repair, Dartnell et al 2007). The absence of plate tectonics (the major potential chemotrophic energy source for growth) and the low water activity at the surface are the other major challenges to life as we know it. If there is any analogue for Martian life on Earth, it is the deep biosphere, with bacteria in low cell numbers, estimated doubling times of thousands of years in substrate depleted, anaerobic environments. It is interesting to note that the deep biosphere may contain up to 30% of the living biomass on Earth (Whitman et al 1998) and from that perspective the two planets may not be so different than their completely different planetary surfaces suggest. Therefore, an analysis of the potential biohazard imposed by Martian microbes is warranted.

3. WHAT IS THE PATHOGENIC POTENTIAL OF PUTATIVE MARTIAN LIFE FORMS?

The first aspect when one asks of the potential of existence of life on Mars regards the life forms that would be most likely to be encountered. The most obvious conclusion based on the observations made until now is that Martian life forms will be necessarily microscopic. Microscopic life on Earth includes viruses, archaea, bacteria, fungi, unicellular plants, and protozoa. The more complex forms of unicellular life are unlikely to be present in the hostile environment of Mars (such as fungi, unicellular plants and protozoa).

Viruses are obligate intracellular organisms, needing a host to replicate and transmit genetic information, which make them additionally vulnerable to extreme environments. However, the example of bacteriophage viruses of extremophile bacteria provides the example of virus presence in extreme environments (Prigent, et al., 2005; Sawstrom, et al., 2008). Thus the possibility of Martian viruses (in case of the presence of bacteria) cannot be completely ruled out.

Therefore, the most likely life form on Mars would be bacterial life. Bacteria are not only the most resilient and adaptable of known life forms but they are also the only class of organisms that can survive "stand alone" and do not rely on other life forms. In addition, the microfossils present in the ALH84001 meteorite are most plausible bacterial (nanobacteria), if they are indeed of biologic origin (McKay, et al., 1996).

How likely is that a Martian bacteria would be pathogenic for humans, or disruptive for an Earth ecosystem? An excellent analysis of these aspects have been provided by Schuerger, based on a plant infection model of pathogenicity (Schuerger, 1998). Terrestrial plant-microbe interaction can be classified by non-interactive, saprotrophism, necrotrophism, biotrophism, symbiosis and commen-salism (Schuerger, 1998). The last four categories imply a certain degree of adaptation and co-evolution between the microbial life and the multicellular organism, and this is absent between microbial Martian life forms and humans (or other Earth organisms). This would imply that the most likely interaction between microbes on Mars and astronauts would be non-interactive or saprotrophic, and hence most likely nonpathogenic. The chance of a Martian microbe, adapted to extremely slow, cold and anaerobic conditions having the ability to attach to cells of a terrestrial host and invade its cells or tissues, and hence produce infection, in full competition with terrestrial microbes, is very small. Less likely is even transmission to a second 'vulnerable' host.

However, a pathogenic potential of Martian microbes cannot be excluded either. Even if they were not capable of directly invading the host and causing infection, Martian microbes could still have pathogenic potential by secreting toxins that could indirectly harm the astronauts (e.g. through wounds, contaminated food). Examples of powerful microbial toxins secreted by terrestrial bacteria indeed abound, e.g. clostridial toxins (Lebrun, et al., 2009). Still, one has to recognize that the majority of such toxins of terrestrial bacteria are proteins, which in turn are recognized by specific cellular receptors, again requiring a history of previous interaction between the pathogenic agent and the host. Would such putative toxins of Martian microbes also be proteins, would they have similar biochemistry, would they even be made of the same aminoacids? Although it is possible that through mechanisms know as panspermia (Joseph and Schild 2010a,b) that microbes from Earth could be transported to Mars (and vice versa) thereby providing opportunities for horizontal gene transfer and thus giving Martian microbes human-infective properties (Joseph and Wickramasinghe 2010), at present there is no hard evidence to substantiate these theories. Thus, these are all questions that cannot be answered at present. Still, how minimal the chance that there may be pathogenic microorganisms on Mars, one cannot completely rule it out (Rummel et al. 2010).

A different aspect of the biohazard potential of Martian microbes is the capacity of such microorganisms to disrupt Earth ecosystems, should contaminated material from a Mars mission reach the environment upon return (Rummel et al. 2010). This risk is most likely also small, as environmental conditions such as temperature, humidity, chemistry, atmospheric pressure, and nutrients fundamentally differ between Earth and Mars. From an evolutionary point of view, it is highly unlikely that a Martian microbe that in Earth terms would be characterized as an extremophile would be able to compete successfully with terrestrial microorganisms, which are optimally adapted to the environment through millions years of evolution. However, long-term subtle influences on terrestrial ecosystems might be induced by introduction of Martian microorganisms and thereby represent a potential hazard. As a conclusion, the National Research Council Space Studies Board who assessed the biohazard posed by Martian microorganisms considered the risk of "back-contamination" as small, but not inexistent, and recommended that spacecraft and samples returning from Mars to be treated as potentially hazardous (National Research Council Space Studies Board, 1997; see also Rummel et al. 2010).

The reverse should also be considered, i.e., the potential contamination of Mars with Earth microorganisms. Although it is not likely that hitchhiking microbes can survive on the harsh surface of Mars, - likewise they should behave as extremophiles - it could be that during drilling missions in the search for water or during accidents such as a vehicle crash during landing, terrestrial microbes gain access to spots where they could potentially survive and multiply. Although this will not immediately result in a risk for human explorers, time and adaptation may result in a biohazard during returning missions to Mars.

4. PREVENTIVE MEASURES AGAINST BIOHAZARDS DURING A MARS MISSION

Despite the low probability of pathogenic microorganisms as indicated above, it cannot be excluded that Mars harbors microscopic life, and the possibility that astronauts would come in contact with it necessitate precautionary measures to insure safety of the crew and Earth habitats upon return of the mission. Based on the chances of an encounter with pathogenic life forms on Mars, a set of recommendations to insure the biosafety of a Mars mission has been recently proposed by Warmflash and colleagues (Warmflash, et al., 2007) (Table 1). Firstly, crew extravehicular activity (EVA) suits should be decontaminated upon habitat return of the astronauts from field activities. Secondly, laboratory facilities on Mars working with Martian samples should be equipped with a minimal biosafely level (BSL) 2 equipment. Thirdly, sterilization methods for the spacecraft upon return should be considered (Trofimov, et al., 1996). Finally, a quarantine program for crews and material brought from Mars also seems to be prudent. Protocols based on the *Apollo Quarantine Program* should be developed and followed strictly. The return from Mars to Earth has a long duration, which may allow for a disease to manifest itself in the astronauts during the flight back, but the certainty of such a scenario cannot be assumed. While the Apollo astronauts were quarantined for 21 days, breaches in the protocol occurred (National Research Council Space Studies Board, 1997), which should be strictly avoided upon the return of a Mars mission.

5. MEASURES IN CASE OF SUSPICION OF INFECTION OR CONTAMINATION

The facts and arguments assembled above present a situation in which the infection of humans with a pathogenic Martian microorganism is highly unlikely. This chance should be even further diminished by strict implementation of preventive biosafety measures. However, the possibility of infection can never be completely excluded. What measures should be taken in case of suspicion of infection with Martian microorganisms?

As with any other infection, one has to think at diagnostic and therapeutic measures, neither of which will be trivial in such a case. In order to diagnose an infection with a putative Mars pathogen, one will initially rely on microbiological culture techniques or direct detection of components or structures of extraterrestrial microbes (Table 1). Very little will be known about the microbiological culture techniques to be used, but one has to consider growth conditions close to those of the original Martian habitat from which the microbe originated.

Direct detection of chemical signatures of Martian microbes could be both challenging, as well as providing a possible window of opportunity. On the one hand, PCR-based diagnostics as used for detection of novel pathogens on Earth will be challenging, as indeed one has to assume the presence of nucleic acids as prerequisites of Martian life. These amplification techniques are based on common sequences within the 16s ribosomal RNA of terrestrial bacteria, and such sequences may be totally absent in Martian microbes. An additional challenge would be to exclude any contamination with terrestrial microbial sequences as the source of a potentially positive PCR-based test (Poole and Willerslev, 2007). We should be aware that the putative Martian microorganisms are likely to use as building blocks at least some aberrant materials such as aminoacids or sugars of varieties inexistent or rare on Earth. For example, the abundance of the aminoacids α–aminoisobutiric acid and isovaline in carbonaceous meteorites (Engel, et al., 1990) may indicate the possibility that such aminoacids are used as building blocks by extraterrestrial microbes. These aminoacids have a structure clearly different from terrestrial aminoacids are practically not found on Earth, and their presence in biologic material may indicate infection with extraterrestrial microbes.

However, Martian organisms may not have a DNA-based genome, or they may use other sugars or even of very different structure (such as PNA's – peptide nucleic acids). If life did not arise on Earth or Mars by mechanisms of panspermia (Joseph and Schild 2010a,b), and unless life has been transferred back and forth between the planets, thereby exchanging DNA, then it must be recognized that if there is Martian life, it may not have arisen in the same way or completed all the same steps as life on Earth. Martian life could have stopped at the pre-RNA or RNA world levels, with replicating macromolecules different from the current one in terrestrial life, in which case, PCR will not work at all.

Treatment of such infection would be a tremendous challenge. Extensive testing capabilities of substances with antibiotic properties should be in place once a Mars mission returns. Such testing of antibiotics should start as soon as a Martian microorganism is identified and cultured, and even if thought non-pathogenic. In addition, the possibility exists that the human (or other animal) immune system does not recognize a pathogen with which it has not shared billions of years of evolution. In that case, it is not possible to raise antibodies naturally and thus biotechnical methods to manufacture synthetic therapeutic tools (e.g., synthetic antibodies against Martian bacteria) should be considered (Table 1).

Common aminoacids

Valine

Alanine

Exotic aminoacids

Isovaline

α-Aminoisobutyric acid

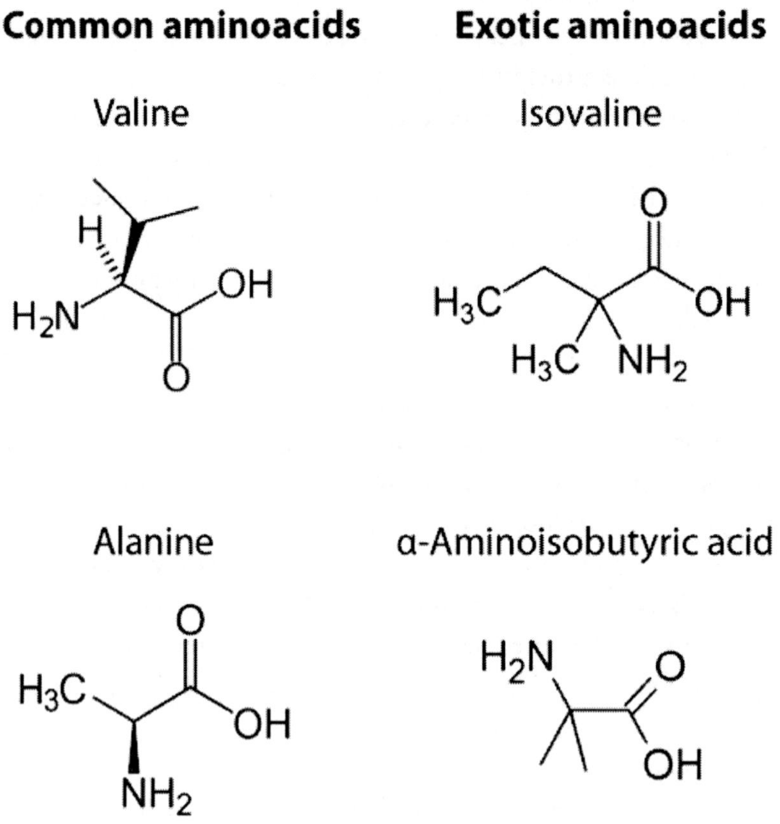

FIGURE 1

Structure of the abundant amino acids α–aminoisobutiric acid and isovaline found in carbonaceous meteorites, and putative building blocks of extraterrestrial microorganisms.

6. CONCLUSIONS

The chance of a human mission to Mars to encounter pathogenic micro-organisms is small, but not zero. A set of safety measures to prevent, diagnose and eventually treat infections with Martian microorganisms should be considered (Table 1), and these may further diminish potential biohazards. Therefore, it may be concluded that the benefits of a mission to Mars, with all the scientific, technological and economical progress that is envisaged, heavily outweigh the low probability of an encounter with a pathogenic microbe, and therefore this should not be an impediment for pursuing human exploration of Mars.

TABLE 1

RECCOMENDATIONS FOR PREVENTION, DIAGNOSIS AND TREATMENT OF AN INFECTION WITH A PUTATIVE MARTIAN PATHOGENIC MICROORGANISM

Situation encountered	Reccomendations
Preventive measures on Mars	- Decontamination of EVA suits
	- BSL2 laboratory for handling Martian samples
Preventive measures upon return	- decontamination of spacecraft
	- quarantine of astronauts and samples
Diagnostic measures	- high standard facilities for microbiological and biochemical diagnostics available upon return of the mission
Treatment measures	- immediate and extensive program of antibiotic testing upon identification of a Martian microorganism
	- biotechnical capacity available for production of synthetic therapeutic tools

Acknowledgements: M.G.N. was supported by a Vici Grant of the Netherlands Organization for Scientific Research. M. S. is supported by ERC grant "MASEM" (242635).

REFERENCES

National Research Council Space Studies Board. (1997). Mars sample return: issues and reccomendations. Washington D.C.: National Academy Press.

Brack A. (1999). Life in the solar system. Adv Space Res 24(4):417-33.

Dartnell, LR, Desorgher L, Ward JM, Coates AJ. (2007). Modelling the surface and subsurface Martian radiation environment: Implications for Astrobiology. Geophysical Research Letters 34, L02207.

Engel MH, Macko SA, Silfer JA. (1990). Carbon isotope composition of individual amino acids in the Murchison meteorite. Nature 348(6296):47-9.

Formisano V, Atreya S, Encrenaz T, Ignatiev N, Giuranna M. (2004). Detection of methane in the atmosphere of Mars. Science 306(5702):1758-61.

Joseph R., and Schild, R. (2010a). Biological Cosmology and the Origins of Life in the Universe. Journal of Cosmology, 5, 1040-1090.

Joseph R., and Schild, R. (2010b). Origins, Evolution, and Distribution of Life in the Cosmos: Panspermia, Genetics, Microbes, and Viral Visitors From the Stars. Journal of Cosmology, 7, 1616-1670

Joseph, R. and Wickramasinghe, N. C. (2010). Comets and Contagion: Evolution and Diseases From Space. Journal of Cosmology, 7, 1750-1770.

Jull AJT, Beck JW, Burr GS. (2000. Isotopic evidence for extraterrestrial organic material in the Martian meteorite, Nakhla. Geochimica Geophys Acta 64:3463.

Houtkooper, J. M and Schulze-Makuch (2010). The Possible Role of Perchlorates for Martian Life. Journal of Cosmology, 5, 930-939.

Klein HP, Horowitz NH, Levin GV, Oyama VI, Lederberg J, Rich A, Hubbard JS, Hobby GL, Straat PA, Berdahl BJ and others. (1976). The viking biological investigation: preliminary results. Science 194(4260):99-105.

Lebrun I, Marques-Porto R, Pereira AS, Pereira A, Perpetuo EA. (2009). Bacterial toxins: an overview on bacterial proteases and their action as virulence factors. Mini Rev Med Chem 9(7):820-938.

Leuko, S., Rothschild, L. J. and Burns, B. P. (2010). Halophilic Archaea and the Search for Extinct and Extant Life on Mars. Journal of Cosmology, 5, 940-950.

Levin, G. L., (2010). Extant Life on Mars: Resolving the Issues, Journal of Cosmology, 5, 920-929.

Malin MC, Edgett KS. (2000a). Sedimentary rocks of early Mars. Science 290(5498):1927-37.

Malin MC, Edgett KS. (2000b). Evidence for recent groundwater seepage and surface runoff on Mars. Science 288(5475):2330-2335.

McKay DS, Gibson EK, Thomas-Keprta KL, Vali H, Romanek CS, Clemett SJ, Chillier XD, Maechling CR, Zare RN. (1996). Search for past life on Mars: possible relic biogenic activity in Martian meteorite ALH84001. Science 273:924-930.

Mitchell, E. D., Staretz, R. (2010). Our Destiny – A Space Faring Civilization? Journal of Cosmology, 12, 3500-3505.

Picardi G, Plaut JJ, Biccari D, Bombaci O, Calabrese D, Cartacci M, Cicchetti A, Clifford SM, Edenhofer P, Farrell WM and others. 2005. Radar soundings of the subsurface of Mars. Science 310(5756):1925-1928.

Poole AM, Willerslev E. (2007). Can identification of a fourth domain of life be made from sequence data alone, and could it be done on Mars? Astrobiology 7(5):801-814.

Prigent M, Leroy M, Confalonieri F, Dutertre M, DuBow MS. (2005). A diversity of bacteriophage forms and genomes can be isolated from the surface sands of the Sahara Desert. Extremophiles 9(4):289-296.

Rummel, J. D., Race, M. S., Conley, C. A., Liskowksy, D. R. (2010). The integration of planetary protection requirements and medical Support on a mission to Mars, Journal of Cosmology, 12. 3834-3841.

Russell, M.J., and Kanik, I. (2010). Why does life start, What does It do, where might it be, how might we find it? Journal of Cosmology, 5, 1008-1039.

Sawstrom C, Lisle J, Anesio AM, Priscu JC, Laybourn-Parry J. (2008). Bacteriophage in polar inland waters. Extremophiles 12(2):167-175.

Schuerger AC. (1998). Application of basic concepts in plant pathogenesis suggests minimal risk for return of extraterrestrial samples from Mars. Lunar and Planetrary Science Conference Procedures, Houston 1998 29:1312.

Scott ER. 1999. Origin of carbonate-magnetite-sulfide assemblages in Martian meteorite ALH84001. J Geophys Res 104(E2):3803-3813.

Sephton, M. A. (2010). Organic Geochemistry and the Exploration of Mars. Journal of Cosmology, 5, 1141-1149.

Squyres SW, Grotzinger JP, Arvidson RE, Bell JF, 3rd, Calvin W, Christensen PR, Clark BC, Crisp JA, Farrand WH, Herkenhoff KE and others. (2004). In situ evidence for an ancient aqueous environment at Meridiani Planum, Mars. Science 306(5702):1709-14.

Straume, T. et al., (2010). Toward Colonizing Mars. Perspectives on Radiation Hazards: Brain, Body, Pregnancy, In-Utero Development, Cardio, Cancer, Degeneration, Journal of Cosmology, 12, 3992-4033.

Stuster, J. (2010). Acceptable Risk: The Human Mission to Mars, Journal of Cosmology, 12, 3566-3577.

Thomas-Keprta KL, Clemett SJ, Bazylinski DA, Kirschvink JL, McKay DS, Wentworth SJ, Vali H, Gibson Jr EK, Jr., Romanek CS. (2002). Magnetofossils from ancient Mars: a robust biosignature in the martian meteorite ALH84001. Appl Environ Microbiol 68(8):3663-72.

Titus TN, Kieffer HH, Christensen PR. (2003). Exposed water ice discovered near the south pole of Mars. Science 299(5609):1048-51.

Trofimov VI, Victorov A, Ivanov M. (1996). Selection of sterilization methods for planetary return missions. Adv Space Res 18(1-2):333-7.

Yung, Y. L., et al., (2010). The Search for Life on Mars. Journal of Cosmology, 5, 1121-1130.

Warmflash D, Larios-Sanz M, Jones J, Fox GE, McKay DS. (2007). Biohazard potential of putative Martian organisms during missions to Mars. Aviat Space Environ Med 78(4 Suppl):A79-88.

Whitman WB, Coleman DC, Wiebe WJ. (1989). Organic materials in a Martian meteorite. Nature 340:220-222.

Zubrin, R. (2010). Human Mars Exploration: The Time Is Now, Journal of Cosmology, 12, 3549-3557.

EXPLORING MARTIAN MYSTERIES

Journal of Cosmology, 2010, Vol 12, 3912-3927.
JournalofCosmology.com, October-November, 2010

- 10 -

Destination Mars: Human Exploration of Martian Mysteries

Markus Hotakainen

Ursa Astronomical Association, Helsinki, Finland

ABSTRACT

Human missions to Mars should adopt as some of their targets the surface anomalies of the red planet. The possibility that some surface features on Mars are exceptionally anomalous could lend additional interest to exploration of any areas of the planet where such features are present in conjunction with other factors such as indications of water or of a past shoreline. Studying them an insight into several aspects of the geological and climatological evolution of Mars could be gained. Some of them could also hold keys into finding Martian life and understanding its history.

Key Words: Mars, surface anomalies, water, oceans, ice, life

1. INTRODUCTION

UNTIL NOW MARS PROBES, ORBITAL or vehicular, have been controlled by a combination of onboard software and radio communication with Earth-based humans. The main focus of interest has been the geological character of the planet (Levine et al., 2010a,b). However, there are many unexplained anomalies which are also deserving of attention and the close inspection of trained teams of astronauts.

The geological processes which characterize Mars are similar to and familiar from Earth (Levine 2010a,b). The principle of the past being the key to the present is applicable on the large scale. With a surface area similar to that of all the continents of Earth and a readily observable geological record of billions of years, Mars is a treasure-trove for planetary research, both of the planet itself, with relation to Earth, and in connection with the origin and evolution of the Solar System as a whole.

However, the exploration of Mars has to take into account the fact that during most of its evolution the conditions have been crucially different from those of the ancient and present-day Earth, which means similar processes could have led to different outcomes. Indeed, it is not just geology and biology, but the fact there are many unexplained Martian anomalies which require trained field scientists. For example, there is a wealth of enigmatic features on Mars defying any simple and straight-forward explanations. Many of these features will keep their secrets until human presence. Even the most sophisticated robotic probes with most advanced artificial intelligence lack – at least for the moment – the ingenuity and creativity of the human mind.

Thus the next step in the exploration of our planetary neighbor, should be a human mission to Mars and the landing a human crew, whose observations will certainly go beyond questions of geology but will also seek to determine whether life in any form exists on the planet today, or has existed in the past. Recent discoveries in the biological sciences have conclusively demonstrated that living organisms are capable of surviving in extreme conditions, and this means we cannot rule out the possibility that some form of organic life exists on Mars even in the conditions that prevail today. In terms of a search for evidence, there are a number of intriguing candidates for a landing site and base camp. Such sites would include proximity to water in the form of ice, which exists at the poles and also likely within the confines of some craters; all of which may harbor life.

Consequently exploring the anomalies of Mars could potentially give new insight into the geological and biological history of Mars *and* Earth. However, this would require human presence, as advocated by astronauts Dr. Edgar Mitchell (Mitchell 2010) and Dr. Harrison Schmitt (2010). Astronaut Schmitt is a trained geologist and the last man to step upon the surface of the Moon and he argues that it is imperative that field geologists accompany the first astronauts to Mars (Schmitt 2010). It is because of this astronaut geologist, and the human presence on the Moon, that the Apollo program contributed to many important discoveries giving clues to the birth and evolution of our cosmic companion.

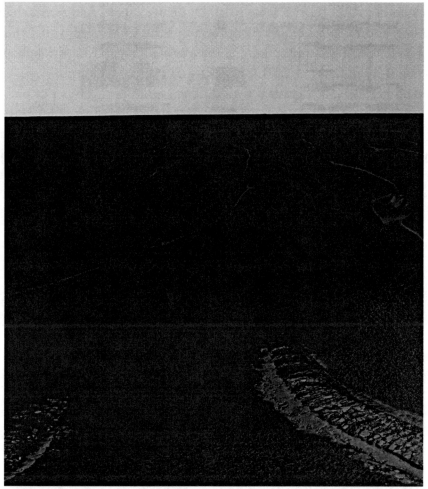

FIGURE 1
Tracks to be followed. Robotic rovers – Sojourner, Opportunity, and Spirit – have paved a way to manned exploration of Mars. Credit: NASA/JPL/Cornell

2. WET MARS

One of the most intriguing questions in the study of Mars has been and still is the existence of water. It is the most important substance making the difference as well as the similarity between Mars and Earth. The currently cold and arid Mars has experienced a period (or periods) of warmer and wetter climate with a considerably thicker atmosphere (Morris et al. 2010).

The images acquired with the high-resolution cameras of several probes launched during the past decade and a half – Mars Global Surveyor, Mars Odyssey, and Mars Reconnaissance Orbiter of NASA, and Mars Express of ESA – and the research conducted on the surface with the rovers – Sojourner, Opportunity, and Spirit – have proved that there has been water on the surface of Mars: lots of water (Carter et al. 2010; Di Achille 2010). And recent studies of the carbon and oxygen isotopes of the atmospheric CO_2 suggests that there has been low-temperature interaction between water and rocks throughout the Martian history (Niles et al. 2010).

The surface of Mars is covered with different kinds of signatures of past water from narrow gullies on the walls of craters and slopes of hills and dunes through dried river beds and valleys carved by water to vast, smooth areas reminiscent of ocean floors. However, whether all of these formations are due to water (or ice) is still a matter of dispute, and other explanations like cryoclastic phenomena and gas-supported density flows have been brought forth (Hoffman 2000).

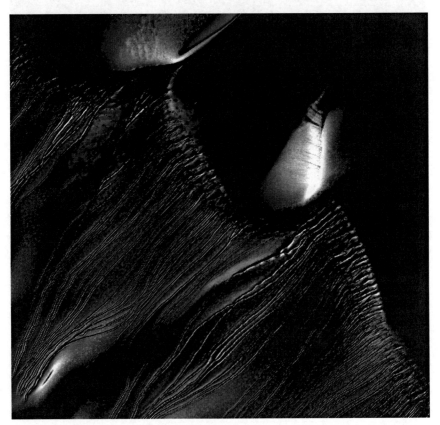

FIGURE 2

On the pole-facing slopes of the dunes in Russell Crater on the southern hemisphere of Mars there are narrow gullies side by side. Credit: NASA/JPL/University of Arizona

In many cases the surface anomalies – just like the more ordinary features – are seemingly the result of flowing water or melting ice, coupled with perhaps tricks of light and shadow and the propensity of the human imagination to conjure forth not just what is, but what might be. However, dismissing these anomalies as illusions or to assume they are the result of natural geological forces, is not in the spirit of science. Assumptions, be they pro or con, are not the same as facts. Rather, these surface anomalies require serious consideration and investigation which in turn would lead to the advancement of science, and lead to a greater understanding of planetary geology and the cause and origins of these anomalies.

3. MARTIAN ARCTIC OCEAN

The most prominent anomaly of Mars is the dichotomy between the northern and southern hemispheres. The southern highlands are covered with craters of all sizes, the northern lowlands are exceptionally flat and smooth. There is also an average difference of 4 kilometers in their elevations (Kiefer, 2008).

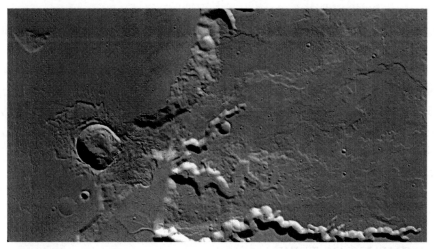

FIGURE 3
The highland-lowland dichotomy boundary of Mars is at most places – like in this area near Medusa Fossae – very apparent. Credit: ESA/DLR/FU Berlin (G. Neukum)

One of the theories is that there had been a vast Martian ocean covering most of the northern hemisphere. An ocean of Martian water would explain the smoothness. However, if there had been a Northern ocean, then where are the ancient shorelines? Shorelines should be evident at the borderline between the highlands and the lowlands. Although many have claimed to have detected these palaeoshorelines (Parker et al. 1993), these claims are in most cases controversial to say the least. If there had been ancient oceans, perhaps the shorelines have disappeared after being blanketed by later sedimentation. Or perhaps the failure to find conclusive evidence is due to the insufficient resolution of the imaging instrumentation of the probes. Trained field observers, could provide the answers.

The existence of an ancient ocean is further implied by the distribution of ancient deltas dilineating the margins of the northern lowlands. The level of these deltas is consistent with the possible palaeoshorelines indicated by the analysis of other morphological and topographical features, as well as the distribution and age of valley networks on the surface of Mars (Di Achille 2010).

Also the recent detection of hydrated silicates in the northern plains indicates a past presence of large amounts of water. The silicates have been known to exist on the southern highlands implying wet conditions early in the history of Mars, but now the same goes with the northern highlands. The ancient deposits of these hydrated silicates, being mostly phyllosilicates, excavated at places by impact cratering, are covered with hundreds of metres of later sedimentation (Carter et al. 2010).

Perhaps the answers concerning the existence of the ancient ocean could be obtained by sophisticated high-resolution cameras and deep-range radars onboard an orbiting craft. However, these answers may not be obtained without in-situ exploration, including deep drilling, which would be difficult to accomplish without human presence.

Finding a landing site with the highest potential to give an answer to this question on the existence of an ancient sea might also prove difficult. The borderline between the highlands and lowlands is at many places a very steep slope making it unsuitable for landing a manned craft. However, there is an area with not only an appropriate location, but a wealth of enigmatic features.

4. CYDONIAN COMPLEX

The first feature to arouse the interest of both the scientific community and the general public was the detection of a "face-like" anomaly in the area known as Cydonia, a plain in a transitional region between heavily cratered southern highlands and the smooth northern lowlands; the so called "Face of Mars". It was photographed by the Viking 1 Orbiter in July 1976 while making detailed imaging for the selection of the landing site for the Viking 1 Lander.

The image of the "Face" gained immediately great publicity and a keen interest in, along with various speculations on, the origin of the familiar looking formation. There are strong advocates for the interpretation that this feature with a length of 2,5 kilometres and a height of 250 metres is artificial, being some kind of a vast monument or perhaps a cenotaph. Later images taken by various orbiters having instruments with higher resolution than those onboard Viking 1 has lent credence to the view of the Face being a natural formation, a rocky hill or mesa with crevices on top simulating the features of a giant face staring upward from the Martian surface.

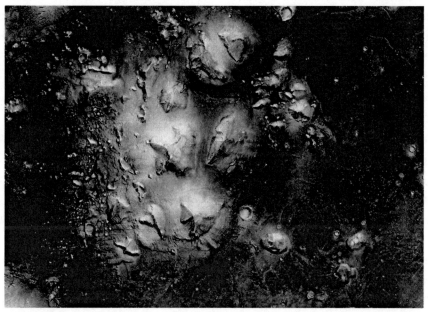

FIGURE 4
The unusual mesas of the Cydonia region has led to wild speculation on their origin. The infamous "Face of Mars" is near the lower right-hand corner of this image. Credit: Mars Express/ESA

The "Face of Mars" is not the only formation on the Cydonia plain having created both interest and controversy. Some of them have been seen as further evidence for the existence of an ancient, intelligent civilization on Mars. There are numerous formations reminiscent of eroded pyramids, ruined castles, and other structures claimed to have an artificial origin. However, the evidence for the artificiality of these formations, along with the "Face of Mars", is not nearly conclusive enough, and the general view is that they are natural products of geologic processes. Nevertheless, the structural anomalies on the Cydonia plain are well worth a thorough investigation because of both their fame and the contribution their study would give to our knowledge on Mars and various processes shaping its surface - whatever the outcome.

The exploration of Cydonia would also give information on the nature of the borderline between the rugged south and smooth north. The area is possibly an ancient coastline with several processes having affected its evolution either concurrently or consecutively – or both. Landing a manned mission on the Cydonia plain would – in addition to the obvious impact on the imagination of the public and media – enable extensive studies of the ancient shoreline and its alterations. This could also shed light on the climate change which caused the transformation of a temperate planet into a frozen celestial object, a kind of "Museum of Water". Situated on the mid-latitudes (40°N) Cydonia would also be an ideal target for the studies on the effects of the variability in the obliquity of the rotational axis of Mars.

At the moment the inclination is very close to that of Earth – 25°11' of Mars compared with 23°59' of Earth – but it has been known a long time that there is a variability of at least 20 degrees over a period of approximately 100 000 years (Ward 1973). However, the variability seems to be chaotic, and during periods of tens of millions of years the inclination could have changed several tens of degrees (Touma et al. 1993). Because of the chaotic nature of the variability it is impossible to track it precisely for more than a few million years into the past, but with statistical methods it is possible to analyze the history of these variations. The value for the maximum of the obliquity is 82°, and the probability for the obliquity having reached 60° during the past 1 billion years is 63 %, and during the past 3 billion years as high as 89 % (Laskar et al. 2004).

This kind of variation has had a dramatic effect on the climate of Mars in the past, and most probably is still having: the variability of the inclination is continuous. While the inclination is small and the planet rotates in an upright position, the polar areas receive much less solar energy than the lower latitudes, and are thus much colder. When the inclination increases, the poles receiving more solar energy warm up, but the equatorial regions cool down.

This variability is the probable cause for the presence of the large reservoirs of near-surface ice in high latitudes of both northern and southern hemispheres of Mars. While the obliquity is high, there is an accumulation of ice, most probably even in the form of glaciers, to the equatorial regions, but during low obliquity, the sublimation of these reservoirs results in deposition of ice to the high latitudes and polar regions (Levrard et al. 2004; Forget et al. 2006).

With a thicker atmosphere in the past this has affected the state of water and consecutively the sea level of the Arctic Ocean, but also the migration of water from the polar areas to the mid- and low-latitudes (Haberle et al., 2004). This in turn could be inferred from the current depth and thickness of permafrost present also beneath the surface of Cydonia.

5. NOCTIC LABYRINTHUS

Noctis Labyrinthus just south of the equator of Mars is a place of major upheaval – or "downfall" to be precise. It is an area of large mesas, broad and flat-topped mountains and hills with steep clifflike slopes, called "chaotic terrain". According to early theories this chaotic terrain formed when large amounts of subsurface ice suddenly melted with the water flowing off (Carr and Schaber 1977) or groundwater being released from aquifers to create large outflows (Carr 1979), and the terrain collapsing as a result.

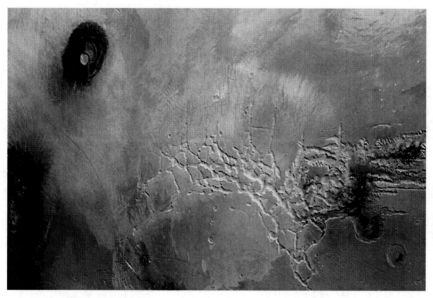

FIGURE 5
Noctis Labyrinthus east of the Tharsis region is a large area of "chaotic terrain".
Credit: Viking/NASA/JPL/USGS

However, there is still no final word on the process creating these features, but water has certainly played a key role in the evolution of the area as shown by the hydrated minerals found in the area (Thollot et al. 2010). Presumably the melting of ice was caused by volcanic activity: Noctis Labyrinthus is adjacent to the Tharsis highlands having many giant volcanoes. An examination of this vast and varied area would yield information on the similar features observed in a smaller scale, like Reull Vallis close to the Hadriaca Patera volcano and several others.

Landing a manned mission in Noctis Labyrinthus would offer an opportunity to study the origin and evolution of the chaotic terrain characteristic of the area. It would be a challenge to make a pin-point landing to avoid the hazards of the rough surface. However, in 1969, Apollo 12 landed right on target, less than 200 meters from the Surveyor 3 probe sent to the Moon 2,5 years earlier. Pin-point landings on a distant celestial body are not impossible.

The average depth of Noctis Labyrinthus is about 5 km (Bistachi et al. 2004), so the atmospheric pressure, albeit still very low, would be above average. Whether the net effect of it would be positive or negative as to the manned mission, is uncertain. Some areas of exploration like meteorology would benefit from the local climate of the valleys – condensate clouds made of water-ice crystals form in this region rather regularly – but this is also an area of dust storm activity.

FIGURE 6
In between the flat-topped mesas of Noctis Labyrinthus there are deep depressions formed by outflow of water and collapse of the ground. Credit: Mars Express/ESA

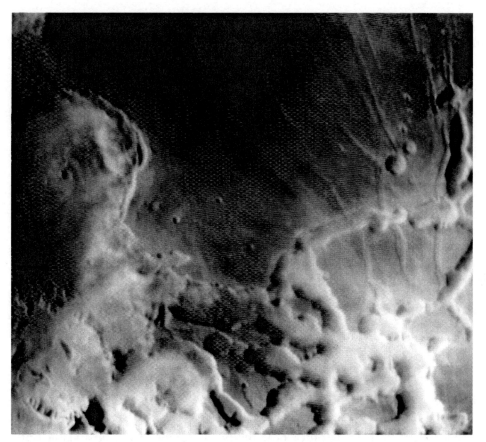

FIGURE 7
Early morning water clouds in and around the canyons of Noctis Labyrinthus. Credit:
Viking/USGS/JPL/NASA

Noctis Labyrinthus could have a definite benefit for a manned mission. There is a possibility for an existence of caves, carved either by the flash floods caused by the sudden melting of subsurface ice, or volcanic flows forming lava tunnels. If existent, they would offer a natural shelter against the ultraviolet radiation of the Sun and the bombardment of the cosmic rays. Otherwise they both would require either a heavily shielded landing craft or a base camp dug beneath the surface. Because of increased mass of the craft these alternatives would make a manned mission more challenging both technically and economically.

Noctis Labyrinthus would be interesting also because of its proximity to two extensive surface features with a history still largely unknown: Tharsis and Valles Marineris. The former is evidently of volcanic origin – there are four giant volcanoes on top of Tharsis – and Valles Marineris was formed by rift faults in the crust of Mars, thus being similar to the East African Rift Valley (Hauber et al. 2010), which has later been expanded because of erosion and massive landslides: it is a sign of a kind of "failed" tectonism. It is still unknown whether they are related to each other, and if so, how; and whether either is caused by the other, or are they both caused by some process even more global than either of them.

6. POLAR EXPEDITION

Landing on the polar regions would offer an excellent opportunity to explore the enigmatic and simultaneously variable surface features on the high latitudes of Mars. There are theories on how the polar formations like polygons, "spiders", "forests", and "swiss cheese", related to the sublimation of either water or CO_2 ice or both, and also larger, but still continuously evolving features like dune fields are formed, but to organize geological field trips to the most interesting of these areas might answer the question of the origin of these formations.

The polar regions of Mars are under continuous change. With the seasons the CO_2 migrates through the tenuous atmosphere from one pole to the other giving rise to a regular variation in the appearance of the polar caps (Smith et al. 2009). While the southern cap completely disappears during the spring of the southern hemisphere, the residual water ice of the northern one holds through the somewhat colder spring and summer of the northern hemisphere.

The retreating ice is annually leaving behind strange formations the nature and especially the birth processes of which are still largely unknown. The future missions to Mars will – and most probably has to – "live off the land" to the largest possible extent, so landing on or near the polar regions would be a viable option. Water in the form of permafrost is found more or less everywhere on Mars, but on the polar regions, as the Phoenix lander in 2008 proved, it is only few centimeters below the surface thus making it easy to reach and utilize (Smith et al 2009).

Figure 8

Geysers of CO_2 carrying dust from the subsurface create curious patterns in the polar areas. Credit: MRO/NASA/JPL-Caltech/University of Arizona

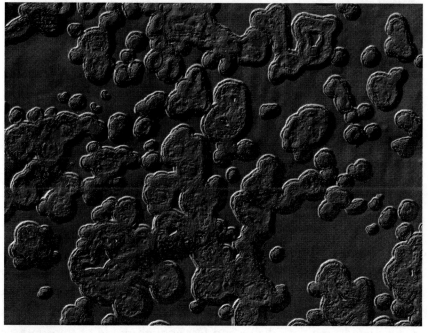

Figure 9

The cause of "Swiss cheese terrain" found only near the southern polar cap are believed to be the differences in the rate of the seasonal changes of the CO_2 and water ices. Credit: NASA/JPL/University of Arizona

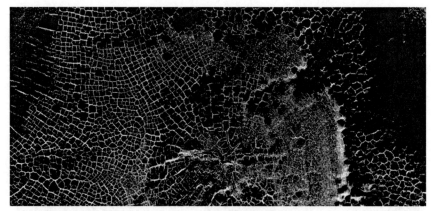

FIGURE 10
Polygons on the "patterned ground" are reminiscent of the phenomena in the Arctic regions of Earth originating from the changes in the subsurface ice. Credit: NASA/JPL/University of Arizona

FIGURE 11
Phoenix probe landed on water ice covered with only a thin layer of dust and sand. Credit: Kenneth Kramer, Marco Di Lorenzo, NASA/JPL/UA/Max Planck Institute

However, landing on the polar regions might prove too hazardous because of potential dust storms, landslides, and the CO_2 ice covering the ground in winter. To avoid any unnecessary and foreseeable risks it might be more practical to establish a base camp on a safer area, but still within easy reach of the places of interest. This would require reliable and fast enough mode of transportation for the astronauts to be able to explore the area within the time-limit set by the duration of both the stay on the surface of Mars and the changing seasons.

7. ACIDALIAN MUD

A question even more intriguing than the past and present existence of water on Mars – but closely related to it – is the one about life on Mars. Despite promising, yet controversial findings in meteorites originated from Mars (Thomas-Keprta 2009), there are no definite proofs neither for the existence nor non-existence of life.

The surface conditions of present-day Mars are very hostile to life or any organic matter, but in the warmer and wetter past there might have been abodes for life to emerge. Whether the conditions have been favorable for long enough time for life to evolve is another matter of dispute, but if there has been life of any kind, based on the current knowlegde it must have been related with the existence of water. One of the potentially promising areas for finding signs of life is Acidalia Planitia. Along with several other areas on the northern lowlands like Utopia, Isidis, and Chryse Planitia, Acidalia is covered in many places with formations resembling very much the mud volcanoes found on Earth. The latest research implies that the number of these potential mud volcanoes is almost 20 000, perhaps even 40 000 (Oehler and Allen 2010).

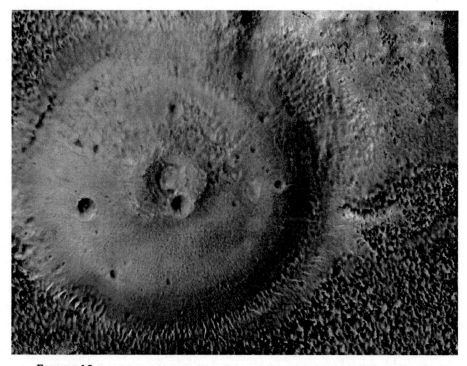

FIGURE 12

Possible mud volcanoes found on Mars could give clues to the ancient Martian life – if there was any. Credit: HiRISE/MRO/LPL (U. Arizona)/NASA

The mud volcanoes are formed by pressurized subsurface gas or liquid erupting to the surface carrying with it material from depths of up the several kilometres. It is possible that there has been – or even today could be – reservoirs of liquid water beneath the thick layer of permafrost; they could have been favorable for bacterial life similar to the extremophiles on Earth. If this is the case, there might still be organic material to be found in the mud and gravel transported to the surface by the eruptions.

Albeit being relatively young, perhaps late Hesperian or early Amazonian, deduced from the fact that they overlap older formations (Tanaka et al. 2005), the mud volcanoes of Acidalia are still two to three billion years old. The strong radiation and surface chemistry would have destroyed any evidence of organic matter, but inside the flat domes with diameters of one kilometre and heights of some 200 metres on average (Oehler and Allen 2010), there might still be some evidence of life to be found. However, reaching it would require, as in the case of the ancient shoreline of the Arctic Ocean, an expedition equipped with drilling machinery.

8. CLAYS OF MAWRTH VALLIS

One of the oldest valleys on Mars, Mawrth Vallis, could also be a potential site for finding signs of ancient Martian life. Recently, it has been observed to bear evidence for a prolonged existence of aquatic environment very early in the history of Mars, when the climate of the planet was more temperate, and potentially suitable for life (Poulet et al. 2005).

A stable and long-term presence of large quantities of liquid water is implied by the deposits of hydrated minerals the formation of which requires wet conditions. However, there are different kinds of hydrated minerals, namely phyllosilicates and hydrated sulphates, which have their origins in different kinds of processes. Phyllosilicates, for example clay, are formed by alteration of minerals of magmatic origin having been in contact with water for a considerable length of time. On the other hand, hydrated sulphates are deposits formed from salted water, and their formation requires an acidic environment, but not extended periods of contact with water.

The detection of phyllosilicates in Mawrth Vallis strongly implies that there has been water on the surface of Mars for extended periods of time in the earliest, the Noachian era, of its evolution, long before the global climate change making Mars a less habitable planet. On Earth the clay minerals are capable of preserving microscopic life, so a thorough examination of the phyllosilicates on Mars could give an answer to the question concerning ancient life on the red planet.

FIGURE 13
Mawrth Vallis has Noachian deposits of phyllosilicate minerals with a possibility to find evidence on Martian life. Credit: HiRISE/MRO/LPL (U. Arizona)/NASA

9. OLYMPUS MONS

Another possible landing site for a manned mission with a great potential to capture the imagination of the general public, but not in anyway without scientific benefits would be Olympus Mons, the highest mountain in the Solar System. It could be a key to several questions concerning both the ancient and recent history of Mars.

Olympus Mons is the greatest of the Tharsis volcanoes not only by height, but by the volume of erupted lava and by the time span there has been eruptions. The formation of the mountain is very similar with those on the Hawaiian islands, but because of the lack of plate tectonics on Mars, the volcanic hot spot has been immobile, and the lava has accumulated into just one gigantic mountain.

Based on research made on the aureole surrounding the mountain the origins of Olympus Mons have been dated to the Hesperian era (Fuller and Head 2003), so it is at least 2 billion years old, but there are volcanic flows with an age of just tens of millions, perhaps only few million years (Neukum et al. 2004). In addition to being active until geologically very recently, the volcanic activity has been episodic.

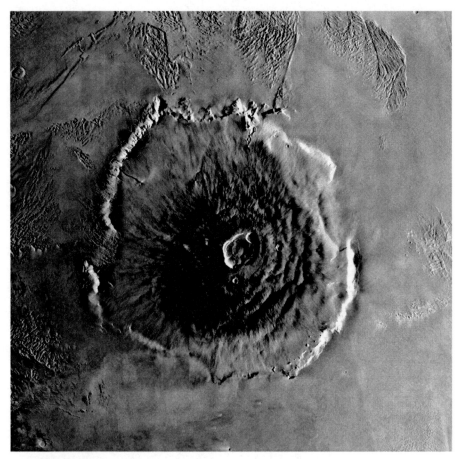

FIGURE 14
*Olympus Mons, the highest mountain in the Solar System, has possibly been active
until rather recently, only few million years ago. Credit: Viking/USGS/JPL/NASA*

Making detailed studies on the different lava layers on Olympus Mons would
enable dating the different stages in the evolution of the mountain with a better
accuracy than the current estimates based on crater counts, and trying to find the
cause for the periodical behaviour – and possibly for the formation of the Tharsis
region as a whole.

Olympus Mons is often shrouded with CO_2 ice clouds making it an ideal site
also for meteorological observations. The nature of the clouds – composition,
density, temperature – could be measured readily with a possibility to take
samples of the ice crystals making up the clouds. The size of the mountain is
such, that it has a major effect on the air currents and wind patterns in the
atmosphere of Mars, and consequently on the Martian weather (Wolkenberg
2008). To make direct observations on the spot of origin of these effects would
help develop more detailed models for both global and local weather phenomena.

All this could be done with robotic probes, but Olympus Mons might offer an extra benefit for a future base camp. The slopes of the mountain could hide a number of caves formed by collapsed lava tunnels. Just like in the case of Noctic Labyrinthus these would offer a natural shelter against the harsh radiation environment on the surface of Mars. The great height of Olympus Mons would also keep a base camp well above all but the largest, global dust storms.

A mountain with a summit rising 25 kilometres higher than the surrounding landscape might appear as an extremely challenging site for any, especially manned, mission and the activities related with it. In reality the slopes of the mountain – apart from the steep cliffs especially on the eastern flank of the mountain – are very gentle, with a gradient of only few degrees. Together with the low gravity of Mars of some 40 % that of Earth this would result in an easy traversing on Olympus Mons, not so much going uphill or downhill, but only walking or driving around.

The obvious downside of Olympus Mons as a potential landing site for a manned mission is the very probable lack of subsurface ice to be exploited by the expedition. If there were any ice, it would be very deep in the ground and practically unattainable without very heavy machinery.

10. SELECTING THE TARGET

Making the selection between various options for the landing site of a manned mission to Mars aimed at exploring the surface anomalies will not be a simple task. The arguments to be taken into account include both the safety and practicality of the mission, and the possibility to make in-situ studies with scientific results increasing our knowledge on the anomalies specifically and on the evolution of Mars in general. However, trying to meet these criteria is an effort worth making.

11. CONCLUSIONS

There are many reasons why future manned missions to Mars must include among their primary objectives the exploration and examination of these surface anomalies: to explore their true nature, to solve their origin, to find out their importance in the geologic evolution of the planet, and – not the least important of the reasons – to feed the imagination of general public and keep up public interest in continued exploration of our planetary neighbor.

REFERENCES

Bistacchi, N., Massironi, M., Baggio, P. (2004). Large-scale fault kinematic analysis in Noctis Labyrinthus (Mars). Planetary and Space Science, 52, 215–222.

Cass, M.H., Schaber, G.G. (1977). Martian permafrost features. J. Geophys. Res., 82, 4039–4054.

Cass, M.H. (1979). Formation of Martian flood features by release of water from confined aquifers. J. Geophys. Res., 84, 2995–3007.

Di Achille, G., Hynek, B.M. (2010). Ancient ocean on Mars supported by global distribution of deltas and valleys. Nature Geoscience, 3, 459–463.

Forget, F., Haberle, R.M., Montmessin, F., Levrard, B., Head, J.W. (2006). Formation of Glaciers on Mars by Atmospheric Precipitation at High Obliquity. Science, 311, 368-371.

Frey, H. (1979). Martian canyons and African rifts: Structural comparisons and implications. Icarus, 37, 142–155.

Fuller, E.R., Head, J.W. (2003). Olympus Mons, Mars: Detection of extensive preaureole volcanism and implications for initial mantle plume behavior. Geology, 31, 175–178.

Hauber, E., Grott, M., Kronberg, P. (2010). Martian rifts: Structural geology and geophysics. Earth and Planetary Science Letters, 294, 393–410.

Hoffman, N. (2000). White Mars: A new model for Mars' surface and atmosphere based on CO_2. Icarus, 146, 326–342.

Kiefer, W.S. (2008). Planetary science: Forming the martian great divide. Nature, 453, 1191–1192.

Laskar, J., Correia, A.C.M., Gastineau, M., Joutel, F., Levrard, B., Robutel, P. (2004). Long term evolution and chaotic diffusion of the insolation quantities of Mars. Icarus, 170, 343–364.

Levrard, B., Forget, F., Montmessin, F., Laskar, J. (2004). Recent ice-rich deposits formed at high latitudes on Mars by sublimation of unstable equatorial ice during low obliquity. Nature, 431, 1072–1075.

Moore, H. et al.(1998) Ice Found in Craters in Cydonia", HYPERLINK "http://spsr.utsi.edu/" \o "http://spsr.utsi.edu/" http://spsr.utsi.edu/

Morris, R.V., Ruff, S.W., Gellert, R., Ming, D.W., Arvidson, R.E., Clark, B.C., Golden, D.C., Siebach, K., Klingelhöfer, G., Schröder, C., Fleischer, I., Yen, A.S., Squyres, S.W. (2010). Identification of Carbonate-Rich Outcrops on Mars by the Spirit Rover. Science, 329, 421–424

Neukum, G., Jaumann, R., Hoffmann, H., Hauber, E., Head J. W., Basilevsky, A. T., Ivanov, B. A., Werner, S. C., van Gasselt, S., Murray, J. B., McCord, T., and the HRSC Co-investigator team (2004). Recent and episodic volcanic and glacial activity on Mars revealed by the High Resolution Stereo Camera. Nature, 432, 971–979.

Niles, P.B., Boynton, W.V., Hoffman, J.H., Ming, D.W., Hamara, D. (2010). Stable Isotope Measurements of Martian Atmospheric CO_2 at the Phoenix Landing Site. Science, 329, 1334.

Oehler, D.Z., Allen, C.A. (2010). Evidence for pervasive mud volcanism in Acidalia Planitia, Mars. Icarus, 208, 636–657.

Parker, T.J., Gorcine, D.S., Saunders, R.S., Pieri, D.C., Schneeberger, D.M. (1993). Coastal geomorphology of the Martian northern plains. J. Geophys. Res., 98, 11061–11078.

Poulet, F., Bibring, J.-P., Mustard, J.F., Gendrin, A., Mangold, N., Langevin, Y., Arvidson, R.E., Gondet, B., Gomez, C. (2005). Phyllosilicates on Mars and implications for early Martian climate. Nature, 438, 623– 627.

Smith, D.E., Zuber, M.T., Torrence, M.H., Dunn, P.J., Neumann, G.A., Lemoine, F.G., Fricke, S.K. (2009). Time variations of Mars' gravitational field and seasonal changes in the masses of the polar ice caps. J. Geophys. Res., 114, E05002.

Smith, P.H., Tamppari, L.K., Arvidson, R.E., Bass, D., Blaney, D., Boynton, W.V., Carswell, A., Catling, D.C., Clark, B.C., Duck, T., DeJong, E., Fisher, D., Goetz, W., Gunnlaugsson, H.P., Hecht, M.H., Hipkin, V., Hoffman, J., Hviid, S.F., Keller, H.U., Kounaves, S.P., Lange, C.F., Lemmon, M.T., Madsen, M.B., Markiewicz, W.J., Marshall, J., McKay, C.P., Mellon, M.T., Ming, D.W., Morris, R.V., Pike, W.T., Renno, N., Staufer, U., Stoker, C., Taylor, P., Whiteway, J.A., Zent, A.P. (2009). H_2O at the Phoenix Landing Site. Science, 325, 58–61.

Tanaka, K.L. (2005). Geology and insolation-driven climatic history of Amazonian north polar materials on Mars. Nature, 437, 991–994.

Thollot, P., Mangold, N., Le Mouélic, S., Milliken, R.E., Roach, L.H., Mustard, F. (2010). Recent hydrated minerals in Noctis Labyrinthus Chasmata, Mars. 41st Lunar and Planetary Science Conference, 1873.

Thomas-Keprta, K.L., Clemett, S.J., McKay, D.S., Gibson, E.K., Wentworth, S.J. (2009). Origins of magnetite nanocrystals in martian meteorite ALH84001. Geochimica et Cosmochimica Acta, 73, 6631–6677.

Touma, J., Touma, J., Wisdom, J. (1993). The Chaotic Obliquity of Mars. Science, 259, 1294–1297.

Ward, W.R. (1973). Large-Scale Variations in the Obliquity of Mars. Science, 181, 260–262.

Wolkenberg, P.M., Formisano, V., Rinaldi, G., D'Amore, M., Geminale, A., Montabone, L., Spiga, A. (2008). An atmospheric hot ring around Olympus Mons – Comparison with mesoscale models (LMD and MRAMS). LPI Contribution, 1447, 9126.

VI

COLLONIZING & TERRAFORMING MARS

Journal of Cosmology, 2010, Vol 12, 3904-3911.
JournalofCosmology.com, October-November, 2010

- 11 -

Mars Base First:
A Program-level Optimization for Human Mars Exploration

Douglas W. Gage, Ph.D.
XPM Technologies, Arlington, VA 22201-4637

ABSTRACT

Much effort has been spent over the past two decades on the development of various plans for the human exploration of Mars, and, while the allocation of functions to specific vehicles has differed, consensus has formed around a conjunction mission (nominally 18 months on the surface) with pre-emplacement of some assets and some level of in-situ resource utilization (ISRU). The exploration program as a whole is envisioned as a sequence of several independent missions, each sent to a different area of interest on Mars, analogous to the Apollo approach to lunar exploration. Creation of a quasi-permanent Mars base is envisioned as following later, at a site to be identified during the initial exploration campaign. This paper argues that a better initial human exploration campaign – i.e., one which will ultimately maximize scientific payoffs and minimize costs and risks, both to crew safety and mission success – is one in which multiple (nominally two, perhaps three) crews land at a single site and live and work in a small underground base habitat constructed in advance by robots. They stay on Mars for an extended period – nominally 96 months, returning not on the first, but on the fourth Hohmann opportunity. They explore an extended area using pressurized rovers, perhaps driving to additional robotically-constructed underground habitats. Such a program-level architecture affords a continuous human presence on Mars, provides better shielding from radiation, reduces the number of crew transits from and to Earth, greatly reduces the maximum mass requirement for EDL (Entry, Descent, and Landing – we would land a crew capsule, not a full habitat), permits deferred development of the return vehicle (which could otherwise be a schedule-limiting element), and allows an initial unmanned return vehicle test supported by "ground crew" to return samples to Earth. Adoption of a "base-first" exploration program will require us to acknowledge and engage the real challenges to the human exploration and colonization of Mars – maintaining the safety, health, productivity, and happiness of a very small population of humans on the surface of Mars for an extended period of time. Apollo/Saturn proved that powerful rocket systems can be developed in less than a decade – but the Mars surface stay presents many specific technical and non-technical challenges that have nothing to do with "rocket science." Now is the time to start thinking seriously about these issues.

Key Words: Mars base, human spaceflight

1. INTRODUCTION

IN THE EARLY 1990S, IN RESPONSE to the demise of the first President Bush's proposed Space Exploration Initiative, or SEI (Synthesis Group, 1991), Martin Marietta engineers Bob Zubrin and David Baker developed a mission concept that came to be called "Mars Direct" (Zubrin, 1996; Gill, 2005). Several of its key features have since been incorporated into NASA and other Design Reference Missions, or DRMs (Rapp & Andringa, 2005; NASA, 1997, 1998, 2001), including the recently released Design Reference Architecture (DRA) 5.0 (NASA, 2009): (1) a conjunction mission comprised of a 6+ month transit to Mars, approximately 18 months spent on the surface of Mars, and a 6+ month return; (2) pre-emplacement on Mars of unmanned assets, including an Earth Return Vehicle (ERV) or Mars Ascent Vehicle (MAV); and (3) in-situ resource utilization (ISRU) to generate fuel (methane) and/or oxidizer (liquid oxygen) from Martian atmospheric carbon dioxide and hydrogen possibly extracted from Martian water. While the partitioning of functional elements into specific vehicles differs among the various proposals, this nominally 30+ month conjunction mission profile leveraging pre-emplacement of resources and ISRU now represents the consensus of thinking both inside and outside NASA.

Since expected high costs make it unlikely that we will invest in developing a system capable of taking humans to Mars and then use it for just one 18 month surface-stay mission – NASA's DRA 5.0 assumes three consecutive missions (NASA, 2009, p. 20) – we should consider how a sequence of manned Mars flights should be configured to create a rational program – one which maximizes scientific payoff consistent with minimizing costs and risks (maximizing crew safety). The Apollo program provides a baseline for comparison: a series of six independent sorties were made to different locations on the lunar surface in the three and a half years between July 1969 and December 1972, with surface stays ranging from 22 to 75 hours and total mission durations from launch to splashdown of between 8 and 13 days. Launch windows occurred every month when the lighting at the planned landing site was correct, so the factor limiting the pace of launches was the assembly and checkout of the required Saturn launch vehicle stages and Apollo command, service, and lunar modules.

We cannot fly our Mars missions this way, however, since the possible launch windows occur only every 26 months. Moreover, in order to maintain effective and efficient operations of the hardware production lines and launch and mission control facilities, it will be necessary to launch during every conjunction launch window. This means (1) that the second mission crew will launch from Earth two months after the first has begun its return trip from Mars, (2) that for a period of four months we will have two crews in space, and (3) that the second crew will not arrive on Mars until eight months after the first crew has departed. This suggests a fairly obvious opportunity: if we want to place a crew at point B on the surface of Mars, and already have a crew at point A, why should we fly the first crew hundreds of millions of km back to Earth, even as we send a second crew all the way from Earth? Why not send a capable ground transport vehicle and have the first crew drive it from point A to a robotically prepared site at point B? This, of course, would require that the two points be within driving range of each other, perhaps 500-1,000 km.

Manned spaceflight is inherently dangerous, especially in its launch and reentry phases – witness the fates of Challenger, Columbia, Soyuz 1, and Soyuz 11 (NASA, 1997, p. 3-13). While proper operation of a life support system is critical on the surface of Mars as well as in space, a spacecraft requires many more active systems than a ground habitat (e.g., vehicle stabilization). Moreover, radiation and meteoroids pose additional risks in space, and zero-gravity is a major complicating factor. On a per-hour basis, humans inside a habitat on the surface of Mars are likely to be far safer than humans in space – and it should be relatively easy to make them safer still. Since even the basic conjunction mission calls for more time on the surface of Mars (18 months) than in transit (12-13 months), it will pay to invest heavily in making the surface stay as safe as possible. In fact, we have a "virtuous cycle": (1) the longer we are planning to stay on the surface of Mars, the safer we can and should make it, and (2) the safer it is on the surface of Mars, the longer we can and should plan to stay.

2. THE BASE FIRST PROGRAM

Consider the following program plan: an initial crew (of 4 to 6 people) is launched to a carefully prepared site, and remains on the surface of Mars for a full 96 months, returning on the fourth minimum-energy Hohmann opportunity. A second crew follows to the same site 26 months later, and also stays for 96 months. Thus, we have 8-12 people at the base for a period of 70 months, we have no gaps in crew presence on Mars, and the operations team never has to deal with two crews in transit at the same time. We continue to build assets at the initial landing site, expanding to a second site only when the first base has achieved true critical mass. Instead of launching an ERV or MAV to Mars 26 months before the first crew, we send it 26 months after the second crew. Moreover, the first return launch of the ERV can be an unmanned test flight, carrying Mars samples back to Earth, and we will have the advantage of having "ground crew" to service the launch. Finally, because we are delaying the launch of the first ERV, we can also defer its development, thus employing a smaller team of rocket developers for a longer period of time (Gage, 2009).

	Mars Base First Program (2-96 month surface stays)	5 Conjunction missions (18 month surface stays)
Time from first crew arrival to last departure	122 mo	122 mo
Time 2 crews on Mars	70 mo (1@70 mo)	--
Time 1 crew on Mars	52 mo (2@26 mo)	90 mo (5@ 18 mo)
Gap: 0 crews on Mars	--	32 mo (4@8 mo)
# crews launched	2	5
EDL mass requirement	6-10 tonnes	40 tonnes
First ERV/MAV launch	52 months AFTER first crew launch	26 months BEFORE first crew launch

FIGURE 1

A capsule comparison of the "Base First" plan proposed here with a program consisting of a sequence of five independent conjunction missions.

2.1 Extended missions dictate early base establishment - It is clear that supporting a crew of 8-12 people on the surface of Mars for 10 full years is a very different proposition from supporting 4-6 people for 18 months. NASA's DRA 5.0 and other DRMs all the way back to the Mars Direct plan envision the crew living inside a habitat sitting on the surface. EDL to get this nominally 10-meter diameter 40-tonne "tuna can" unit to the surface of Mars in one piece represents a major technological challenge. The alternative approach proposed here for the "Base First" plan would have the crew live and work in underground tunnels constructed by robots before the crew arrives, creating a true Mars Base. Living underground would provide much better protection than a surface habitat against radiation, which, while much less intense on the Martian surface than in interplanetary space, is much more intense than on Earth or in low earth orbit (LEO) (Rapp, 2006). This approach also carries the additional advantage that a crew-landing vehicle could be much smaller than the 40 tonne tuna can habitat, greatly reducing the structural mass that would have to be brought from Earth and radically simplifying the EDL problem (Gage, 2009). For comparison, the Apollo Command Module (CM) weighed 6 tonnes, while the Apollo Lunar Module (LM, nee LEM), including both descent and ascent stages, was 15 tonnes.

The underground space would be quickly and continually expanded to eventually include living, sleeping, and dining areas, galley, pantry, and garden; medical/dental clinic with mini-intensive care unit, exercise facilities (gym and track), spa, and swimming pool; medical, biological, chemical, and geological laboratories; manufacturing and repair shops; storage for food, other supplies, spare parts, and collections of samples. Many critical systems will be required to support human life and mission operations, including thermal control, air, water, waste, computing, and communications, but these various subsystems can be installed in the constructed underground base in a much more loosely coupled manner than would be possible in a tightly integrated habitat transported from Earth, thus simplifying component repair and replacement.

3. THE CHALLENGES OF THE BASE FIRST PROGRAM

The concept of a base on Mars is obviously not a new one: Google searches for "Mars Base" or "Mars Colonization" yield a plethora of diverse links. The "Mars Homestead" project (Mars Foundation, 2010) seems to be pursuing (at the "enthusiastic volunteer" level) a fairly reasonable broad-based exploration of many of the issues involved in establishing a permanent human presence on Mars. However, a common thread among mainstream efforts is the (usually implicit) assumption that the establishment of a base will occur only after a sequence of manned sorties to identify the best location for the base or colony. The key arguments in this paper are that (1) we should plan for a Mars base beginning with the first humans we send to Mars: we should plan to send fewer people to Mars, but have each of them stay much longer, and (2) we should invest heavily up-front in developing and refining the surface segment of the human Mars mission, because travelers to Mars will spend (much) more time on the surface of Mars than in space, and this is where the critical challenges and payoffs lie. This will require either that NASA move well beyond its traditional focus on space transportation systems, or that one or more other entities assume a leadership role in the Mars exploration enterprise.

3.1 Can humans survive and succeed on a ten-year mission? - Some may object that a mission profile calling for an eight-year stay on the surface of Mars (and ten years away from Earth) is unreasonable – that the psychological stresses of living in such a small isolated group for so long would put the success of the mission, if not the crew's survival, at unacceptable risk. However, the history (and especially the prehistory) of humanity is one of many small groups of people migrating into the unknown with no intention of returning, and, in fact, informal surveys suggest that many people would be willing to sign up for a one-way trip to Mars (Krauss, 2009). We find many examples of small groups that have successfully lived in nearly constant isolation, including bands of hunter-gatherers, Inuit family groups, pre-20th century ship crews, castaways, and some soldiers and prisoners.

However, while humans on Mars will be physically isolated from Earth, they will have high bandwidth connectivity to the rest of the humanity (albeit with a 6-44 minute round trip latency). They need not be lonely; the World Wide Web will grow into the Solar System Wide Web. But we must thoroughly explore the full range of issues associated with long-term connected-but-physically-isolated living, including understanding how and how well high-bandwidth network communications can compensate for the lack of physical contact, and develop an experience base on Earth before we dare send people on such a mission. Since it is likely that the success of the mission may depend on the "chemistry" of the specific personalities involved, it may be that a crew should begin living together as a coherent group (if not in full isolation) well before their launch. The psychological and psychiatric issues associated with spaceflight have been studied since the beginning of the space age; see, for example, Kanas & Manzey, 2003; Kanas & Ritsher, 2005, and Johns, 2004.

Since living beneath five meters of regolith will mitigate the radiation hazard on the surface, the principal physiological challenge posed by the base-first mission (beyond those posed by a 30-month conjunction mission) is the loss of bone density and strength associated with the outward and return 6+ month zero gravity transits and eight years of 0.38 g Mars gravity. A focused exercise regimen, possibly combined with dietary modification, should at least partially mitigate these effects (Keyak et al. 2009), and at some point it might be possible to install a one-g centrifuge in the base. Long-term exposure to a low-pressure high-oxygen atmosphere in the base habitat, which could be adopted in order to reduce EVA prebreathe time (Gage, 2006; NASA, 2001, p. 20), would constitute a second physiological risk factor – but this is a risk which can be evaluated by experimentation on Earth.

An advantage of the base-first exploration strategy is that it will allow people to extend their stay on Mars, which would be absolutely necessary if the ERV or MAV could not be made ready during the return launch window, and might be desirable in other cases – imagine that the crew exobiologist on the first conjunction mission were to discover living Martian life just a few weeks before she is scheduled to return to Earth. And, of course, one of the classical planetary exploration science fiction tropes (e.g., Landis, 2000; Varley, 2005) – is that, when it is time to return to Earth, one or two characters (usually a couple) announce "we're going to stay."

3.2 Required technologies and tools - Viewed from an engineering perspective, it is clear that a Mars base will constitute a complex "system of systems", one whose development will involve a large number of technical disciplines, and this fact must be explicitly acknowledged if we are to succeed. Here is a listing of some of the technologies and tools we will need to develop in order to create a human base on Mars:

- Surface nuclear power plant (nominally 150 kW electrical, plus thermal energy)
- Cryogenic storage and handling tools/systems
- Thermal control systems (including insulation) – different on Mars than in space
- Methane (and/or propane?)-oxygen power sources (electrical, thermal, motive; very small to very large)
- Vehicles (manned and unmanned, ground and air, pressurized and unpressurized, all sizes)
- Construction technologies and equipment (including robots, autonomous or supervised)
- Communications and navigation systems (intra-base, off-base, and off-planet; supporting systems, vehicles, and people)
- Ultra-reliable computing and other IT support (redundant, radiation-hard; wearable systems, etc)
- Medical strategies/tools: auto-medicine (taking care of yourself), para-medicine (taking care of each other), and tele-medicine (accessing medical resources back on Earth)

Not only is this not "rocket science", it's not even just technology. We need to think about construction, physiology, and robotics; and psychology and sociology; and nutrition, gardening, and medicine; and architecture, history, insulation, and HVAC; and power distribution, IT, sensors, and AI; and biology, chemistry, geology, and seismology; and...

In fact, the successful development of an effective base on Mars will require more than a solid systems-centric engineering perspective; it will also require a human-centric perspective, involving numerous social as well as technical disciplines. In essence, we are attempting to design the smallest-scale possible viable human economy and supporting ecology, and we don't know in advance what this "nano-society" will look like. But NASA as an organization is focused on the "rocket science." To understate the case considerably, "studies of surface activities and related systems have not always been carried out to the same breadth or depth as those focused on the space transportation and entry or ascent systems needed for a Mars mission" (NASA, 2001, p. 1). Perhaps the National Science Foundation (NSF), with its broad scientific purview and experience managing U.S. Antarctic bases, might effectively participate in the development of the Mars base.

3.3 Base development process and technology context - Developing all the pieces for an effective base on Mars will be a complex undertaking, one quite distinct from the development of the system that will be required to transport humans to and from Mars. What is required is the development of an overall plan, starting from the physiological and psychological needs of a human crew, defining their task-oriented and other activities, leading to system and subsystem models, assessing and adopting/adapting technologies to implement them, and eventually validating the various subsystems through extensive testing and simulations here on Earth (Gage, 2006). This should be a "spiral" process that will be iterated until it is time to go, with multiple agencies/entities involved (e.g., development of a surface-sited nuclear power plant by DOE, healthcare systems by NIH). The initiation of this activity need not and should not wait for a specific commitment to build the Mars transportation system. Perhaps the most difficult challenge will be to manage a complicated program with a relatively small budget (as compared to rocket development), across multiple agencies, over a period of many years.

Many technologies and systems developed for Earth will be carried unchanged to Mars; others will have to be adapted to the particular situation of our Mars base. Rapidly changing technology complicates the development process: at what point do we decide to adopt or adapt a given product or system for inclusion in our long-term Earth-based Mars base prototyping/simulation enterprise? We can freely experiment with COTS elements, but the decision to embark on a costly program to modify existing products for use on Mars must not be taken too early, or we will, like the U.S. military with its communications systems, be trapped in an expensive web of obsolete proprietary systems even as the rest of the world adopts technologies with much higher performance and much lower cost.

The rapid evolution of technology also carries short-term challenges with respect to what we actually send to Mars: given the 26 month synodic period between launch windows, an assembly-test-launch (ATL) time that is not much shorter, and the 12-18 month COTS electronics product innovation cycle, we will have to decide whether to introduce a new generation of IT for each successive mission, and it will clearly be impossible to perform a full-mission duration test of new subsystems as they are deployed. Fortunately, the loose coupling of subsystems in the Mars base environment will allow easy module upgrades and the use of redundant units to ensure system-level reliability.

4. CONCLUSIONS

Because of the limitations placed by orbital mechanics on energy-affordable transits between Earth and Mars (transits that last 6+ months, and are possible only every 26 months), it would be suboptimal to execute the initial human exploration of Mars as a sequence of independent sorties, analogous to the Apollo program. Costs and risks can be significantly reduced by pursuing a program in which the first humans we send to Mars remain there for many (nominally eight) years, living and working in a safe and productive underground base, constructed in advance of their arrival by robots (Gage, 2010). Twenty-six months after the first crew's arrival, a second crew will land at the same base, and other sites of interest can be visited using ground vehicles. Human presence on Mars will be continuous for the (nominally ten year) extent of the program, but fewer crews will be exposed to the risks of spaceflight than in a sequence of independent conjunction missions.

The enthusiasm of wannabe Martians (Gill, 2005) notwithstanding, and independent of the cleverness of NASA's rocket scientists in developing the vehicles and systems needed to transport humans to Mars, many years of preparation involving many other technologies and disciplines will be required before humans will actually set foot on the Martian surface. A major challenge will be the management of this extended process of preparation, with a relatively small budget (as compared to rocket development), across multiple agencies, and over a period of many years.

REFERENCES

Gage, D.W. (2006). Begin High Fidelity Mars Simulations Now. Ninth International Mars Society Convention, Washington, DC, 3-6 August.

Gage, D.W. (2009). Prepare Now for the Long Stay on Mars. Twelfth International Mars Society Convention, College Park, MD, 30 July - 2 August.

Gage, D.W. (2010). Unmanned systems to support the human exploration of Mars. Proc. SPIE Vol 7692, 7692M.

Gill, S. (2005). The Mars Underground. DVD available online at http://store.oculefilms.com/buy-dvds, Ocule Films, Venice, CA.

Johns, W. (2004). Psychological Factors for Mars Settlement. Online at http://www.marshome.org/files2/Psychological%20Factors%20for%20Mars%20Settlement.ppt.

Krauss, L. (2009). A One-way Ticket to Mars. New York Times Op-Ed. August 31. Online at http://www.nytimes.com/2009/09/01/opinion/01krauss.html.

Kanas, N., Ritsher, J. (2005). Psychosocial issues during a Mars mission. AIAA 1st Space Exploration Conference, Orlando, FL, January 30-February 1.

Kanas, N., Manzey, D. (2003). Space Psychology and Psychiatry. Microcosm Press, El Segundo CA.

Keyak, J.H., Koyama, A.K., LeBlanc, A., Lu, Y., Lang, T.F. (2009). Reduction in proximal femoral strength due to long-duration spaceflight. Bone, vol 44, issue 3, pp 449-453.

Landis, G.A. (2000). Mars Crossing. Tom Doherty Associates, NewYork.

Mars Foundation. (2010). The Mars Homestead Project. http://www.marshome.org/

NASA. (1997). Human Exploration of Mars: The Reference Mission of the NASA Mars Exploration Study Team, NASA SP-6107, NASA Johnson Space Center, Houston TX.

NASA. (1998). Reference Mission 3.0; Addendum to the Human Exploration of Mars: The Reference Mission of the NASA Mars Exploration Study Team, NASA SP-6107-ADD, also EX13-98-036, NASA Johnson Space Center, Houston TX.

NASA. (2001). The Mars Surface Reference Mission: A Description of Human and Robotic Surface Activities, NASA TP-2001-209371, NASA Johnson Space Center, Houston TX.

NASA. (2002). Analysis of Synthesis Group Architectures. ExPO Document XE-92-004.

NASA. (2009). Human Exploration of Mars Design Reference Architecture 5.0. Online at http://www.nasa.gov/pdf/373665main_NASA-SP-2009-566.pdf.

Rapp, D., Andringa, J. (2005). Design Reference Missions for Human Exploration of Mars. JPL Report D- 31340, also presented at ISDC, Arlington, VA, May, 2005.

Rapp, D. (2006). Radiation effects and shielding requirements in human missions to the moon and Mars, Mars 2, 46-71 (2006), online at http://marsjournal.org/contents/2006/0004/files/rapp_mars_2006_0004.pdf.

Synthesis Group. (1991). America at the Threshold: America's Space Exploration Initiative. Online at http://history.nasa.gov/staffordrep/main_toc.pdf.

Varley, J. (2005). In the Hall of the Mountain Kings, in the anthology Fourth Planet from the Sun. Thunder's Mouth Press, New York.

Zubrin, R. (1996). The Case for Mars. Simon & Schuster, New York.

Zubrin, R. (2003). Mars on Earth. Tarcher/Penguin, New York.

Journal of Cosmology, 2010, Vol 12, 3946-3956.
JournalofCosmology.com, October-November, 2010

- 12 -

Expedition to Mars. The Establishment of a Human Settlement

V. Adimurthy, Ph.D., Priyankar Bandyopadhyay, G. Madhavan Nair, Ph.D.

Indian Space Research Organization (ISRO) Bangalore, India

ABSTRACT

The long-term future of the human race must be in space. Mars, like Earth, is a rocky planet with ranges of surface temperature that man has managed to deal with on Earth. Most likely, Mars will be the first place in the solar system where humans can colonize outside Earth. During the last forty-five years numerous unmanned spacecrafts have been probing Mars through fly-by, orbiter and lander missions, and sample return missions are on the anvil. These robotic explorations will pave the way for manned missions to Mars. The opportunities for Mars expeditions are chosen through assessment of the variability across Earth-Mars synodic cycles. Mars expedition is undertaken through execution of split mission architecture. Portions of assets required for Mars expedition are sent to Mars prior to the crew. For an expedition of a crew of six to Mars, nine launch vehicles each having Low Earth Orbit (LEO) payload capacity of about 100 tonne are required, in addition to a flight of man-rated vehicle to transfer the crew to interplanetary module. Nuclear power is a key element for propulsion as well as for sustaining crew on Mars.

Key Words: Mars, Interplanetary Travel, Mission Architecture, Mars Expedition, Human Settlement, Long Duration Spaceflight

1. MARS EXPEDITION OVERVIEW

MARS IS A ROCKY PLANET, LIKE EARTH. It is formed around the same time, yet with only half the diameter of Earth. The levels of mean surface temperature that exist on Mars are also present on Earth in some extreme places where humans have successfully ventured. The corresponding lengths of a day and the obliquities of Earth and Mars are also similar. Both planets experience seasons owing to their eccentricities and obliquities (see Levine et al., 2010a,b,c,d). Among the planets and moons in the solar system, where man can colonize, Mars will be the first. As the resources of Earth are finite, it is likely, that the long-term future of the human race is in space (Mitchell and Staretz 2010; Zubrin 2010). Humans, as a species, should start thinking towards freeing themselves from the constraint of a singularity called Earth.

Within a decade of the first launch of artificial satellite in 1957, Mariner- 4 flew past Mars in 1965, providing the first close-up photographs of Mars. In the early 1970, Mars- 2 and Mars- 3 from the erstwhile USSR were the first two space probes to enter into orbits around Mars. In 1976, the two American Viking probes entered Mars orbit and released static lander modules that made a successful soft landing on the planet's surface. The next leap in Mars exploration came, in proving the feasibility of autonomous control over a robot, the rover, *Sojourner*, as part of the Mars Pathfinder mission in 1997. In the second decade of the 21st century, plans are already afoot for a Mars Sample Return mission that would use robotic systems and a Mars ascent rocket to collect and send samples of Martian soils to Earth for detailed physical and chemical analysis. These robotic explorations will pave the way for manned mission to Mars (Gage 2010a; Podnar et al., 2010). In the spirit of human endeavor and exploration, a human mission to Mars could be likened to the expedition by Vasco da Gama and Christopher Columbus, to discover the sea route between Europe and India. Of course, technical challenges for human mission to Mars are several orders more complex.

A single human expedition to Mars with a crew of six, would require at least nine launch vehicles with a Low Earth Orbit (LEO) payload capacity of about 100 metric tonne. More complex is the scenario, in the context of human settlement on Mars. To establish a vibrant colony, we estimate that an initial population of 150-180 would be required to allow normal propagation for 60-80 generations, which is equivalent to 2000 years. To promote the variability in gene pool, it is advantageous to choose settlers from diverse backgrounds and with different skill. As a consequence, Mars expedition and colonization would entail international cooperation on an unprecedented scale (Joseph 2010).

2. MISSION OPPORTUNITIES

The dates for a Mars expedition are chosen through assessment of the variability of mission opportunities across Earth-Mars synodic cycles. Mission opportunities occur approximately every 2.1 years in a cycle that repeats roughly every 15 to 17 years (the synodic cycle). Within this time span mission opportunity characteristics are similar but not the same. Along with human missions, one-way cargo delivery trajectories would also be required and which would depart during each opportunity preceding each crewed mission.

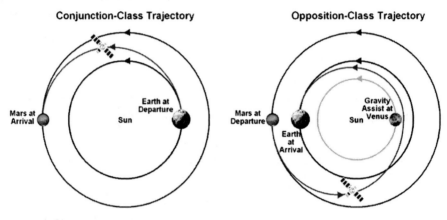

FIGURE 1

Trajectories for the expedition to Mars

Two types of mission trajectories, namely, conjunction-class and opposition-class, have been discussed in the context of human expedition to Mars (Rapp, 2008) and are depicted in Figure 1. Crew in conjunction class trajectory from Earth would spend about 6-9 months in space before arriving at Mars to spend another 14-20 months before the initiation of return trip, again through conjunction class trajectory. However, there can be unforeseen circumstances such as severe dust storms, radiation hazards, alien pathogens (Levine et al., 2010a,b,c,d; Rummel et al., 2010; Straume et al., 2010; Stuster 2010), which may compel them to leave Mars early for safe return. Under these emergency contingencies, they can lift-off from Mars after spending 1-3 months on the surface and follow the opposition class of trajectory to Earth through gravity assist by Venus.

The return to Earth through opposition-class trajectory will take much longer (10-15 months). Conjunction-class trajectories are the main focus for expedition to Mars. However, opposition-class trajectories, notwithstanding having higher propulsive, shielding and life support requirements, can serve as a mission salvage option. But even before mission salvage, there must be requirements for mission abort.

A planned option to provide an abort-return trajectory may bolster the capability to handle any unforeseen eventualities. In such a transfer, the spacecraft is placed on a trajectory from Earth to Mars, which also returns to Earth at a specified point in the future, without the need for any major additional action except for small mid-course corrections. Typically such free-return trajectories employ a resonance in the orbital periods of the transfer trajectory and Earth about the Sun. Abort options clearly place a requirement on the outbound spacecraft to support the crew during the lengthy return to Earth.

3. PREPARATION FOR MARS EXPEDITION

The Mars expedition should be undertaken through execution of split mission architecture. A portion of assets required for Mars expedition (e.g. food, supplies, fuel, robots) would be sent to Mars prior to the arrival of the crew (Gage 2010a,b; Podnar, et al., 2010). This mission architecture allows a lower-energy/longer-duration trajectory to be utilized for these pre-deployed assets. The flexibility to pre-position some of the mission infrastructure also allows for the preparation of the ascent propellant, while on Mars, using the Martian environment as the source for raw materials (Zubrin 2010). The pre-deployment of a propellant processing facility and implementation of in-situ resource utilization results in a net decrease in the total mass that is needed to complete a mission. It also results in a significant reduction in the size of landers.

It is envisaged that at least two major components are required: propellant and consumables. Fortunately, abundant resource of water on Mars has the potential to provide propellants and consumables. Water can be electrolyzed to produce hydrogen, which can be used as propellant. A surface nuclear power reactor can provide energy required to carry out electrolysis. Furthermore, this power system would be adequate to meet the needs of the human crew when they arrive.

The separation of the mission elements into pre-deployed cargo and crew vehicles allows the crew to fly on a higher-energy/shorter-duration trajectory, thus minimizing their exposure to the hazards associated with inter-planetary travel (Hawley 2010; Straume et al., 2010; Stuster 2010).

Owing to the significant amount of mass required for a human mission to Mars, numerous heavy-lift launches which can put a LEO payload of the order of 100 metric ton, would be required. The mission architecture conceived for the expedition to Mars for human settlement is described in Figure 2.

Of the ten launches required for a six member crew mission to Mars, the first five launches would ensue two years prior to the actual crew launch. These are represented in the left side of the Figure 2. The pre-deployment of the first two cargo modules, namely, the descent/ascent vehicle and the surface habitat, takes

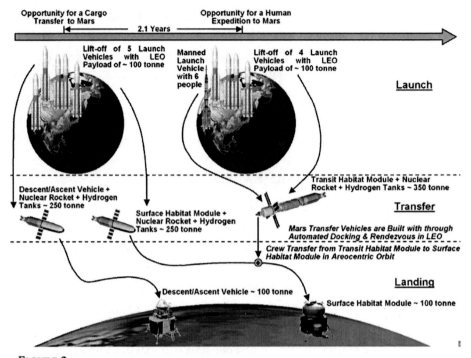

FIGURE 2
The mission architecture for the expedition to Mars

place through the first five launches. These two vehicle sets to Mars are first assembled (via rendezvous and docking), and checked out in LEO. After all of the systems have been verified and are operational, transfer phase is initiated and they are injected into a minimum energy transfer from geocentric orbit to Mars just over 2 years prior to the launch of the crew. The descent/ascent vehicle module lands on Mars at a pre-selected location and its nuclear reactor starts transforming water from Mars environment into propellant to be used for Mars ascent as well as surface operations. The surface habitat module enters into an eccentric areocentric orbit and waits for the arrival of the crew.

The second batch of launches for Mars expedition begins during the next injection opportunity. These include the launch, assembly, and checkout of the Mars Transfer Vehicle which contains Transit Habitat Module, Nuclear Rocket and Hydrogen Tanks. Mars Transfer Vehicle serves as the interplanetary support vehicle for the crew for a round-trip mission to Mars, though it must be recognized that this may be a one way journey (Schulze-Makuch and Davies 2010). Assembly of these modules would require four launches from Earth.

After the Mars Transfer Vehicle is operational, the flight crew embarks on the journey to Mars along an appropriate fast-transit trajectory towards Mars. These are depicted on the right side of the Figure 2. The length of this outbound transfer to Mars is dependent on the mission date, and ranges from 175 days to 225 days. Upon arrival at Mars, the crew vehicle enters into an elliptic parking orbit before performing a rendezvous with the Surface Habitat Module, which is already in Mars orbit.

There are many who envision a human mission to Mars before the end of the next decade (Joseph 2010; Mitchell and Staretz 2010; Zubrin 2010). However, if we were to wait until 2035, this window of opportunity would shorten the trip considerably, to a little over 200 days, which makes it more practical for settlement and resupply. Thus, the the initial plans for Mars settlement may take place 25 years into the future, beginning in 2035. This 2035 opportunity offers a trajectory to Mars, from a circular Earth orbit of 400 km or so, and is achieved through a velocity addition of around 3.7 km/s. Upon reaching Mars, another velocity reduction of a little less than 1 km/s is employed in conjunction with aero-capture to enter into an elliptic Mars Orbit that is an eccentric areocentric orbit of 250 km • 34000 km.

The 2035 opportunity can be utilized to send necessary infrastructure and other materials to Mars for eventual human settlement. During the very next opportunity in 2037, the crew can experience a typical transit time of a little less than 180 days to Mars. Crew can stay on Mars for about 540 days and can return to Earth in 2040. During any of these opportunities, it is also possible to achieve Trans Mars Insertion (TMI) using Moon and Earth gravity assists (Penzo, 1998).

Nuclear thermal rocket propulsion is preferred as the interplanetary propulsion system for both cargo and crew vehicles. In nuclear thermal rockets, a fluid with low molecular weight like liquid hydrogen is heated using energy liberated by fissile material and expelled through a nozzle to create thrust. Propellants, other than liquid hydrogen, such as ammonia or water, would provide reduced exhaust velocity but their greater availability in places like Mars can increase operational flexibility. Nuclear power systems are also best suited for surface operations where sunlight is limited and reliability is paramount. On Mars, though the night period is only 12 hours, sunlight at the surface is reduced to about 20% that of Earth. Martian dust storms and missions to the higher latitudes, further decrease the availability of sunlight for solar power (see Levine et al., 2010a,b,c,d).

A key advantage of surface fission reactor systems is that they produce constant power to allow continuous day and night surface operations. Relative to comparable solar power systems with energy storage, a fission reactor offers significant mass and volume savings. This large number of launches necessitates a significant launch campaign with international participation that must begin several months prior to the opening of Mars departure window. The reference strategy that is adopted eliminates on-orbit component-level assembly of the mission elements by segmenting the systems into discrete modules and using automated rendezvous and docking of the major modules in LEO. Launches occur 30 days apart and are completed several months before the opening of Mars departure window to provide a margin for technical delays and other unforeseen problems. This strategy requires that the in-space transportation systems and payloads stay in LEO for several months prior to departure for Mars. In this regard, the risk due to derelict space objects must be mitigated through Space Object Proximity Analysis (SOPA).

4. MARS TRANSFER TRAJECTORY

A Mars transfer trajectory is of the order of 200 days. During this time, the crew prepare for the tasks ahead. Prior to and during the journey precautions must be taken to ensure the physical and psychological health of astronauts as the perils are many (Bishop 2010; Fiedler and Harrison 2010; Harrison, and Fiedler 2010; Hawley 2010; Suedfeld 2010). Creating a home-like environment that mirrors life on Earth is important for Mars expedition. Family members can contribute in engaging the crew in two-way communications. With such support the crew in space could still be an integral part of the family back home (Pletser 2010; Johnson, 2010). However, in this context, it may be recalled that the maximum distance between Earth and Mars is roughly 400,000,000 km, 1,000 times the distance between Earth and Moon. Because of the loss of signal strength due to the increased distance, communications from Mars are more challenging than communications from the Moon. The large distance to Mars also implies long signal transit times, with a round-trip time of up to 44 minutes. An interplanetary electronic mail service can be the choicest mode of communication in the future.

Radiation in space is a major issue to contend with (Straume et al., 2010). From the standpoint of humans in interplanetary space, the two important sources of radiation for Mars expedition are, the heavy ions (atomic nuclei with all electrons removed) of galactic cosmic ray and sporadic production of energetic protons from large solar particle events (Straume et al., 2010). The constant bombardment of high-energy galactic cosmic ray particles delivers a lower steady dose rate compared with large solar proton flares which on occasion deliver a very high dose in a short period of time (of the order of hours to days). Various active and passive shielding options to protect astronauts from space radiation are described by Seedhouse (2009) and Straume et al., (2010).

Another important factor for Mars expedition is absence of gravity during Mars transfer trajectory (Moore, et al. 2010). One of the major effects of prolonged weightlessness seen in long-duration space flights has been a reduction in bone mineral density (Harrison and Fiedler 2010; Moore, et al. 2010). In this context, alternate medicinal systems like ayurveda hold promise in alleviating ailments resulting from weightlessness. Formulations comprising *Terminalia arjuna, Withania somnifera* and *Commiphora mukul* are well known for their bone remineralization (Mitra et al., 2001). At the same time, the efficacy of ayurvedic formulations in zero gravity conditions needs to be ascertained. Similarly, physical exercise and yoga can help maintain physiological and psychological health of the crew. Some aspects of yoga exercises are the theme of experiments conducted during the joint Indo-Soviet manned space program in Salyut-7 (Wadhawan et al. 1985: Harland, 2005). The results suggest that yoga exercises have definite beneficial effects in preventing muscular atrophy and on the psycho-physiological well being during space flight.

5. ENTRY, DESCENT AND LANDING (EDL)

One of the most important strategies needed to enable human mission to Mars is efficient aero-assist technology for descent. Mars entry, descent and landing bring forth many engineering challenges (Drake 2010). Challenges emanate from an atmosphere, which is thick enough for substantial heating, but not sufficiently dense for low terminal descent velocity. Also the surface environment is made of complex terrain patterns with rocks, craters, and dust. Relative to Earth, Mars atmosphere is thin, less than 1% in surface pressure (see Levine et al., 2010a,b,c,d). As a result, Mars entry vehicles tend to decelerate at much lower altitudes and, depending upon their mass, may never reach the subsonic terminal descent velocity of Earth aerodynamic vehicles. Because hypersonic deceleration occurs at much lower altitudes on Mars than on Earth, the time remaining for subsequent entry-descent-landing events is often a concern. On Mars, by the time the velocity is low enough to deploy decelerators, the vehicle may be near the ground with insufficient time to prepare for landing.

As the spacecraft approaches the planet Mars it will enter into a hyperbolic path. Mars can be reached and a successful landing ensured entirely through expending propulsive energy or through direct entry, aero-capture or aero-braking, which utilize the Martian atmosphere to a substantial advantage (see Figure 3). In direct entry mode, the module soft lands on the Martian surface after reducing the velocity using on board propulsive energy and by continuously maneuver the module in such way that it lands on the surface vertically with velocity not larger that a few meters per second. Direct entry is associated with intense heating and high deceleration load (Drake 2010).

The Viking 1 and 2 missions employed direct entry to place the landers on the Martian surface. Aero-capture is a method that is employed to directly capture into a planet's orbit from a hyperbolic arrival trajectory using a single,

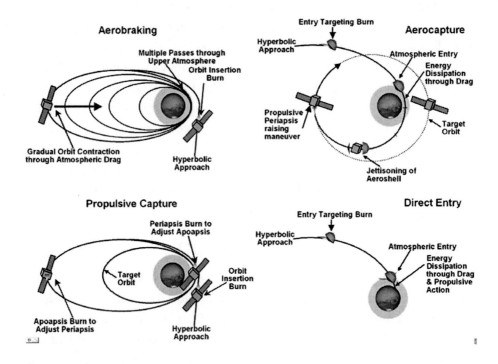

FIGURE 3

Various planetary capture and entry strategies

atmospheric aerodynamic drag pass, thereby reducing the propellant required for orbit insertion. In aero-capture a spacecraft dips into the mid-atmosphere with hyperbolic entry velocity so as to achieve the capture velocity and orbital insertion in a single drag pass. Subsequently, a propulsive maneuver is performed to raise the periapsis. In contrast, when aero-braking is employed, a spacecraft is first captured into an elliptic orbit around the celestial body through propulsive maneuver. Then the orbit size is reduced gradually over a period of time through successive upper-atmospheric drag passes. Though aero-capture uses less propellant as compared to aero-braking, additionally an aeroshell is a required to protect the payload from aero-thermal load during the atmospheric passage of spacecraft. The landing site is chosen at a relatively flat area where cargo elements can land safely. Also the landing site must be suitable to allow for in situ resource utilization (Gage 2010b).

6. THE NEW BEGINNING ON MARS

Mars is a unique and universal place, not for one group, not for one creed, not for one nation but for the entire humanity (Mitchell and Staretz 2010). The human mission to Mars will be a major step towards fulfilling the age old dream and human desire to explore the planets. It may take a large number of missions to Mars to perfect the technology associated with human travel to deep space. It may take the crew a few weeks or months to acclimatize to the gravity of Mars (about 0.38 g). After the crew has acclimated, the initial surface activities would focus on transitioning from a lander mode to a fully functional surface habitat mode (Boston 2010; Gage 2010b; Schulze-Makuch and Davies, 2010; Zubrin 2010). This would include performing all remaining setup and checkout operations that could not be performed prior to landing, as well as transfer of hardware and critical items from the pre-deployed Mars-Ascent-Descent Vehicle System. Thus, a new beginning for the humanity is ushered in. The journey to Mars will be a first step toward colonizing the entire cosmos.

Acknowledgements: Authors would like to acknowledge S. Swaminathan, Ph.D. for his valuable contributions in the preparation of this paper.

REFERENCES

Bishop, S. L. (2010). Moving to Mars: There and Back Again. The Psychology and Culture of Crew and Astronaut, Journal of Cosmology, 12, 3711-3722.

Boston, P.J. (2010). Location, Location, Location! Lava Caves on Mars for Habitat, Resources and Science, Journal of Cosmology, 12, 3957-3979.

Drake, B. G. (2010). Human Exploration of Mars: Challenges and Design Reference Architecture 5.0, Journal of Cosmology, 12, 3578-3587.

Fiedler, E. R., and Harrison, A. A. (2010). Psychosocial Adaptation to a Mars Mission, Journal of Cosmology, 12, 3685-3693.

Gage. D. W. (2010a). Robots on Mars: From Exploration to Base Operations, Journal of Cosmology, 12, 4051-4057.

Gage, D. W. (2010b). Mars Base First: A Program-level Optimization for Human Mars Exploration, Journal of Cosmology, 12, 3904-3911.

Harland, D.M. (2005). The Story of Space Station Mir. Springer, New York.

Harrison, A. A., and Fiedler, E. R., (2010). Mars, Human Factors and Behavioral Health, Journal of Cosmology, 12, 3685-3693.

Hawley, S. A. (2010). Mission to Mars: Risks, Challenges, Sacrifices and Privileges. One Astronaut's Perspective, Journal of Cosmology, 12, 3517-3528.

Johnson, M.P.J. (2010). The roles of NASA, U.S Astronauts and their Families in Long Duration Missions. Acta Astronautica, 67, 561-571.

Joseph, R. (2010). Marketing Mars. Financing the Human Mission to Mars and the Colonization of the Red Planet, Journal of Cosmology, 12, 4068-4080.

Levine, J.S., Garvin, J.B. and Beaty, D.W. (2010a). Humans on Mars: Why Mars? Why Humans? Journal of Cosmology, 12, 3627-3635.

Levine, J.S., Garvin, J.B. and Head III, J.W. (2010b). Martian Geology Investigations. Journal of Cosmology, 12, 3636-3646.

Levine, J.S., Garvin, J.B. and Elphic, R.C. (2010c). Martian Geophysics Investigations. Journal of Cosmology, 12, 3647-3657.

Levine, J.S., Garvin, J.B. and Hipkin, V. (2010d). Martian Atmosphere and Climate Investigations. Journal of Cosmology, 12, 3658-3670.

Mitchell, E. D., Staretz, R. (2010). Our Destiny – A Space Faring Civilization? Journal of Cosmology, 12, 3500-3505.

Moore, S. T. et al. (2010). Journey to Mars: Physiological Effects and Operational Consequences of Long-Duration Microgravity Exposure, Journal of Cosmology, 12, 3781-3793.

Mitra, S. K., Venkataranganna, M.V., Gopumadhavan, S., Mohamed Rafiq, Anturlikar, S.D, Venkatesha Udupa, U, Seshadri, S.J., Sundaram, R and Madi Tripathi (2001). The Beneficial Effect of OST-6 (OsteoCare), a Herbomineral Formulation, in Experimental Osteoporosis. Phytomedicine, 3, 195- 201.

Penzo, P.A. (1998). Planetary Missions from GTO using Earth and Moon Gravity Assists. AIAA-98-4393.

Pletser, V. (2010). A Mars Human Habitat: Recommendations on Crew Time Utilization, and Habitat Interfaces, Journal of Cosmology, 12, 3928-3945.

Podnar, G. et al., (2010). Telesupervised Robotic Systems and the Human Exploration of Mars, Journal of Cosmology, 12, 4058-4067.

Rapp, D. (2008). Human Missions to Mars, Springer, New York.

Rummel et al., (2010, The Integration of Planetary Protection Requirements and Medical Support on a Mission to Mars, Journal of Cosmology, 12, 3834-3841.

Schulze-Makuch, D., and Davies, P. (2010). To Boldly Go: A One-Way Human Mission to Mars, Journal of Cosmology, 12, 3619-3626..

Seedhouse, E. (2009). Martian Outpost. Springer, New York.

Straume, T. et al., (2010). Toward Colonizing Mars. Perspectives on Radiation Hazards: Brain, Body, Pregnancy, In-Utero Development, Cardio, Cancer, Degeneration, Journal of Cosmology, 12, 3992-4033.

Stuster, J. (2010). Acceptable Risk: The Human Mission to Mars, Journal of Cosmology, 12, 3566-3577.

Suedfeld, P. (2010). Mars: Anticipating the Next Great Exploration. Psychology, Culture and Camaraderie, Journal of Cosmology, 12, 3723-3740.

Wadhawan, J., Dikshit, M.B., Chatterjee, P.C. (1985). Some Aspects of Yogic Exercise during Space Flight. Aviation Medicine, 29, 31-37.

Zubrin, R. (2010). Human Mars Exploration: The Time Is Now, Journal of Cosmology, 12, 3549-3557.

Journal of Cosmology, 2010, Vol 12, 3957-3979.
JournalofCosmology.com, October-November, 2010

- 13 -

Location, Location, Location! Lava Caves on Mars for Habitat, Resources, and the Search for Life

Penelope J. Boston, Ph.D.
New Mexico Institute of Mining & Technology (NMT) and
The National Cave and Karst Research Institute (NCKRI)

ABSTRACT

Over the course of humanity's history as a species, the use of caves, rock shelters, and other natural geological features has played an important role in our survival and cultural development. We suggest that the use of such natural features in future human exploration of Mars and Earth's moon could be a timely and practical solution to a number of potential dilemmas presented by the extreme and challenging nature of the environments on these bodies. Lava tubes, other caves, cavities, and canyon overhangs that are being identified on other planets are sites of intense scientific interest for geology, atmospheric climate records, and potentially biology. They may offer easier subsurface access for direct exploration and drilling, and could provide extractable minerals, gases, and ices. In the past few years, examples of such structures on Mars, the Moon, and potentially other bodies have increasingly come to light. Thus, the real estate is out there waiting for us to modify it for our exploration missions. The present Martian surface environment is extremely cold, dry, chemically active, and high in both ultraviolet and ionizing radiation. Galactic Cosmic Radiation (GCR) and episodic waves of high energy particles from solar proton events (SPE) necessitate the provision of robust radiation protection for habitats, workspaces, vehicles, and personal space suits. The mass penalty of providing this is a major driver in our consideration of the use of natural rock mass for radiation protection for habitats and workspaces, arguably the most massive components of an integrated human exploration equipment suite. Planetary protection considerations emerging from recent studies advocate a localization and zoning of degrees of human impact, much like that being implemented in the Antarctic as *Special Regions*. Containment of the primary human habitation and work activities within the confines of a subsurface habitat are highly consistent with these new approaches to Planetary Protection forward contamination. To begin to think about caves in the extraterrestrial exploration context, we have developed the notion of a complete, functioning subsurface habitat system. A suite of relatively low technology modifications to caves to improve habitability and safety are suggested. This system can integrate a spectrum of missions from both robotic precursors to human expeditionary missions and ultimately colonization.

Key Words: cave, lava tube, lava cave, habitat, geomicrobiology, astrobiology, habitability, planetary protection, radiation protection, pressure shell

1. INTRODUCTION

OVER THE COURSE OF HUMANITY'S HISTORY AS a species, the use of caves, rock shelters, and other natural geological features has played an important role (e.g. Bonsall, 1997; Leroi-Gourhan, 1982; Lommel, 1966; Redondo et al., 2010; Valladas et al., 2001; Whitley, 2009). In fact, much of what we know about early hominids, including *Homo neanderthalensis*, and early *Homo sapiens* comes from remains and artifacts that have been preserved for us in caves around the world (Bahn, 2007; Chazan et al., 2008; Dye, 2008; Knudson et al., 2005). These remains include precious DNA-containing tissues (eg. Lalueza-Fox et al, 2007) and apparent artifacts of extreme antiquity in the 60,000-100,000 year range (e.g. Henshilwood et al., 2009; Texier et al., 2010; Vanhaeren et al, 2006). Giving us a picture of more recent millennia, textiles, mummified humans, and other artifacts have shown us aspects of caves in human life ranging from the domestication of cotton to the trade goods that passed through the Silk Road across Asia (e.g. King, 1974; Whitfield, 1996).

The present Martian surface environment is extremely cold, dry, chemically active, and high in both ultraviolet and ionizing radiation (e.g. McKay and Stoker, 1989; Carr, 1981; Jakosky et al., 1997). Indeed, the consensus has been that even organic materials may not be able to survive on the surface (Klein, 1978) although a recent new look at the subject has suggested that perchlorates found by the Phoenix Mission (Kounaves et al. 2010) may have interfered with the Viking organic results (ten Kate, 2010) and necessitates a new look at the question of Martian habitability (Stoker et al., 2010). Nevertheless, the surface is unarguably a very hostile place for our type of life. With all the truly compelling sites on Mars that have been championed as the "best" sites for future robotic and human Mars missions, why have we chosen to focus on caves? Is it just because we are looking for a new twist on the old theme of planetary exploration, life on other planets, or human colonization of extraterrestrial surfaces? No, we have championed this idea because caves provide both unique scientific targets especially for astrobiology (Boston et al., 1992, 2001; Grin et al., 1998; Le'veille' & Datta, 2010) and critical practical human support functions in an extremely challenging radiation and physical environment (Boston, 2000; Boston et al., 2003, 2004 a & b).

2. WHY CAVES?

Caves in general are poorly understood and underappreciated by the vast majority of the population. Scientists and engineers are no exception to this generalization. Because we are surface-inhabiting creatures, we bring a certain amount of *surface chauvinism* to our perception of caves as well as the oceans and the upper atmosphere. However, many modern and ancient indigenous peoples have been well acquainted with the properties of the caves in their environments. They made extensive use of them for shelter, materials acquisition, water and ice repositories, burial chambers, ritual sites, protection from temperature extremes, and refuge from human enemies (e.g. Adler, 1996; Arnold, 1971; Arnold and Bohor, 1975; Hatt et al., 1953; Sieveking, 1979; Tankersley, 1997; Watson, 1986; Wright, 1971). Interestingly, a tent-shaped hut was constructed inside a French cave (the Grotte du Lazaret near Nice) about half a million years ago by members of some early human groups, perhaps *Homo heidelbergensis,* one of many possible progenitors of both *Homo sapiens* and *H. neanderthalensis* (Jelinek, 1975). Evidently, early hominid inhabitants built a fireplace and other amenities at this site. Obviously, the benefits of construction within a cave were clear to many who have preceded us!

Conditions in cave interiors are typically radically different from and in many ways more benign than the surface environment (Boston et al., 2001). This has enabled microorganisms, larger organisms and even humans to gain protection by using caves as habitat. Many microbial forms are unique to the subsurface and have developed into countless novel strains (Northup and Lavoie, 2001; Boston et al., 2001, Barton & Northup, 2007). Caves also function as sole habitats for a wide array of highly specialized and unique native cave animals (Culver & Pipin, 2009).

The tendency of the uninitiated to imagine all caves as nasty, dank, smelly and rather creepy places akin to the dungeons of fairy tales has made them seem unappealing to some. Indeed, some caves are like that, but many caves, especially in arid environments, are relatively dry, easily accessible, geologically and tectonically stable, and quite "homey". People as diverse as the mushroom growing epicures of the Loire Valley, the Dogon people of Mali, and the miners of Naica, Mexico, respectively, use caves routinely for highly specific economic purposes, to shelter from an otherwise unsupportable surface environment, and sometimes as a source of immense wealth. In recent times, the use of caves on extraterrestrial bodies for human habitation has been suggested by several groups. Lunar lava tube bases received much of the early attention (Horz, 1985; Walden, 1988; Kokh, 1996; Taylor,1998) because lavatube-like structures were clearly visible in early lunar image data (Greeley,1969). Mars lava tubes have also been considered to have great potential as habitat and greenhouse structures (Walden et al., 1988; Frederick, 1999; Boston, 2000; Boston et al., 2003, 2004 a & b).

3. LAVA CAVES AND OTHER CAVE TYPES ON EARTH AND SOLAR SYSTEM BODIES

Contrary to popular belief and the advertisements of tourist caverns, caves on Earth are not rare! They are a globally distributed geological phenomenon that occurs in every major rock type (including salt!) and even in polar and high altitude ices. We have argued that this is not surprising because on any planet with a surface that has an internal or external source of energy (including impact cratering), there will be cracks in that surface. Such cracks form the basis for cave formation by a variety of terrestrial and non-terrestrial mechanisms from simple tectonic caves to highly complex solutional structures (Boston, 2004). Additional cave types are produced by melting within a solid as in the case of ices and lavas. Table 1 shows the plethora of different physical and chemical mechanisms currently known that produce cavities in Earth's crust and may be even more common on other Solar System bodies.

Recent image datasets from MOC (the Mars Onboard Camera aboard the Mars Global Surveyor mission, 1996-2006, MGS) were used to claim the presence of lavatubes on Mars (Boston (2004), and the HiRise camera on the Mars Reconnaissance Orbiter (MRO, inserted into Mars orbit in 2006) now shows clear evidence of large tubes visible in a number of volcanic regions on Mars (Figure 1). More extensive analyses of Martian tubes have been conducted by Cushing and colleagues (2007) and Wynne et al (2008) and a tube system has even been identified by a middle school student group in June 2010 from HiRise image data! In the lunar case, years of speculation about lunar lavatubes has been

TABLE 1

A systematic physics and chemistry based view of cave formation mechanisms derived from Earth examples and extrapolated to Martian conditions. Lavatubes are the only confirmed cave type on Mars to date, however, numerous plausible mechanisms may have produced other varieties of cave that are more difficult to detect remotely. Revised and expanded from Boston, 2004.

Site Type	Possible Parent Materials	Possible Formation Mechanisms	Possible Unique Martian Mechanisms
Solutional Caves	Soluble rock types	Dissolution of rock by fluid	
Acidic solution	Limestone, dolomite, other carbonates	Weak carbonic acid (aqueously dissolved CO_2); Strong sulfuric acid (subsurface H_2S dissolving at the water table or other fluid repository)	Status of carbonates on Mars poorly constrained. Sulfur-rich crustal materials may make sulfuric acid created caves more common than they are on Earth.
Water solution	Quartzite, sandstone, opalinized silicates, salt (NaCl), gypsum (sulfates)	Water dissolution of highly soluble rocks, organic chelation of silicates making them more soluble	Liquid carbon dioxide as a solvent?
Melt Caves	Solid, meltable rock & ice types	Melting to the liquid or plastic state and refreezing	
Lava tubes	Basalt, andesite, etc.	Molten rock with differential cooling	Scale of known tubes much larger than Earth
Glacier or "true" ice caves	Ice masses	Thermal and pressure-induced localized melting in water ice	Melting in carbon dioxide ice and super-cooled water ice
	Subice volcanic caves	Lava/ice or lava/permafrost interactions	Lava interactions with CO2 ice or ice-clathrate interactions, sublimation of ground ices
Sublimation caves	Subsidence caves	None known for Earth	Ground ice sapping or sublimation and subsequent collapse in permafrost
Fracture Caves	Solid rock of many types		
Tectonic or talus caves	Any brittle crustal rock or ice	Faulting, Mass wasting	Cratering causing massive instantaneous fracturing with subsequent release of fluids & dissolution
Elastic decompression	Granite or other rock that forms from intruded batholith masses below planetary crusts	Decompression of large rock masses due to removal by erosion of massive overlying facies	Martian subsurface structure poorly constrained, unknown whether batholiths or other upwelling hot rock melt structures are present
Erosional Caves			
Wind-scoured caves	Rock, especially sedimentary, welded tuff	Wind blasting abrasive particles	Global or regional dust storms?
Canyon rock shelters	Solid rock, sedimentary rock, welded tuff	Spring sapping at an impermeable rock layer weakens at that point	Weeping cliffs with periodic subsurface ice-melting or sublimation events
Coastal caves	Solid rock, welded tuff	Water action (waves, floods)	Massive flood events in Mars' past
Suffosional Caves	Unconsolidated materials, regolith, alteration clays, fault gouge	Suspension of fine particles in fluid and sapping from below	On Earth, such caves are rare and geologically very short-lived. On Mars, such structures may persist much longer in the tectonically quiescent, hydrologically episodic, and reduced weathering environment.

confirmed by the finding of a clear lavatube entrance (Figure 2) by the recent SELENE mission (Haruyama et al. 2009). Interestingly, the dimensions of this tube are enormous (130m diam), consistent with the large sizes of Martian tubes, possibly a result of the lower surface gravity on both bodies with respect to Earth (Boston, 2004). Of course, significant differences in lava chemistry, temperature, and resulting rheology may also play a part in the production of such enormous tube structures. A study in 1995 by Keszthelyi has quantified heat loss during lavatube flow and concluded that basaltic volcanic flows that are tube-fed can produce tubes hundreds of kilometers long with effusion rates of only a few tens of cubic meters per second, thus Martian tubes could be produced by low to moderate effusion rates.

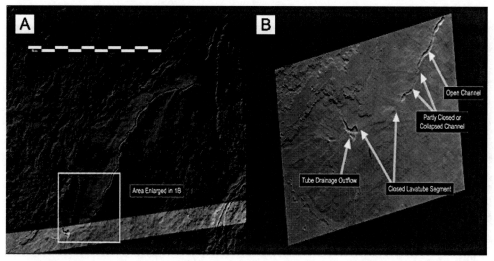

FIGURE 1

Mars lavatube, on the East flank of Olympus Mons. A - Context image showing the entire extent of a long lava channel with segments of closed tube. B. High resolution closeup of open and closed channels and the probable drainage point through which lava exited the tube structure. HiRISE Camera System, NASA/JPL/University of Arizona.

Besides lavatube caves, there exists the potential for other cave types to be present on Mars. Abundant evidence of water-created features exists on Mars (e.g. Malin and Edgett, 2000; Carr and Wanke, 1992; McKay et al., 1992) and recent detection of subsurface hydrogen probably associated with water has been reported (Boynton et al., 2002; Feldman et al., 2002). Ice-created features have also been explored (Squyres et al., 1987 and 1992). Where cracks have been formed in the Martian surface, fluid flow may dissolve at least some of the material depending upon chemistries of the solids and liquids involved potentially even produced by a type of catastrophic speleogenesis (cave formation) from impact cratering (Boston et al., 2006). Mechanisms for cave formation on Mars may differ from Earth providing a variety of new features not found here (Grin et al., 1998, 1999; Boston et al., 2004 a, b, 2006).

Although the presence of significant carbonate deposits has been the subject of speculation and prediction (Nedell et al., 1987; McKay and Nedell, 1988), presence of carbonate in Martian dust (Hamilton et al., 2005; Bandfield et al., 2006), and recent large scale detection in Nili Fossae from orbit (Ehlman et al., 2008), confirmation of carbonate outcrops has been slow in coming. Finally, an apparent positive detection of magnesium iron carbonates has been claimed for Gusev Crater in the Columbia Hills (Morris et al., 2010). However, the precise geological circumstances of these carbonate detections is unclear, thus, no predictions can be made as to whether carbonate dissolutional caves may be likely on Mars.

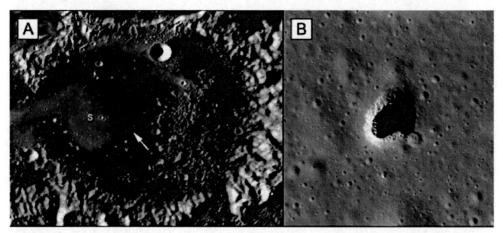

FIGURE 2

Lunar lavatube skylight. A – Context image showing the Mare Ingenii (Sea of Cleverness) region of the Moon where a lavatube skylight has been found. B - This lavatube skylight pit is ~130 meters in diameter. This size dwarfs Earth lavatubes and is similar to tube dimensions identified on Mars. The total image width is ~550 meters. Illumination is from the upper right. LROC Frame: NAC M128202846LE. Credit: NASA, Goddard, Arizona State University.

Recent mineralogical results from both MER Opportunity at Meridiani Planum (Christensen et al., 2005; Squyres et al., 2005; Yen et al., 2005) and from the Mars Express OMEGA instrument (Arvidson et al., 2005; Gendrin et al., 2005; Bibring et al., 2005) indicate that evaporite basin settings rich in various sulfates may be a major environmental type on Mars. Gypsum caves on Earth occur in arid and even some non-arid environments, some of great size and length (eg. Klimchouk, 2009; Klimchouk & Aksem, 2005). Fracturing and subsequent dissolution by surface or upwelling subsurface waters are the potential cave-forming mechanisms involved on Earth that may also be present at least episodically or in the past on Mars.

4. RADIATION ENVIRONMENT ON MARS

We believe that the radiation environment on Mars is the single most challenging aspect of the planet for future human exploration and even for life detection missions. The radiation environment on Mars has long been thought to be very extreme and antithetical to the survival of both indigenous microbial life and to human life, however, no direct measurements have yet been successfully taken on landed missions. The MARIE instrument aboard the Mars Odyssey Mission (launched 2001) was intended to characterize the radiation environment during transit from Earth to Mars and in Mars orbit. However, the instrument malfunctioned during the cruise phase and was not able to take measurements in Mars orbit. The lack of direct measurements has prompted investigators to try to deal with radiation issues using a combination of calculations and modeling efforts. However, the radiation situation is highly complex with a plethora of relevant particles including gamma radiation, heavy iron nuclei, protons and electrons in the solar wind, neutrons and more! The MARIE instrument did take some data and this was used to create a cosmic ray environment map of Mars (Figure 3).

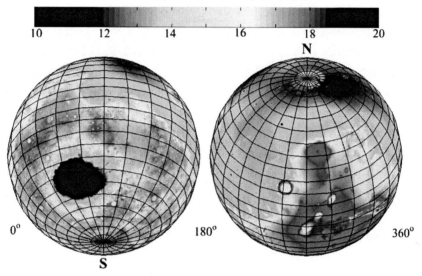

FIGURE 3

Cosmic Ray Environment map – This map was produced based on cosmic radiation data by the Mars radiation environment experiment (MARIE), an instrument on NASA's Mars 2001 Odyssey spacecraft. Those data were combined with Mars altimetry from the MOLA instrument aboard MGS (Mars Global Surveyor. Lowest elevations are anticipated to have lowest radiation levels because more atmosphere exists above them to prevent penetration to the surface. The color scale is 10 rems (dark blue) to 20 rems (dark red). Astronauts on the International Space Station in Earth orbit encounter radiation equivalent to an annual rate of 20-40 rems. JPL Press Release, March 1, 2002. Credit: NASA/JPL/JSC

Local environmental radiation doses in the Martian crustal material are anticipated to be smaller than on Earth due to trace quantities of U, Th and K. Data from Martian meteorites (Shubert et al., 1992) is used to bolster this argument. Thus the extrinsic sources of GCR and solar particle (or proton) events (SPE) are the focus of attention and are both anticipated to be vastly more intense than on Earth.

The sun ejects primarily protons and electrons which are major constituents of the ordinary solar wind. These particles typically have velocities of around 400 km s-1 by the time they get to the Earth (Barnes, 1992). The energy range for these particles ranges from eV to keV (Hundhausen, 1995). According to Skidmore and Ambrosi (2009), the thin Martian atmosphere is capable of attenuating the proton flux below energies of 100 MeV, thus only a few solar wind protons ever get to the Martian surface. However, as a result of these high energy protons, a cascade of neutrons and other high energy particles are created at high altitudes in the Martian atmosphere and unfortunately these secondary products DO reach the surface. As these authors point out, the GCR protons ARE able to directly reach the Martian surface because their energies range above 100 MeV. Furthermore, the protons accelerated by SPE are of energies above this 100MeV threshold and thus, also are expected to directly reach the Martian surface. However, direct information about the interaction of gamma radiation with the Martian surface during SPE is not currently available.

A combination of calculations based on assumptions and modeling has been used to try to approximate both GCR and solar intensities likely from SPE, i.e. solar particle events (Dartnell et al., 2007 a & b; Morthekai et al., 2007; Banerjee & Dewangan, 2008; McKeever et al., 2003). Table 2 compares the results of several of these studies and this has provided at least some constraint for our thinking about what the magnitude of the radiation challenge might be and how deeply it may penetrate the Martian surface. From these authors, we see that attenuation to very low levels has occurred by the 3 to 4 meter depth. These models are mostly using "soil", i.e. unconsolidated regolith, as the crustal material. Rock materials may have greater ability than regolith to attenuate radiation in some cases. Basalt and andesite are likely surface lithologies to be found on Mars and their presence has been confirmed (Mustard et al., 1997; Salvatore et al., 2010). The primary relevant constituent in both rock types appears to be silicon content (Goldstein & Wilkins, 1954).

In a study of the radiation attenuation properties and scattering coefficients of andesite (a volcanic rock) and diorite (a comparable igneous intrusive rock) with respect to gamma radiation (El Taher et al. 2007), these authors show that small differences in precise chemical compositions of the rock types (especially of heavier minerals, e.g. iron and manganese oxides) have consequences for their attenuation behaviors. Penetration and diffusion of gamma radiation in material is expressed by the attenuation coefficient μ. In turn, μ is a function of photon energy (E) and atomic number of the impacted material (z). The density (rho) of target materials is critical. The sidescattering coefficient (phi) which captures the magnitude of the secondary daughter product production, is of particular interest

TABLE 2

Comparison of radiation values from several prior modeling studies. Environmental Radiation pertains to the radiation emitted from radioactive elements in the Martian crustal material.

Source	Martian Surface GCR & SPE* (unless otherwise specified)	Depth	Material	Method
Bannerjee & Dewangan, 2008	~ 65mGya−1 (minimum solar activity M=300 MeV)	Surface	Soil	Monte Carlo code GEANT4
	~ 43mGya−1 (average solar activity M=600 MeV)	Surface	Soil	"
	~ 0.5mGya-1 (average solar activity M=600 MeV)	3 meters depth	Soil	"
	~ 30mGya−1 (peak solar activity M=900 MeV)	Surface	Soil	"
	negligible	4 meters depth	Soil	"
Environmental	~ 0.7mGya−1 (environmental dose rate, 5% water)	Surface	(From U, Th, K radioactive sources)	Lodders & Fegley, 1998
McKeever et al, 2003	~ 78mGya−1 (minimum solar activity)	Surface less than 2 m	Regolith	HZETRN code
	~ 54mGya−1 (average solar activity)	Surface, less than 2 m	Regolith	
SPE**	~2.7 mGya−1 (averaged over 11 yr solar cycle)	Surface	Regolith	
	~ 25mGya−1 (peak solar activity)	Surface less than 2 m	Regolith	
	~ 12mGya−1 (peak solar activity)	>2 meters depth	Regolith	
Environmental	0.4 mGya-1 (from Martian meteorites)		(From U, Th, K radioactive sources)	Shubert et al., 1992
Morthekai et al, 2007	(minimum solar activity M=300 MeV)	Surface	Soil	Monte Carlo code GEANT4
	(peak solar activity M=900 MeV)	Surface	Soil	
		40 cm depth	Soil	
Dartnell et al, 2007a,b	6.2 cGy/year, building to a peak of 6.6 cGy/year at 40 g/cm2 shielding	Surface to 4.5 m in 5 cm thick layers	Regolith, water ice, and rock versions	Monte Carlo code GEANT4

* GCR = Galactic Cosmic Radiation; SPE = Solar Particle Event
Most of the dose is deposited below the surface as protons (Banerjee & Dewangan, 2008). Dartnell and colleagues (2007 a, or b) however, show that for protons with energies less than 1 GeV, no dose peak occurs and they decay as a simple exponential curve. No secondaries are generated, thus shielding material attenuates them.

for our purposes. These authors show that the higher density materials contain a greater number of scattering centers which causes more elimination of photons, thus attenuate better in direct proportion to density. They tested rock materials with densities ranging from ~ 2.4 to 2.7 gm cm-³. Their explanation for this effect relies on the decrease of intramolecular voids and hence the potential for increased multiple scattering events. The side scattering photons, phi, also attenuate in direct proportion to increasing density as well. So the message for our purposes is that these rock types can significantly attenuate gamma radiation especially when endowed with enough heavy materials, including silicon and iron. Note that basalt has an even higher density (3 gm cm-³) than andesite and diorite, but its silica content is somewhat lower by percentage (45-52 wt%)

compared to 57-63 wt% (andesite) and 55- 65 wt% (diorite) (Le Maitre et al., 2002).

The results of models like those presented in Table 2 are assumption dependent and the Martian radiation problem is still not well constrained. Because of this situation, a major study of astronaut safety issues vis a vis Mars has strongly advocated a radiation measurement precursor mission to definitively measure radiation at the Martian surface, characterize some relevant aspects of that surface, and then calculate the human dose that would be experienced at the measured sites (NRC, 2002). Further, they recommend a mission that will *"measure the absorbed dose in a tissue-equivalent material on Mars at a location representative of the expected landing site...."* Such information will be critical in formulating an optimal strategy for future human missions because there are significant trade-offs that have to be made. For example, Cuccinotta and colleagues (2001) have pointed out that a long-duration stay on the surface that is coupled with short transit time launch opportunities may actually be better from the radiation exposure standpoint than a short-duration surface stay coupled with longer interplanetary transit times because the planet itself and even its tenuous atmosphere provide better radiation shielding than the interplanetary exposure.

The effective radiation dose is particularly sensitive to the quantity of secondary neutrons given off when the primary radiation influx interacts with matter (NRC 2002), in our case, a particular lithology like basalt or carbonates or evaporites. Neutrons are absorbed better when there is hydrogen present in the material. When a material contains heavy nuclei, this results in secondary neutron generation (daughter products) and this means that attenuation of both primary and secondary radiation must be accounted for in any assessment of radiation attenuation by rock material. Depending on the precise composition of the rock, e.g. basalt, there may or may not be hydrogen containing minerals in the mix. For example, on Earth basalts may contain mica or amphibole and both of those are hydrated minerals. We know from the complex sulfate composition of Mars (e.g. Chipera & Vaniman, 2007; Bonello et al., 2005; Clark et al., 2005; Squyres et al., 2004; Larsen et al., 2000) that at least those mineral suites are capable of being highly hydrated on the Martian surface and we speculate that the basalts may also have avoided the dewatering that may or may not have been the fate of lunar basalts (see recent debate on the hydration state of the moon in McCubbin et al., 2010 and Sharp et al., 2010).

Amidst all this complexity and uncertainty what can be said at present about the radiation environment on Mars? In a study modeling the effects of hydrogen and iron concentrations on absorbed dose, the investigators varied amounts of hydrogen and iron (Clowdsley et al. 2001, cited in NRC 2002). This preliminary study concluded that the current understanding of the elemental composition of Martian soil is adequate for radiation transport calculations through bulk Martian regolith. The authors even suggest that significant variations in the concentration of hydrogen and iron have little effect on the absorbed dose. Thus, even our relatively crude approximation of radiation conditions on Mars is adequate to state that radiation will be one of the primary environmental conditions of concern for humans and other organisms on the Martian surface. Further, the attenuation properties of rock materials are adequate to provide significant shielding with a few meters of overlying rock, precisely the condition found in most caves at least here on Earth.

5. CAVES AS SCIENCE SITES

Although the search for life or life's past traces on Mars is a high priority within the astrobiology community, we have argued that any present Martian life is more likely to be analogous to Earth's subsurface biosphere than anything currently found on Earth's surface (Boston et al., 1992, 2001; Boston, 1999, 2000b). Even traces of ancient life will have been preserved more successfully in the subsurface than the surface. In this capacity, we have been strong advocates of science missions to detect and then investigate extraterrestrial caves, especially those on Mars, and with particular emphasis on life detection missions (see Figure 4 for examples of potential Martian subsurface habitats). However, the high premium placed on Mars life detection missions and the extreme delicacy of the balance between investigation and potential damage or contamination of such life requires extraordinary protocols analogous to those of biohazard containment of the most virulent Earth viruses (Rummel, 2001). This requirement for complete containment must be coupled with the demands of field science, active exploration, robotic functionality, and ultimately astronaut survival. Nothing of this magnitude has ever been attempted in human history. Pristine Earth caves that contain numerous novel species of microorganisms provide genuine biologically sensitive sites for developing and practicing the operations required for application to Mars life detection sites and ultimately extraterrestrial human habitat. In this capacity, the work that our team and other investigators do in the subsurface is of great importance in informing us about the potential for such future missions.

Caves of scientific interest have a high probability of interesting geological and biological features, e.g. extreme age, isolation from the surface, evidence of gases or water. They must be big enough for instrumentation, microrobotic access, or drilling into but not necessarily for humans. Caves of all depths may be scientifically interesting, but very deep caves may be the most interesting for stratigraphy, mineralogy, geomorphology, and life detection. For scientific purposes caves without natural openings or with very limited openings are highly desirable because of superior preservation of the contents. Certainly caves in or

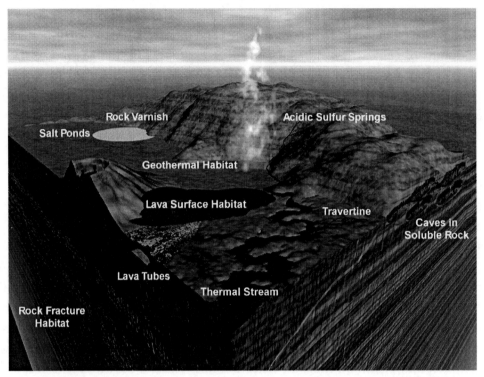

FIGURE 4

Artist's conception of an astrobiologically hectic corner of Mars. This illustration demonstrates a plethora of possible extremophile habitats on Mars for which there are analogs on Earth that have been studied by the astrobiological community. Included are three subsurface examples: lavatube caves, caves in soluble rock, and the rock fracture habitat itself. Lavatube caves, visible on the surface, have been confirmed for Mars (see text). Dissolutional caves and a potential rock fracture environment have yet to be confirmed and are more challenging to discover using orbital data. Art courtesy of R.D. Frederick.

near geologically or hydrothermally active areas would be of great interest to many scientific disciplines.

6. EARTH CAVE ANALOG VALUE

Caves on Earth can provide significant value as analogs of some aspects of the Martian environment. The unusual gases (H_2S, CO_2, CO, NH_3, and others) contained in the air of some Earth caves present the opportunity to practice protection from and management of poisonous or deleterious atmospheres. The condition of no or little atmosphere is not easily simulated in Earth caves, of course. Conceivably, any caves existing at extremely high altitudes in the Andes or the Himalayas could provide a partial simulation but other logistics probably make it too difficult to be worth the effort involved. However, some caves are depleted in oxygen or heavily laden with toxic gases, thus requiring full breathing gear for investigators (e.g. Hose et al., 2000).

The greatest similarity between terrestrial caves and those of other planets exists in the realm of operational considerations within the confines and topography of caves. That is, the very experience of living, working, doing science, and extracting resources in the lightless and potentially hazardous cave environment is the primary value of the Earth cave analog for human exploration. This particularly extends to matters of planetary protection. The minute a human investigator enters a cave on Earth, the potential exists for deleterious impact on the indigenous biota (Moser et al., 2001; Boston et al., 2006b). But unlike the way that we deal with extreme biocontainment in the laboratory, in the cave environment investigators must deal with a difficult and dangerous environment and still accomplish their goals while simultaneously protecting the biota.

7. PLANETARY PROTECTION

Planetary protection considerations emerging from recent studies (e.g. MEPAG, 2006; Criswell et al., 2005; Hogan et al., 2006) advocate a localization and zoning of degrees of human impact, much like that being implemented in the Antarctic as *Special Regions.*

The July 2008 COSPAR document (COSPAR Panel on Planetary Protection, 2008) attempted to define "Special Regions" by way of measurable parameters so that such metrics can be applied to specific areas on Mars that may need particular protection. Such an idea has already been put forward by MEPAG (2006), especially for temperature and water activity (biological availability of water to enter into biochemical reactions). Although Conley and Rummel (2010) point out that there is a consensus that future human exploration of Mars will inevitably result in some forward contamination, they further advocate monitoring and minimization as essential steps in controlling such contamination. These authors specifically advocate *"selecting landing sites so that release of contamination will remain local".* We suggest that containment of the primary human habitation and work activities within the confines of a subsurface habitat are highly consistent with this specific recommendation and with the new approaches to Planetary Protection forward contamination discussed here.

Further, we have suggested that Earth caves with sensitive biological (especially microbiological) contents can serve as a particularly cogent testbed for the development of minimal impact or non-impact methodologies and protocols for extraterrestrial use because the same operating principles are also essential to our work with indigenous cave and subsurface microorganisms (Boston et al., 2006; Moser et al. 2001).

8. CAVES AS POTENTIAL MARTIAN HABITAT

We have argued that the intense radiation environment on the surface of Mars makes the subsurface the best place to look for any extant Martian life (Boston et al., 1992). In our view, it is also clearly the only currently practicable human habitat choice. Whether natural caves, artificial tunnels, or bermed structures are used, the problems of living in a subsurface environment may be more easily overcome than developing methods to ameliorate the effects of radiation as experienced on the surface. The problem of protecting humans on the surface in suits and transportation devices even for limited duration forays is intractable enough (e.g. Carr & Newman, 2008; Zeitlin et al., 2006; Cowing, 2004; Hodgson, 2001).

To explore the idea of the Martian subsurface as habitat, we have studied a number of aspects of the problem (Boston et al., 2004 a & b). This section provides a brief synopsis of the main points of those studies. The primary objective of these studies was to show that relatively simple, easily deployable subsurface habitats are constructible in caves, lavatubes, and other subsurface voids. Further, we wished to demonstrate that they are suitable to sustain small animals, plants, and ultimately humans in an otherwise hostile environment. The secondary objective was to show that humans can do useful work and scientific exploration in a subsurface environment facing some of the constraints that they will meet in the Martian environment including potential biologically sensitive sites. The third objective was to separate those features of a Mars or lunar subsurface mission that can be simulated in an Earth cave (e.g. Mars-derived breathing mixtures) from those that cannot (e.g. lunar or Martian gravity) to provide the basis for future work by other investigators.

We focused particularly on two aspects of the habitat issue: 1) the radiation protection provided by the rock overburden of caves, and 2) the ability of caves to provide a natural "pressure vessel" for the construction of subsurface habitats and their ability to hold breathable air. Secondary considerations included insulation from thermal oscillations, protection from impacting objects, sealability to contain a higher than ambient atmospheric pressure, and possibly access to potentially important subsurface resources, e.g. geothermal energy sources, water, reduced gases, and minerals.

The caves most suitable for human habitation purposes will be shallow, easily accessible, and relatively horizontal. Obviously, large spacious rooms and passages with smooth walls and floors are highly desirable. There are countless such examples on Earth in both lavatube and solutional caves (e.g. see Figure 5 for one that we have used in robotic access simulations, Dubowsky et al., 2004).

Figure 5
Hibashi Cave, Saudi Arabia. This lavatube cave has a smooth floor courtesy of blown in sand. We believe that the frequent dust storms on Mars probably have deposited a similar blanket of material on the floors of open lavatubes that may be there. Such a flat floor would be a valuable asset for a potential Martian habitat cave. Image courtesy of J. Pint.

Geological stability (very likely in the long history of relatively tectonically quiet Mars), and formation in relatively impermeable materials will provide safety and a degree of sealing against gas losses from habitat leaks. However, all rock (including human concrete structures!) fractures over time and we anticipate that such fractures must be sealed. We developed the idea of an inflatable liner that can conform to the natural shape of the cave. After inflation, a secondary step to rigidify the structure with foamed in place insulation, much as is done in conventional construction, is then conducted. An interior finish can be applied over these structural elements to provide a more human acceptable living space.

Compatible with this construction methodology is the idea of foamed in place airlocks (Figure 6). We envision standardized rigid airlock assemblies that can be custom tailored to the precise topology of a cave cross section by a combination of rough cutting a skirting area surrounding the airlock, combined with another use for foamed in place technology to support and seal the airlock assembly.

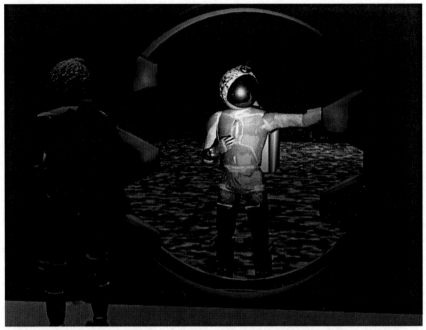

FIGURE 6
Artist's conception of a foamed in place airlock assembly. A rigid airlock mechanism can be mounted in a cuttable frame. This can be trimmed to the approximate shape of a cave cross section, then filled in with rigidifying foam to seal the unit. Art courtesy of R.D. Frederick.

Natural openings are convenient but not essential as drilling can provide access to cavities through a few meters of rock. Location within a lower area like a canyon or a crater could provide additional protection from some surface conditions like large dust storms and a slightly higher ambient atmospheric pressure. Natural gas blowout features of lavatubes, known as *hornitos*, occur in Earth caves and may be present in Martian caves. Such natural entry holes are convenient and relatively easily modifiable. Resource-providing caves obviously must contain minerals, volatiles, or other assets. To be useful, they must provide nearer proximity to the resources than surface drilling and mining would afford. Additionally, they could be used as storage space for geothermal fluids, volatiles, and raw materials. Caves that contain volatiles will likely be naturally sealed from the outside low Martian air pressure.

Any residual geothermal activity within caves could be a potential source of power as can solar arrays deployed on the surface. Geothermal activity occurs in some caves on Earth, but on Mars we have little grasp of the planet's geothermal gradient, nor any current knowledge about active geothermal hotspots. For lighting, a combination of light piping of natural surface light with optical fibers from the exterior, and interior artificially powered lighting and temperature control will probably be necessary due to atmospheric opacity during dust storm events. Optical fiber or mirror light piping is a very convenient methodology pioneered in the mid 1990's (e.g. Swift & Smith, 1995). When fibers are used, they are now available for a multitude of applications and come in flexible or rigid form. Many of them can select specific wavelengths, and be designed to deliver all of their light to the end point or to attenuate the light along a length.

Lastly, not all of the cave interior must be given over to containment of breathable air and interior habitat. We also envision that low permeability sealed parent material on a cave interior would allow for filling parts of a cave with inert gases compressed (but not separated out) from the Mars atmosphere, thus allowing a "shirt-sleeve" environment in terms of pressure, but requiring thermal protection and oxygen breathing gear much as we currently use to work in Earth caves that contain poisonous gases like CO_2, H_2S, and others.

9. CONCLUSION

Mission planners cannot take particular options seriously unless the necessary background work is in place. While we have simply scratched the surface on many of the issues that we have consider in our studies (Boston et al., 2003, 2004a,b) , we hope that we have helped to lay a solid foundation that can be built upon further by ourselves and other interested colleagues. The Solar System is an amazing place filled with many spectacular natural features. We believe that it is time that caves took their place alongside the other natural phenomena worthy of exploration, study, and human utilization.

Acknowledgements: Thanks go to Gus (R.D.) Frederick for Figures 4 and 6. Appreciation to long time collaborators and cave partners Mike Spilde, Diana Northup, Kenneth Ingham, Jim Werker, and Val Hildreth-Werker. The author thanks NCKRI (the National Cave and Karst Research Institute) for support of various relevant field activities. Part of the work reported here was supported by the NASA Institute for Advanced Concepts (NIAC) grants #CP-01-01, and #CP-99-01 to the author, #CP 02- 02 to S. Dubowsky, and Subcontract 07605-003-020 to P. Todd. Fundamental geomicrobial studies of cave and rock varnish organisms and how we deal with them with minimally contaminating methodologies have been funded by the National Science Foundation EAR 0719669 and EAR 0311990 (P. Boston, PI), and LExEn DEB 9809096 (C. Dahm, PI).

REFERENCES

Adler, M.A., ed. (1996). The Prehistoric Pueblo World, A.D. 1150-1350 University of Arizona Press, Tucson, AZ.

Arnold, D.E. (1971). Ethnomineralogy of the Ticul, Yucatan potters: Etics and emics. Amer. Antiquities 36:20-40.

Arnold, D.E. and Bohor, B.F. (1975). Attapulgite and maya blue: An ancient mine comes to light. Archaeology 28:23-29.

Arvidson, R.E., Poulet, F., Bibring, J.P., Wolff, M., Gendrin, A., Morris, R.V., Freeman, J.J., Langevin, Y., Mangold, N., and Bellucci, G. (2005). Spectral reflectance and morphologic correlations in eastern Terra Meridiani, Mars. Science 307(5715):1591- 4.

Bahn, P.G. (2007). Cave Art: A Guide to the Decorated Ice Age Caves of Europe. 3rd Ed. Frances Lincoln Pub. 224 pp.

Bandfield, J.L., Glotch, T.D., and Christensen, P.R. (2003). Spectroscopic identification of carbonate minerals in the Martian dust. Science 301(5636):1084-1087.

Banerjee, D., and Dewangan, A. (2008). Simulation of the cosmic-ray induced dose-rate within a Martian soil profile. Radiat. Msmts 43(2-6):797-801.

Barnes, A. (1992). Acceleration of the solar wind. Rev. Geophys. 30, 43–55.

Barton, H.A., and Northup, D.E. (2007). Geomicrobiology in cave environments: Past, current and future perspectives. J Cave Karst Stud 69:163-178.

Bibring, J.P., Langevin, Y.,Gendrin, A., Gondet, B., Poulet, F., Berthe, M., Soufflot, A., Arvidson, R., Mangold, N., Mustard, J., and Drossart, P. (2005). Mars surface diversity as revealed by the OMEGA/Mars Express observations. Science 307, no.5715, p.1576-81.

Bonello, G., Berthet, P., d'Hendecourt, L., (2005). Identification of magnesium sulfate hydration state derived from NIR reflectance spectroscopy. In: Lunar and Planetary Science, vol. XXXVI, Abstract #1996, Lunar and Planetary Institute, Houston (CD-ROM).

Bonsall, C. (1997. The Human Use of Caves. British Archaeological Reports, Intl. London, UK. 218 pp.

Boston, P.J. (1999). The Search for Extremophiles on Earth and Beyond: What is extreme here may be just business-as-usual elsewhere. Ad Astra Magazine Vol. 11, No.1, January/February (1999).

Boston, P.J. (2000a). Bubbles in the rocks: Natural and artificial caves and cavities as life support structures. In, R.M. Wheeler and C. Martin- Brennan, eds. Mars Greenhouses: Concepts and Challenges: Proceedings from a (1999) Workshop. NASA Tech. Mem (2000-208577.

Boston, P.J. (2000b). Life Below and Life 'Out There'. Geotimes 45(8):14-17.

Boston, P.J. (2004). Extraterrestrial Caves. Encyclopedia of Cave and Karst Science. Fitzroy-Dearborn Publishers, Ltd., London, UK. Pp. 355-358.

Boston, P.J., Frederick, R.D., Welch, S., Werker, J., Meyer, T.R., Sprungman, B., Hildreth-Werker, V., Murphy, D., and Thompson, S.L., (2004)a. Human Utilization of Subsurface Extraterrestrial Environments. Final report for NIAC CP 01-01, Phase I I. http://www.lib.usf.edu/karst-test/docs/NIAC_Cave_II.pdf.

Boston, P.J., Frederick, R.D., Welch, S.M., Werker, J., Meyer, T.R., Sprungman,B., Hildreth-Werker, V., Thompson, S.L. (2004b). Extraterrestrial SubsurfaceTechnology Test Bed: Human Use and Scientific Value of Martian Caves NIAC – Caves of Mars page 64 Space Tech. & Applic. Forum (2003) Proc.. AIP #654. Amer. Inst. of Physics, College Park, MD.

Boston, P.J., Frederick, R.D., Welch, S.M., Werker, J., Meyer, T.R., Sprungman,B., Hildreth-Werker, V., Thompson, S.L., and Murphy, D.L. (2003). "Human utilization of subsurface extraterrestrial environments," Grav. & Space Biol. Bull. 16(2):121- 131.

Boston, P.J., Hose, L.D., Northup, D.E., and Spilde, M.N. (2006a). The microbial communities of sulfur caves: A newly appreciated geologically driven system on Earth and potential model for Mars. In, R. Harmon, ed. Karst Geomorphology, Hydrology, & Geochemistry Geological Soc. Amer. Special Paper 404. Pp. 331- 344.

Boston, P.J., Ivanov, M.V., and McKay, C.P. (1992). On the possibility of chemosynthetic ecosystems in subsurface habitats on Mars. Icarus 95:300-308.

Boston, P.J., Northup, D.E., and Lavoie, K.H. (2006b). Preserving the unseen. Chapter in Cave Conservation Techniques, edited by V.Hildreth-Werker and J. Werker, National Speleological Soc. Pp. 288-300.

Boston, P.J., Spilde, M.N., Northup, D.E., Melim, L.A., Soroka, D.S., Kleina, L.G., Lavoie, K.H., Hose, L.D., Mallory, L.M., Dahm, C.N., Crossey, L.J., and Schelble, R.T. (2001). Cave biosignature suites: Microbes, minerals and Mars. Astrobiology Journal 1(1):25-55.

Boynton, W. V. and others. (2002). Distribution of hydrogen in the near-surface of Mars: Evidence for subsurface ice deposits. Published online May 30 (2002); 10.1126/science.1073722 (Science Express Reports.)

Carr, C.E., and Newman, D.J. (2008). Characterization of a lower-body exoskeleton for simulation of space-suited locomotion. Acta Astronautica 62:308-323. (doi:10.1016/j.actaastro.(2007).11.007.

Carr, M.H. (1981. The Surface of Mars, Yale University Press, New Haven, Conn.

Carr, M.H. and Wanke, H. (1992). Earth and Mars: Water inventories as clues to accretional histories. Icarus 98(1):61-71.

Chazan, M., Ron, H., Matmon, A., Porat, N., Goldberg, P., Yates, R., Avery, M., Sumner, A., and Horwitz. L.K. (2008). Radiometric dating of the Earlier Stone Age sequence in Excavation I at Wonderwerk Cave, South Africa: preliminary results. J. Human Evol. 55(1):1-11.

Chipera, S.J., and Vaniman, D.T. (2007). Experimental stability of magnesium sulfate hydrates that may be present on Mars. Geochim. Cosmochim. ACTA 71(1): 241- 250.

Christensen PR, Wyatt MB, Glotch TD, Rogers AD, Anwar S, Arvidson RE, Bandfield JL, Blaney DL, Budney C, Calvin WM, Fallacaro A, Fergason RL, Gorelick N, Graff TG, Hamilton VE, Hayes AG, Johnson JR, Knudson AT, McSween HY Jr, Mehall GL, Mehall LK, Moersch JE, Morris RV, Smith MD, Squyres SW, Ruff SW, Wolff MJ. (2004). Mineralogy at Meridiani Planum from the Mini-TES Experiment on the Opportunity Rover. Science 306(5702):1733-9.

Clark, B.C., et al., (2005). Results and implications of mineralogical models for chemical sediments at Merididani Planum. In: Lunar and Planetary Science, vol. XXXVI, Abstract #1446, Lunar and Planetary Institute, Houston (CD-ROM).

Clowdsley, M. et al. (2001). Examination of the Sensitivity of Mars Surface Radiation Exposures to Variations in Regolith Composition of Iron and Hydrogen, cited in NRC, (2002).

Conley, C.A. and Rummel, J.D. (2010). Planetary protection for human exploration of Mars. Acta Astronaut. 66:792-797.

COSPAR Panel on Planetary Protection. (2008). Policy on Planetary Protection, (2008). Online at /http://cosparhq.cnes.fr/Scistr/Scistr/ PPPolicy(20-July-08).pdf

Cowing, K.L. (2004). Mountaineering and climbing on Mars. J. Brit. Interplanet. Soc. 57(3-4):113-125.

Criswell, M.E., Race, M.S., Rummel, J.D., and Baker, A. (Eds.) (2005). Planetary protection issues in the human exploration of Mars, Pingree Park Final Workshop Report, NASA/CP-(2005)-213461.

Cucinotta, F., W. Schimmerling, J. Wilson, L. Peterson, G. Badhwar, P. Saganti, and J. Dicello. (2001). Space Radiation Cancer Risk Projections for Exploration Missions: Uncertainty Reduction and Mitigation. JSC 29295. Johnson Space Cent., Houston, TX.

Culver, D.C., and Pipan, T. (2009). The Biology of Caves and Other Subterranean Habitats. Oxford University Press, Oxford, UK. 214 pp.

Cushing, G.E., Titus, T.N., Wynne, J.J., and Christensen, P.R. (2007).THEMIS observes possible cave skylights on Mars. Geophys Res Lett 34(17):L17201.

Dartnell, L.R., Desorgher, L., Ward, J.M., and Coates, A.J. (2007b). Modelling the surface and subsurface Martian radiation environment: Implications for astrobiology. Geophys. Res. Lett. 34(2):L02207.

Dartnell, L.R., Desorgher, L., Ward, J.M., and Coates, A.J. (2007a). Martian sub-surface ionising radiation: biosignatures and geology. Biogeosci. 4(4):545-558.

Dubowsky, S., Iagnemma, K., and Boston, P.J. (2004). Microbots for Large-scale Planetary Surface and Subsurface Exploration. Phase I Final report for NIAC CP. 02-02. http://www.lib.usf.edu/karst-test/docs/Microbot_NIAC_II.pdf

Dye, H.D. (Ed.). (2008). Cave archaeology of the Eastern Woodlands: Papers in honor of Patty Jo Watson, Univ. Tennessee Press. 278 pp.

Ehlmann, B.L., Mustard, J.F., Murchie, S.L., Poulet, F., Bishop, J.L., Brown, A.J., Calvin, W.M., Clark, R.N., Des Marais, D.J., Milliken, R.E., Roach, L.H., Roush, T.L., Swayze, G.A., and Wray, J.J. (2008). Orbital identification of carbonate bearing rocks on Mars. Science 322(5909):1828-1832.

El Taher, A., Mahmoud, H.M., and Abbady, A.G.E. (2007). Comparative study of attenuation and scattering of gamma-rays through two intermediate rocks. J. Pure Appl. Phys. 45:(198-203.

Feldman, W. C. and others (2002). Global distribution of neutrons from Mars: Results from Mars Odyssey. Published online May 30 (2002); 10.1126/science.1073541 (Science Express Reports.)

Frederick, R.D. (1999). Martian lava tube caves as habitats. Second Annual Mars Society Convention, Boulder, CO, Aug. 12-15, (1999.

Gendrin, A., Mangold, N., Bibring, J.P., Langevin, Y., Gondet, B., Poulet, F., Bonello, G., Quantin, C., Mustard, J., Arvidson, R., and LeMouelic, S. (2005). Sulfates in Martian layered terrains: the OMEGA/Mars Express view. Science 307(5715):1587-91.

Goldstein, H., and Wilkins.(1954). USAEC Report NYO, 3075 (NDA-15C-41Nuclear Development Associates. As cited in, El Taher et al. (2007).

Greeley, R. (1969). On the origin of lunar sinuous rilles. Modern Geology 1:75-80.

Grin, E.A., Cabrol, N.A., McKay, C.P., (1998). Caves in the Martian regolith and their significance for exobiology exploration. 29th Lunar and Planetary Science Conference. Abstract # 1012.

Grin, E.A., Cabrol, N.A., and McKay, C.P. (1999). The hypothesis of caves on Mars revisited through MGS data; Their potential as targets for the surveyor program. Workshop on Mars (2001): Integrated Science in Preparation for Sample Return and Human Exploration. 2-4 Oct. (1999(. Lunar and Planetary Inst., Houston, TX.

Hamilton, V.E., McSween Jr., H.Y., and Hapke, B. (2005). Mineralogy of Martian atmospheric dust inferred from thermal infrared spectra of Martian aerosols. J. Geophys. Res. 110, E1(2006).

Haruyama, J., Hioki, K., Shirao, M.,Morota, T., Hiesinger, H., van der Bogert, C.H., Miyamoto, H.,Iwasaki, A.,Yokota, Y., Ohtake, M., Matsunaga, T.,

Hara, S., Nakanotani, S., Pieters, C.M. (2009). Possible lunar lava tube skylight observed by SELENE cameras. Geophys Res Lett 36:L21206.

Hatt, R.T., Fisher, H.I., Langebartel, D.A. and Brainerd, G.W. (1953). Faunal and archaeological researches in Yucatan caves. Cranbrook Inst. Sci. Bull. 33:1-1(19.

Henshilwood, C.S., d'Errico, F., and Watts, I. (2009). Engraved ochres from the Middle Stone Age levels at Blombos Cave, South Africa. J Human Evol 57(1):27-47.

Hodgson, E. (2001). CP 00-02 Phase I – A Chameleon Suit to Liberate Human Exploration of Space Environments. NASA Institute for Advanced Concepts.

Hogan, J.A., Fisher, J.W., Race, M.S., Joshi, J., and Rummel, J.D. (Eds.) (2006). Life Support & Habitation and Planetary Protection Workshop Final Report, NASA/TM-(2006)-213485.

Horz, F. (1985. Lava tubes: Potential shelters for habitats. In, W. Mendell, ed., Lunar Bases and Space Activities of the 21st Century. pp. 405-412. Lunar and Planet. Inst., Houston, TX.

Hose, L.D., Palmer, A.N., Palmer, M.V., Northup, D.E., Boston, P.J., and Duchene, H.R. (2000). Microbiology and geochemistry in a hydrogen sulphide-rich karst environment. Chemical Geology 169, 399-423.

Hundhausen, A.J. The solar wind, in: Kivelson, M.G.,and Russell, C.T., (Eds.) (1995. Introduction to Space Physics. Cambridge University Press, pp.91–128.

Jakosky, B.M., Zent, A.P., and Zurek, R.W. (1997). The Mars water cycle: Determining the role of exchange with the regolith. Icarus 130(1):87-95 .

Jelinek, J. (1975. Pictorial Encyclopedia of the Evolution of Man. Prague, Hamlyn. Keszthelyi, L. (1995). A preliminary thermal budget for lava tubes on the earth and planets. JGR-Solid Earth 100:20411-20420.

King, M.E. (1974). The Salts Cave textiles: A preliminary account. Academic Press, NYNY. 40 pp.

Klein, H.P. (1978). The Viking biological experiments on Mars. Icarus 34:666-674. Klimchouk, A. (2009). Morphogenesis of hypogenic caves. Geomorph.106(1-2 Sp. Iss.):100-117.

Klimchouk, A.B., and Aksem, S.D. (2005). Hydrochemistry and solution rates in gypsum karst: case study from the Western Ukraine. Environmen. Geol. 48(3):307-3(19.

Knudson, K.J., Tung, T.A., Nystrom, K.C., Price, T.D., and Fullagar, P.D. (2005). The origin of the Juch'uypampa Cave mummies: strontium isotope analysis of archaeological human remains from Bolivia. J. Archaeol. Sci. 32(6):903-913.

Kokh, P. (1996. Searching Mars for lavatubes. Moon Miners' Manifesto, #93. Section 6.9.3.2.093, Artemis Data Book.

Kounaves, S.P., Hecht, M.H., Kapit, J., Gospodinova, K., DeFlores, L.,Quinn, R.C., Boynton, W.V.,Clark, B.C., Catling, D.C., Hredzak, P., Ming, D.W., Moore, Q., Shusterman, J., Stroble, S., West, S.J., and Young, S.M.M. (2010). Wet Chemistry experiments on the (2007) Phoenix Mars Scout Lander mission: Data analysis and results. J Geophys Res – Planets 115:E00E10.

Lalueza-Fox, C., Rompler, H., Caramelli, D., Staubert, C., Catalano, G. Hughes, D., Rohland, N., Pilli, E., Longo, L., Condemi, S., de la Rasilla, M., Fortea, J., Rosas, A., Stoneking, M., Schoneberg, T., Bertranpetit, J., and Hofreiter, M. (2007). A melanocortin 1 receptor allele suggests varying pigmentation among Neanderthals. Science 318(5855):1453-1455.

Larsen, K.W., Arvidson, R.E., Jolliff, B.L., Clark, B.C., (2000). Correspondence and leastsquares analyses of soil and rock compositions for the Viking Lander 1 and Pathfinder landing sites. J. Geophys. Res. 105, 29207–29221.

Le Maitre, R.W. (Ed.) (2002). Igneous Rocks: A Classification and Glossary of Terms, Recommendations of the International Union of Geological Sciences, Subcommission of the Systematics of Igneous Rocks. Cambridge University Press.

Leroi-Gourhan, A. (1982). The Dawn of European Art: An Introduction to Palaeolithic Cave Painting (Imprint of Man). Cambridge University Press, Cambridge, UK. 144 pp.

Leveille, R.J., and Datta, S. (2010). Lava tubes and basaltic caves as astrobiological targets on Earth and Mars: A review. Planet Space Sci. Special Issue 58(4):592- 598.

Lodders, K., and Fegley, B. (1998). The Planetary Scientist's Companion. Oxford University Press, Oxford.

Lommel, A. (1966). Prehistoric and primitive man cave paintings, totems, jewellery, masks, pottery, textiles. McGraw-Hill Book Co., NYNY.

Malin, M.C. and Edgett, K.S. (2000). Evidence for recent groundwater seepage and surface runoff on Mars. Science 288:2330-31.

McCubbin, F.M., Steele, A., Hauri, E.H., Nekvasil, H., Yamashita, S., and Hemley, R.J. (2010). Nominally hydrous magmatism on the Moon. Proc. Nat. Acad. Sci. US 107(25):11223-11228.

McKay, C.P., Friedman, E.I., Wharton, R.A., and Davis, W.L. (1992). History of water on Mars: A biological perspective. Adv. Space Res.12:231-38.

McKay, C.P. and Nedell, S.S. (1988) Are there carbonate deposits in Valles Marineris, Mars? Icarus 73, 142–148.

McKay, C.P. and Stoker, C.R. (1989). The early environment and its evolution on Mars: Implications for life. Reviews of Geophysics 27(2):189-214.

McKeever, S.W.S., Banerjee, D., Blair, M., Clifford, S.M., Clowdsley, M.S., Kim, S.S., Leuschen, M., Prather, M., Reust, D., Sears, D.W.G., Wilson, J.W., (2003). Concepts and approaches to in situ luminescence dating of martian sediments. Radiat. Meas. 37, 527–534.

MEPAG Special Regions — Science Analysis Group. (2006). Findings of the Mars special regions science analysis group. Astrobiol. 6:677–732.

Morris, R.V., Ruff, S.W., Gellert, R., Ming, D.W., Arvidson, R.E., Clark, B.C., Golden, D.C., Siebach, K., Klingelhofer, G., Schroder, C., Fleischer, I., Yen, A.S., and Squyres, S.W. (2010). Identification of carbonate-rich outcrops on Mars by the Spirit Rover. Science 329(5990):421-424.

Morthekai, P., Jain, M., Dartnell, L., Murray, A.S., Botter-Jensen, L., and Desorgher, L. (2007). Modelling of the dose-rate variations with depth in the Martian regolith using GEANT4. Nuclear Instruments & Methods In Physics Research Section AAccelerators Spectrometers Detectors and Associated Equipment 580(1):667- 670.

Moser, D., Boston, P.J., and Harris, M. (2001). Sampling in Caves and Mines. In Encyclopedia of Environmental Microbiology. Wiley and Sons, Publ.

Mustard, J.F., S. Murchie, S. Erard, and J.M. Sunshine. (1997). In situ compositions of Martian volcanics: Implications for the mantle." Journal of Geophysical Research 102:25,605-25,615.

Nedell, S.S., Squyres, S.W., and Andersen, D.W. (1987). Origin and evolution of the layered deposits in the Valles Marineris, Mars. Icarus 70(3):409-440.

Northup, D.E. and Lavoie, K.H. (2001). Geomicrobiology of caves: A review. (2001). Geomicrobiol J 18(3):(199-222.

NRC National Research Council, Space Studies Board. (2002). Safe On Mars: Precursor Measurements Necessary to Support Human Operations on the Martian Surface. National Academy Press, Washington, DC. 64 pp.

Redondo, F.A.G., Martin-Loeches, M., and Pobes, E.S. (2010). The prehistory of mathematics and the modern mind: mathematical thought and resourcefulness in the Palaeolithic Franco-Cantabrian region. Dynamis 30:167.

Rummel, J.D. (2001). Planetary exploration in the time of astrobiology: Protecting against biological contamination. Proceedings of the National Academy of Sciences 98(5):2128–2131.

Salvatore, M.R., Mustard, J.F., Wyatt, M.B., and Murchie, S.L. (2010). Definitive evidence of Hesperian basalt in Acidalia and Chryse planitiae. J Geophys Res. – Planets 115:E07005).

Sharp, Z.D., Shearer, C.K., McKeegan, K.D., Barnes, J.D., and Wang, Y.Q. (2010). The chlorine isotope composition of the moon and implications for an anhydrous mantle. Science 329(5995):1050-1053.

Shubert, G.C., Solomon, S.C., Turcotte, D.L., Drake, M.J., Sleep, N.H., (1992). Origin and thermal evolution of Mars. In: Kieer, H.H., Jakosky, B.M., Snyder, C.W., Matthews, M.S. (Eds.), Mars. University of Arizona Press, Tuscon, pp. 147–183.

Sieveking, A. (1979). The Cave Artists. Thames and Hudson, Ltd., London, UK. 221 pp.

Skidmore, M.S., and Ambrosi, R.M. (2009). Gamma-ray spectroscopy in Mars orbit during solar proton events. Adv. Space Res. 44(9):10(19-1029.

Squyres, S.W., Clifford, S.M., Kuzmin, R.O., Zimbelman, J.R. and F.M. Costard. (1992). Ice in the Martian Regolith. In: H.H. Kieffer et al., (eds.) Mars, University of Arizona Press, Tucson, AZ. pp. 523-554.

Squyres, S.W. et al., (2004). In situ evidence for an ancient aqueous environment at Meridiani Planum, Mars. Science 306, 1709–1714.

Squyres, S. W., J. P. Grotzinger, R. E. Arvidson, J. F. Bell, III, W. Calvin, P. R. Christensen, B. C. Clark, J. A. Crisp, W. H. Farrand, K. E. Herkenhoff, J. R. Johnson, G. Klingelho_fer, A. H. Knoll, S. M. McLennan, H. Y. McSween, Jr., R. V. Morris, J. W. Rice, Jr., R. Rieder, and L. A. Soderblom. (2005). In situ evidence for an ancient aqueous environment at Meridiani Planum, Mars, Science, 306, 1709-1714.

Squyres, S.W., D.E. Wilhelms, and A.C. Moosman. (1987). Large-scale volcano-ground ice interactions on Mars. Icarus 70:385-408.

Stoker, C.R., Zent, A., Catling, D.C., Douglas, S., Marshal, J.R., Archer, D., Clark, B., Kounaves, S.P., Lemmon, M.T., Quinn, R., Renno, N., Smith, P.H., Young, S.M.M. (2010). Habitability of the Phoenix landing site. J Geophys Res – Planets 115:E00E20.

Swift, P.D. and Smith, G.B. (1995). Cylindrical mirror light pipes, J Solar Energy Materials and Solar Cells 36:159-168.

Tankersley, K.B., Munson, C.A., Munson, P.J., Watson, P.J., Shaffer N.R. and Frushour, S.S. (1997). Archaeology and speleothems. In, C.Hill and P. Forti, eds., Cave Minerals of the World, Second Edition. Pp. 266-270. Nat. Speleolog. Soc., Huntsville, AL.

Taylor, A.G. and Gibbs, A. (1998). New views of the Moon: Integrated remotely sensed, geophysical, and sample datasets. September 20, (1998) at the Lunar and Planetary Institute in Houston, Texas.

ten Kate, I.L. ()2010). Organics on Mars? Astrobiol 10(6):589-603.

Texier , P.J., Porraz, G., Parkington, J., Rigaud, J.P., Poggenpoel, C., Miller, C., Tribolo, C., Cartwright, C., Coudenneau, A., Klein, R., Steele, T., and Verna, C. (2010). A Howiesons Poort tradition of engraving ostrich eggshell containers dated to 60,000 years ago at Diepkloof Rock Shelter, South Africa. Proc. Nat. Acad. Sci. US (PNAS) 107(14):6180-6185.

Valladas, H., Clottes, J., Geneste, J.M., Garcia, M.A., Arnold, M., Cachier, H., and Tisnérat-Laborde, H. (2001). Palaeolithic paintings. "Evolution of prehistoric cave art," Nature 413(6855):479.

Vanhaeren, M., d'Errico, F., Stringer, C., James, S.L., Todd, J.A., and Mienis, H.K. (2006). Middle Paleolithic shell beads in Israel and Algeria. Science 312(5781):1785-1788.

Walden, B.E. (1988. Lunar base simulations in lava tubes of the Pacific Northwest. Abstr. AAAS Pacific Region Meeting, Corvallis, OR. June, (1988).

Walden, B.E., Billings, T.L., York, C.L., Gillett, S.L., and Herbert, M.V. (1988). Utility of lava tubes on other worlds. Abstr. AAAS Pacific Region Meeting, Corvallis, OR. June, (1988).

Watson, P.J. (1986). Prehistoric cavers of the eastern woodlands. In, C. Faulkner, ed., The Prehistoric Native American Art of Mud Glyph Cave. Pp. 109-116. Univ. Tennesee Press, Knoxville, TN.

Whitfield, R. (1996). Dunhuang: Caves of the Singing Sands, Buddhist Art from the Silk Road. Textile Arts Publications.

Whitley, D.S. (2009). Cave Paintings and the Human Spirit: The Origin of Creativity and Belief. Prometheus Books. 322 pp.

Wright, R.V. (1971). The Archaeology of the Gallus Site, Koonalda Cave. Inst. Aboriginal Stud., Canberra, Australia. 133 pp.

Wynne, J.J., Titus, T.N., and Diaz, G.C. (2008). On developing thermal cave detection techniques for earth, the moon and Mars. Earth Planet. Sci. Lett. 272(1-2):240- 250.

Yen, A. S., and 35 co-authors. (2005). An integrated view of the chemistry and mineralogy of martian soils. Nature, 463, 49-54, (2005).

Zeitlin, C., Guetersloh, S.B., Heilbronn, L.H., Miller, J., and Shavers, M. (2006). Radiation tests of the extravehicular mobility unit space suit for the international space station using energetic protons. Radiat. Msmts. 41(9-10):1158-1172.

- 14 -
The Mars Homestead for an Early Mars Scientific Settlement

Bruce Mackenzie[1], Georgi Petrov,
Bart Leahy, and Anthony Blair
[1]Mars Foundation

ABSTRACT

The Mars Homestead ™ is a major project of The Mars Foundation ™. The goal is to develop a detailed plan for a permanent Mars Scientific Settlement similar to the NASA Design Reference Mission. Emphasis is placed on achieving success as early as possible, and maximizing the use of local resources. Two variations of the master plan are considered. The first is a hillside linear layout. The second, still under development, is a redesign that places the same program in an open plain site. The first crew on Mars would delay major scientific studies to concentrate on producing building materials, habitat construction and food. After the initial base is complete, construction would continue, allowing an additional dozen scientists to arrive every two years, bringing only their personal effects and specialized equipment. The cost of sending the manufacturing equipment can be recouped by eliminating the cost to return later crews to Earth after short rotations. Preliminary estimates show that the launch cost of a permanent scientific base for a dozen scientists (and growing) would be the same as the "Design Reference Mission" for three round-trip missions of six people each, while avoiding the significant risks of multiple return trips to Earth. Given the cyclical history of aerospace funding, this plan to establish an early permanent scientific base and grow it into a vigorous settlement may advance the settlement of space by an entire generation. Participation is welcome to everyone, including contributions by non-aerospace engineers.

Key Words: Mars, Space Settlement, Space Colonization, in-situ resource, construction, life in extreme environments, Astrobiology, space radiation environment

1. INTRODUCTION

THE MARS HOMESTEAD PROJECT is the key endeavor of the Mars Foundation™, a non-profit, 501(c)(3) organization dedicated to establishing permanent, thriving settlements beyond Earth, starting with Mars. The goal of the project is to develop a reference design for the first permanent Mars settlement built primarily from local resources. This early, permanent base can support scientific investigations more cost effectively and safely than a series of round-trip missions. Likewise, instead of focusing on the application of a single technology, the project takes a broad view of all the necessary structures and materials for the constructing the initial 12-person phase of a growing settlement.

Preliminary estimates indicate that the delivered mass to establish a permanent scientific base for the initial dozen scientists would total about 250 tons. This is approximately the same mass as NASA's Design Reference Mission (DRM), but is more cost effective. The DRM proposes three round-trip missions of six people each, totaling less than 24 person-years of time on the surface over a decade. Our alternative delays scientific studies, though it supports 12 person-years of effort each year, quickly growing to 24, 36, and more person-years every other year.

This cost savings is possible by eliminating the need for most return flights, their fuel manufacture, and consumables. Given the cyclical history of aerospace funding, it is critical to establish a reasonably self-sufficient base before enthusiasm and funding wanes. Prior to the crew's arrival, pre-positioned habitats, food, and equipment will be sent to Mars to produce: power, breathing gases (oxygen (O_2), nitrogen (N_2), and argon (Ar), water, fuel, and possibly industrial materials. The initial crew of four would concentrate on establishing long-term life support equipment, and would set up automated equipment to produce building materials, such as fiberglass, polymers, ceramics, glass and brick. Assuming the life-support system is reliable, they would then be joined by two more crews, bringing the total to 12 people. After full production equipment is in service, they would assemble additional manufacturing modules, greenhouses, labs, and living quarters built from local materials. Construction would continue after the crew moves to the permanent living space, allowing additional crews to arrive every two years. The new arrivals need only bring their personal effects and specialized equipment, allowing the scientific base to grow at about one-third the cost of round-trip missions.

2. HILLSIDE SETTLEMENT

A detailed master plan for a "Hillside Settlement" has already been established. It would include living quarters, workshops, greenhouses, maintenance facilities, waste processing, refining, manufacturing, and other areas needed to live and work (Figure 1). Once the team establishes residency in their completed quarters, they will resume construction of additional facilities to hold another dozen persons, mainly scientists who will arrive every subsequence two years. As progress continues, the base will grow into a permanent manufacturing settlement, producing equipment for additional scientific outposts and other permanent settlements.

FIGURE 1
Aerial View Of The Hillside Settlement, At The Stage When First Occupied By 12 People.

2.1 Plains Settlement - Realizing that a convenient hillside might not be available near the desired settlement location, the team is now undertaking a new study to adapt the same program to an open plain area, or most any site which is flat enough to be a good spacecraft landing site. One of the primary differences between the hillside and the plains settlements is the lack of radiation shielding provided by the hillside. Preliminary images are shown in Figure 2.

2.2 Materials - The plains settlement could be located anywhere on Mars where there are available water (as ice in the soil) and silicates suitable for the manufacture of fiberglass. If water-ice is not available, it could extracted from the air, but at an unreasonable cost. If fiberglass is difficult to manufacture from the local soils, alternate materials are possible, including: polymers such as polyester and nylon, spun basalt, recycled aluminum, iron and steel from the iron carbonyl process. As many materials as possible would be produced locally using general purpose equipment sent from Earth.

Interior furnishings could be constructed of locally produced fiberglass, plastic, pressboard panels, extra piping, parachute cloth, and masonry. Once extra greenhouse space is available, bamboo, paper, soybean oil, cloth, and other plant products grown on site could provide additional raw materials for furnishings.

2.3 Cylindrical Habitats - Materials such as fiberglass and polyester would be used for fabricating pressurized horizontal cylindrical habitats. They are arbitrarily dimensioned to be 5 meters in diameter and 20 or 30 meters long. This provides significant width of the floor, and space below the floor and above a ceiling for equipment and storage. Connection rings are 2.5 meters in diameter. Additional spherical connection nodes allow 4 cylindrical modules to be connected together. This avoids the structural problems of connecting to the side of a cylindrical module.

Large inflatable 'assembly tents' house bulky fabrication and fiberglass spinning equipment, and provide valuable factory floor space (left-foreground in Figure 2). Half-sized modules would be spun on a fiber glass lathe, and then joined together in a separate 'finishing tent', which also serves as an air-lock (middle-foreground in Figure 2).

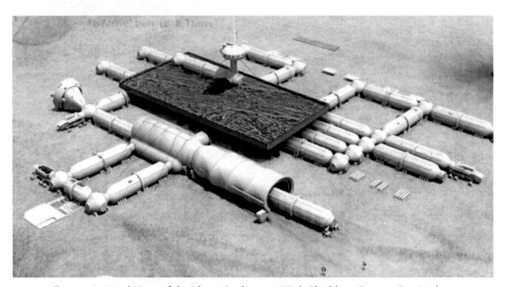

Figure 2 *Aerial View of the Plains Settlement, With Shielding Canopy Cut Back*

2.4 Radiation Shielding Canopy - Sintered bricks would be used for foundations supports and load bearing columns. The columns support an overhead canopy holding regolith for radiation sheilding. Protection against cosmic rays can be improved at a later date by adding a top layer of any material with lightweight atomic nuclei, such as bags of water-ice, waste plant material, or any carbon-based compound; when such materials are available. Gaps below the modules and between the modules and canopy are critical for external maintenance and cooling. This design of raised modules can be adapted for a polar settlement built on permafrost, provided the foundations are kept cool by the wind. In Figure 2, the canopy is partially cut back to show the horizontal cylindrical habitats and spherical connection modules.

A possible alternative is to construct a box frame around the cylindrical module (not shown). This allows more assembly to be performed indoors. The modules may be stacked when the settlement is large enough to justify two story structures. Overhead trays constructed indoors as part of the box frame would later be covered with regolith for radiation shielding.

2.5 Other Features - To allow extra light and overhead radiation shielding, greenhouses would be lit through side windows from external gimbaled mirrors (shown and described later in this paper). The settlement primarily utilizes nuclear power (not shown), not just for electrical generation, but also for industrial heat, such as manufacture of polymers, preheating during sintering of ceramics, manufacture of fuels, and breathable air.

Note the 'observation deck' rising above the canopy in the background of Figure 2. It has water tanks in the ceiling and awning for radiation shielding. The observation deck nominally serves as a supervisory station for construction of settlement and vehicles, much like an airport control tower, but with ubiquitous remote cameras that is not really needed. The real justification for it is psychological; it provides a place with a spacious view for meals, small meetings, private relaxation, and a sense of accomplishment, which are critical for settlers who will be on Mars for many years.

3. MARTIAN TECHNOLOGIES / SOCIAL CONSIDERATIONS

Any task as massive and difficult as exploring and settling Mars requires advance planning, development, and other efforts to literally and figuratively "get off the ground." These steps include feasibility studies, prototype projects, operational and environmental research, identifying funding sources, hardware development, and preliminary exploration. The Foundation has already taken the first step by developing a feasibility study and—like NASA, though with a different emphasis—a design reference mission. The Mars Homestead Project—our feasibility study—began by identifying the basic mission of a Mars settlement and the primary challenges involved in building such a habitat.

3.1 Marian Settlement Challenges - The Mars Homestead Project began with a simple operational premise: establish a permanent base on Mars to perform high-value scientific and engineering research and then to grow the outpost into a larger, permanent settlement. This vision includes:

- Testing manufacturing methods to use Mars resources
- Searching for past and present life on Mars
- Conducting basic science research to gain new knowledge about the history of the solar system
- Performing scientific research to use Martian resources to support human life

3.2 Maximum Flexibility vs. Mass Production - The first Mars settlement will face slow and limited communications with Earth. This will require a great deal of self-sufficiency in developing tools and structures. This situation will require flexible small-scale manufacturing and creative engineers skilled machinists capable of both high-tech and low-tech manufacturing—the equivalent of medieval blacksmiths.

3.3 Worker Productivity - Because the first crews to Mars will most likely be small—growing from 4 to 8 to 12 people initially—the crew must, of necessity comprise individuals capable of multitasking in a variety of disciplines. This would include civil engineering, life support systems, medical science, botany, microbiology, mechanical engineering, laboratory science, and, especially, simple mechanical maintenance. Such multitasking necessarily implies high worker productivity and proficiency across multiple disciplines. Illnesses or deaths will leave the crew short of vital skills and will necessarily result in emotional and social strain among the crew, though obviously they will have support from experts on Earth. This support would include redesigning equipment on Earth, and transmitting the new designs to be manufactured on Mars.

3.4 Limited Earth-Mars Transportation - Many high-technology items will need to be flown to Mars from Earth, such as nuclear reactors and specialized laboratory equipment. The settlers must be able to manufacture low-technology, high-mass materials (e.g. bricks, Fiberglas, or iron) on Mars. We assume no radical advances in propulsion in the near future, and that any incremental advances will be used to reduce the cost rather than to increase the mass delivered to Mars. In addition, the orbital mechanics of Earth and Mars dictate that flights only occur during the launch windows every two and one-seventh Earth years. Such potential delays mean that several fundamental requirements must be met:

Earth-Mars cargo support flights must be restricted to high-cost, high-technology equipment. Mission-critical equipment sent from Earth must be highly reliable. The crew must have the capability to manufacture simple but necessary materials or equipment on-site.

3.5 Unknowns - While any exploration or settlement crew is unlikely to encounter dragons or hostile natives, they will still have to cope with "unknown unknowns" related to interactions between people or equipment and the Martian environment, including the thin atmosphere, soil chemistry, low gravity, sub-arctic temperatures, radiation, low magnetic fields, microbial-natives, dust, and weather. Such unknowns will impact settlers' efforts to adapt Earth-based mining, refining, and manufacturing to Mars, the safety and reliability of equipment, and the overall cost of establishing the settlement.

In addition to the unknowns, one advantage of a one-way long term mission is the possibility of discovering native Martian pathogens. To help alleviate the public fear of transmission of Martians germs to Earth a one-way mission would serve as a method of temporary quarantine to determine if such pathogens pose a serious risk to humans. In the likelihood that such pathogens exist, the initial crew would have to stay on Mars indefinitely.

4. MARTIAN ENGINEERING CONSTRAINTS/SOLUTIONS

Mars presents any future builders with a variety of serious challenges and constraints, from near-vacuum atmospheric conditions to sub-zero temperatures lower than anything experienced in Antarctica.

Table 1 below summarizes some of the highlighted differences between the two planets. Given these extremes and constraints, it is important for any future expedition to identify and budget for the resources it will have available upon arrival. To build a permanent habitation, a settlement expedition will have the tools and equipment aboard the arriving spacecraft (most likely several vehicles, some arriving before the crew), local materials (meaning the land, water, and atmospheric components of the planet Mars), and the human beings and robots making up the crew.

Table 1

Planetary Comparison Between Earth and Mars

	Earth	Mars
Distance from Sun	150 million km	225 million km
Diameter	12,756 km	6,786 km
Axial Tilt	23.5°	25°
Length of Year	365.25 Earth Days	687 Earth Days
Length of Days	24 hours	24 hours 39 minutes
Gravity	1 g	.38 g
Temperature Range	-88°C to 58°C	-127°C to 17°C
Atmospheric Pressure	1013 mb (avg.)	7 mb (avg.)
Atmospheric Gases	78% N_2, 21% O_2	95% CO_2
Polar Ice Caps	Water Ice	Water Ice and Dry Ice (CO_2)
Highest Point	Mount Everest – 8.848 km above sea level	Olympus Mons –27 km above Martian avg.
Lowest Point	Marianas Trench – 11.022 km below sea level	Hellas Basin – 4 km below Martian average

Imported Resources - For the purposes of the original study, the team estimated a total arrival mass of 250 metric tons (tonnes) of cargo. Within this mass would be:

Crew – Increasing crew size from 4 to 8 to 12 people, to make on-site decisions, set up and test and repair equipment, and perform many non-repetitive assembly tasks. Having people on Mars is the whole purpose for going.

Robotic Equipment / Robots – Both general purpose and dedicated robotic systems for the majority of repetitive tasks.

Food and Shelter – Temporary living quarters, life support, and food necessary to support that labor until a permanent habitat is established.

Power System – including three nuclear reactors providing industrial heating and up to 400 kilowatts (kW) of electrical power, and the cables and other components needed to distribute the power throughout the settlement.

Construction Equipment – such as tractors, dump trailers, drag lines, sifting equipment, "fork lifts", etc. Initial Mining, Refining, and Manufacturing Equipment—to make gases, chemicals, metals, plastics, ceramics, masonry, glass out of the local atmospheric and regolith materials at hand. In some cases, the manufacturing equipment can be bootstrapped to make more capable manufacturing equipment once on-site.

High-Tech / Low-Mass Items – Any items that would be difficult to manufacture initially on Mars, such as sensors, computers, cameras, radios, other electronics, bearings, precision tools, small valves, medicines, medical devices, and small high-precision mechanical devices.

Scavenged Equipment – Anything that can be re-used from the Mars landing craft, whether designed to be reused in advance, or cannibalized after the fact. These include avionics, computers, control systems, wiring, actuators, sensors, nylon from parachutes, sheet metals, tanks, piping, etc.

Local Resources - The "mining" equipment for the first Mars settlement must also include air miners that will ingest material from the Martian atmosphere, which consists of 95.3% carbon dioxide (CO_2), 2.7% N_2, 1.6% Ar, 0.15% O_2, and 0.03% water vapor (H_2O) at a pressure around 7 millibars (mbar). After collecting this material, the air miners will use standard, existing chemical processes to create oxygen, habitat buffer gases (N_2/Ar mix), methane (CH_4), and hydrogen (H_2) fuel, longer-chain hydrocarbons, and plastics (including epoxy). Depending on the amount of water available in the local ground or atmosphere, Martian water also could used to nourish plants in the settlement's greenhouse. In addition to air mining, a similar process will be performed with available local resources in the Martian rock and regolith through ore beneficiation, that is, the crushing and separating ore into valuable substances or waste by any of a variety of techniques, (Petrov, 2004).

5. CONSTRUCTION

Mining is only the first step in building the settlement. Once raw materials are obtained, the settlement structure can be built. Due to the extreme cost of importing materials and equipment, relying on habitats brought entirely from Earth is an unsustainable strategy, unless there are revolutionary advances in transportation or manufacturing.

The Mars Homestead Project plan provides efficiency in transportation, infrastructure, safety, and ease of expansion. As a two-avenue structure, the "Linear City" of the Mars settlement will have separately pressurized segments with inflatable modules or regolith-supported masonry to keep the utilities, such as air, water, and power distribution following the "streets" of the city in sub-floor panels beneath the floors.

The team agreed that excavating to build the settlement was much simpler than tunneling for the reasons described in Table 2, below.

TABLE 2

Comparative Advantages Of Excavation Vs. Tunneling

Excavation and Covering	Tunneling
Excavation is optional, as long as firm soil is reached	Excavation and covering necessary for entrance area and extraction of raw materials
Can be located in most any terrain	Prefer a hill or vertical entrance to reach solid rock
Building the cover may be done accurately and safely	
Suitable for many soil types, use sintering for needed strength	Relies on strength of rocks above
Can fragment rock and permafrost using chemical explosives	Requires cutting of solid rock
Lower precision requirement	

The estimated work will require 20 man-weeks of work to excavate, based on conservative estimates. Small domes and vaults can be built in two days on Earth. As a conservative estimate, the project team assumed that it would take three times as long to do the work on Mars, including arches and walls. Each unit would take two to four man-weeks to complete. The large vaults are twice as big, so they would most likely take eight man-weeks each. The large dome will require special construction methods, so the team assumed an assembly time of 50 man-weeks. The total amount of time required to perform masonry work would be 298 man-weeks.

Given the design of the team's settlement, summarized in Table 3, and the constraints inherent in building on Mars, we estimated the total construction time to break down in Table 4.

Given an available building staff of four individuals, the entire settlement construction effort would take 560 man-weeks, with an additional 182 man-weeks for safety and helping the engineers install inflatable modules, airlocks, doors, windows, skylights.

TABLE 3

Summary Of Masonry Structures

Vaults	Radius (m)
8	3.25 x 10
6	2 x 8
3	1.5 x 13
3	1.25 x 8
28 (small domes)	2
1	5

TABLE 4

Construction Time

Construction	Man weeks
Fragmentation	4
Excavation	16
Transport	20
Processing	20
Placement (masonry)	298
Placement (cover)	20
Total	378

As the land is being leveled for development, the settlers will use the excavated regolith (dirt) as raw material to make brick, glass, and metal. The emphasis of this settlement will be on using local materials as much as possible to complete construction of a habitat. Continuing to rely on habitats brought from Earth is an unsustainable strategy. Therefore, settlers must maximize their use of Martian materials and simple, well understood, and tested building techniques.

Antique as well as modern methods of construction would be used to build the first Mars settlement. Ancient methods of construction would include masonry work, while modern methods would rigid cylinders made by winding fiberglass; sheet metal would be used where practical; and inflatable structures with rigid internal floors and partitions could be included as well.

The masonry structures would be based on bricks sintered from Martian regolith, reinforced with glass fibers if needed. If that is not practical, the masonry blocks could be glass blocks, fused basalt, or cut stone. Brick manufacturing would require some 2,200 cubic meters of material, which can be obtained from the regolith excavated from the hill. Heat from the nuclear reactors can pre-heat the kiln needed to fire the brick, with the final temperature reached by electric heat or burning chemical fuel produced from the CO_2 atmosphere. There will be two kilns of 1.5 cubic meters' capacity each, with a firing of time 8 hours. Given two batches per day and 6 cubic meters of brick being processed per day, it will take 370 days to make all of the brick required to build the settlement. Given the simplicity of this automated process, human intervention will be required only to maintain the equipment. A total of 20 man-weeks will be required for brick manufacturing. After the bricks are manufactured, the team must start assembling their new home. This will require a variety of different-sized vaults to achieve the size and comfort level required by the inhabitants.

6. BETTER MARTIAN LIVING SPACES

Given the known and potential unknown constraints on future human explorers, the study members believed that it was not enough to build a survivable Martian habitat; priority was given to building pleasant, permanent habitations on or under the Martian surface. As noted earlier, the settlers will use the temporary shelter of their arriving spacecraft until construction of the first phase is completed. The sooner settlers are able to establish a homelike environment for themselves, the sooner we can save costs and risks of frequent return flights. Also, a permanent base can easily support scientific exploration, while leading to the true goal of settlement for large numbers of people, with animals and plants.

6.1 Vision - The Hillside Settlement - The Mars Homestead Project team designed a partially underground, partially exposed facility built largely from local materials. Using a combination of imported and local resources, the Hillside Settlement will be approximately 90% self-sufficient by mass and will provide the settlers with the industrial capabilities they need to explore and settle the frontier.

The team selected Candor Chasma at reference coordinates 69.95W x 6.36S x -4.4km as the site for the Hillside Settlement. This site is part of the Valles Marineris canyon complex and has been photographed by Mars Global Surveyor. It consists of a number of mesas suitable for providing shelter as well as room for expansion. The site was chosen primarily for its proximity to several dry riverbed-like features. It is also reasonably close to Pavonis Mons, Olympus Mons, and other features of geologic interest. Site selection and response to the site are key design issues for the first settlement. There are a large number of engineering and logistical challenges to be considered in placing the settlement, (James 1998). The settlement must be located in proximity to geologically varied regions of scientific interest. There must be easily accessible in-situ resources. Low elevation is important to take advantage of higher atmospheric pressure, which aids both in harvesting gases from the atmosphere and landing with parachutes. Locations near the equator offer the highest solar intensity for crops, and access to and from any orbit.

There are many architectural and urbanistic issues to consider as well, (Alexander, 1977). The first permanent habitat will be our first cut into the new frontier. This is a very important step that will have an effect on the future development of the planet. A transplant of a pattern from the homeland in a model such as the "Law of the Indies", in which the Spanish crown dictated the masterplan of every city in the New World, will not work. There simply won't be the resources available to impose a predetermined plan that doesn't account for local conditions. The other extreme of a completely unplanned and pragmatic approach like the one taken by North American settlers is also unrealistic. The base must react to the site within a set of guiding rules.

The site must also provide a real and perceived sense of protection. On Earth, the location of a new settlement has traditionally been dominated by considerations for defense. On Mars there are no natives to defend against, but the need for a psychological sense of protection against a hostile environment is still real. Finally, any settlement needs a symbol that the residents can identify with, a landmark that identifies the structure as home, or a sacred place that will symbolize the founding of the community. It should be recognized that eventually the site of the first settlement will become a place of veneration and pilgrimage.

In the years to come, the Foundation will continue to gather more accurate and detailed information about Mars and almost certainly will reach a much more informed decision about the final site. For the purpose of this project, a site was chosen based on the best available information as of autumn 2004.

Candor Chasma - A group of mesas in the middle of Candor Chasma fulfill many of the requirements for a successful site. Candor Chasma is one of the northern branches of Valles Marineris (Figure 3). The elevation of the floor of the canyon is almost 4,800 meters below the planetary mean, offering relatively high atmospheric pressure. The walls of the canyon are about eight kilometers above the base. The panorama of the cliffs and valleys will present a beautiful view for the first settlement. Additionally the mesas themselves present an interesting site, giving the settlement a view in the close range as well. There are relatively flat areas to the north and south of the mesas that can serve as landing areas, where the flight paths to and from orbit will not overfly the settlement.

FIGURE 3
Candor Chasma, The Hillside Settlement Reference Location Shown On An Elevation Map Of Mars

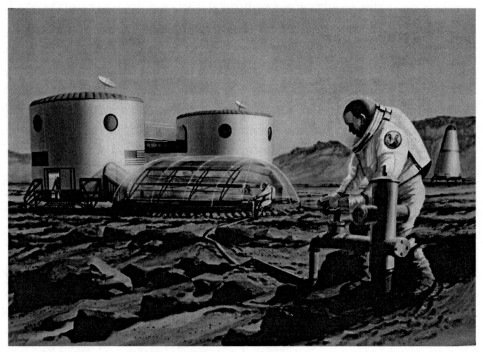

FIGURE 4

Initial, "Tuna-Can" Style Temporary Habitats Brought From Earth, Greenhouse, Ascent Craft, And Water Well To Tap Permafrost

As seen in the site plan, Figure 5, the Hillside Settlement reference design is quite extensive. It provides habitation for 12 people and covers 800 square meters, with greenhouses, fuel, nuclear plants, and manufacturing facilities located outside. Private living spaces, labs, and common areas are situated inside the mesa, with some of the rooms providing views of the outside.

Entrance - The main entrance is composed of two pressurized modules: the main entry and a garage. The main entry module consists of two airlocks with multiple egress options. One of the airlocks will provide direct access to the Martian surface for daily exterior activities, such as construction and repair work. The other airlock will be a docking port for rovers, (Petrov, 2005).

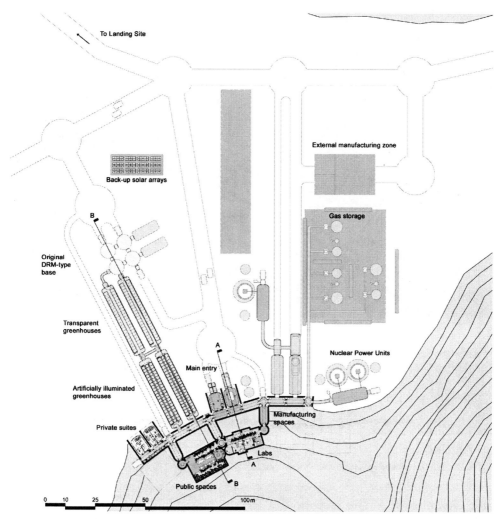

FIGURE 5
Hillside Settlement Reference Design—Site Plan

Social Spaces - The settlement will include areas for the entire group to gather in one place. These spaces are arranged along the infrastructure, with vegetation mediating the spaces between humans in a "Chinese garden" fashion. The human activities are located in the center with the trees surrounding them on two sides. This is where the community comes together on a daily basis (Figure 6 and later Figure 9). The space includes the communal kitchen and dining areas. The two bays above the dining area are covered with two-story high barrel vaults marking this as a special subspace of the segment. The second level includes balconies, catwalks, public work spaces, and also spaces for exercise and entertainment. Figure 6 below depicts how vegetation would be used in the floor plans.

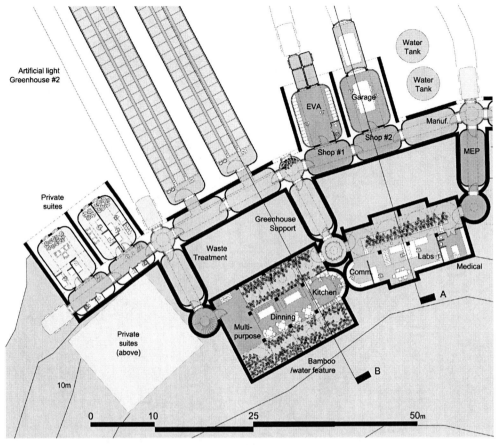

FIGURE 6
Close Up Schematic of Settlement Social Spaces

The Greenhouse - The greenhouses are the largest modules of the settlement and are situated on the flat land extending away from the mesa. The greenhouses are built with redundancy in mind: if one unit fails, demand can be covered by adjacent units. Each module has the complete ability to recycle water, air, and nutrients. The waste handling and water purification units are in the greenhouses and their adjacent modules. Note that 100% recycling is not required, since gases, water, and minerals can be extracted from Mars air and soil. Two greenhouses are transparent to take advantage of natural sunlight, and are supplemented with artificial light as needed. The other two greenhouses are opaque, covered with regolith, and artificially lit (Figure 7).

FIGURE 7
Concept Design for Gravel Bed Hydroponic Greenhouse Interior

Growing food on Mars would be much different from Earth. One extreme difference is the amount of ultraviolet (UV) radiation that reaches the surface of the planet. This is a major problem when trying to grow Martian-based crops. One possible solution would be to create crop plants that are UV tolerant. This would decrease the need for protective UV coating on Martian green houses.

Other photosynthetic producers on Earth have a natural way of resisting UV radiation with no cost to the productivity of photosynthesis. Microalgal groups create mycosporine-like amino acids (MAA) that absorb UV, (Karentz, 1991). This process is very prevalent in the Antarctic regions, where the ozone hole forms annually. The introduction of MAA algae genes into crops might help Martian agro-productivity.

The crops include a variety of vegetables and grains, as well as some non-edible plants grown for their fibers or usefulness in making particle board. Initially, crops will be grown hydroponically, in trays made from local fiberglass or ceramic pots, and filled with gravel. Later, we hope to grow plants in native soils after leaching harmful minerals as needed and adding organic material. Additionally, since some residents plan to spend a considerable number of years, or the rest of their lives on Mars, the greenhouses are deliberately designed for long sight lines, walking, jogging, or for solitude after work hours.

Side-Lit Greenhouse - One possible design for a passively heated and cooled greenhouse is shown in Figure 8. It is designed to use minimal electrical energy for heating and cooling and ventilation, especially during emergency conditions.

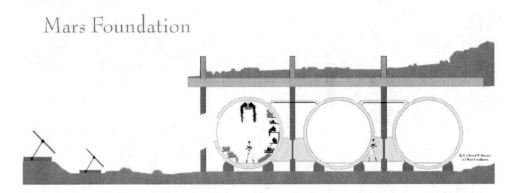

Mars Foundation

FIGURE 8
Side-Lit Greenhouse, With Radiation Shield Canopy

Sunlight is concentrated by moveable mirrors at the left side, and focusing them through side windows. An overhead scaffold holds radiation shielding material (regolith, ice, or any other material) to retain heat. Air ducts and passageways allow warm air convection to other habitat modules to the right, such as workshops, the kitchen, etc. Not shown are curtains which can be closed at night to retain heat when needed, or opened to allow wind cooling, with chimneys for additional convective cooling if needed.

In the event that an extreme emergency makes much of the base unusable and/or limits electrical power, the crew can retreat to these greenhouses and adjacent workshops and kitchen. There they have all the essentials to live and can repair whatever equipment may have caused the problem. This 'Side Lit" greenhouse was not incorporated in the Hillside Settlement design, but is being used in the follow-on designs.

Using Vegetation as a Fundamental Aspect of Settlement Life - Internal vegetation will be an integral part of the Mars settlement's social structure and sustainability, as it will be used as a symbol, life support component, mediator of exterior views, green belt for breaking up social spaces, and overall mediator of social life, (Alexander, 1997).

As a symbol, vegetation will hold a special place immediately between the main entrance and the formal meeting space, as the settlers will plant trees on arrival. These plants will symbolize hope and the permanence of the settlement. The trees will grow as the settlement expands, and when people arrive from

Earth, the first thing they will see as they enter is the grove of trees. These trees could also provide an occasional fruit or nut.

As a life support component, plant-rated greenhouses will optimize atmosphere, light, structure, and safety for specially designed plants. The farmers will plant seedlings and harvest the crops from inside a pressurized area with the aid of robots.

As a mediator of views within the settlement structure, plants will be placed in front of any windows to the outside, allowing settlers to look at the red of Mars through the soothing green of Earth vegetation. This added "touch of home" aids in settler psychology and stability. Additionally, every private suite has a small garden area in front of its window, and connector segments will terminate with small gardens and a window onto Mars. As a green belt, vegetation will separate heavily traveled areas from work spaces.

As a mediator of social life, plant life can be located either at the center of, or on the periphery of, social spaces. Where plants are in the center, the various social spaces are arranged around the periphery and every social space has views through the vegetation to the other social spaces. Those space with plants around the edge are like a "clearing in the woods" or a "Chinese garden", where the social space is surrounded and protected by trees. The edges of the space will be hidden, thus obscuring the limited size of the space. Finally, the lifecycle of the vegetation will provide seasonal visual changes in the living space, (Alexander, 1977).

Other Humanizing Features - In addition to plants, the overall settlement structure will be laid out with human psychology in mind, as it will feature social "spaces" to cover the entire range of human experiences, from the individual to couples to informal subgroups, sub-team meetings, and the entire community.

Figure 9, an image by Phil Smith, depicts large, open, multi-story masonry structures that are buried at the edge of the hillside. Natural sunlight is directed into this area from light shafts above (Figure 10). A Mars settler leans on the bamboo rail and takes in the view, overlooking lush green spaces.

FIGURE 9
Interior of Two Story Common Area

As life in our urban society has shown, artificial lighting is not enough to meet human needs. Therefore, the domes where vegetation is housed will have shafts that reach to the surface, as shown in Figure 10, below. There, light is collected by lenses and brought down to an opening in the top of the dome, where it is spread out again by another lens. This segment encloses a total volume of 1,530 cubic meters and has 125 square meters of floor space.

Rigid cylinders are used in the flat, open areas away from the hillside. The smaller, standard-size rigid modules would be wound fiberglass, constructed inside an inflatable construction tent. The larger ones would be sheet metal welded or riveted on-site, outdoors. Most would be covered with at least 1 meter of regolith (dirt). These modules are at the left of the cut-away Figure 10, and top of the site plan, Figure 5. They include: greenhouses, nuclear power 'balance of plant' in purple, and some manufacturing spaces.

Buried masonry vaults and domes are used for much of the living space. We excavate into the hillside in the loose talus slope, and build the vaults there. To hold the internal pressure, several meters of regolith (overburden) must be placed over them. Obviously, the masonry must be strong enough to hold the weight of the overburdened, and have buttresses shaped to hold the weight whether they are pressurized or depressurized. To keep the air from leaking, the bricks are glazed and caulked on the inside. In addition, alternating layers of sand and vapor barrier are placed outside the masonry to collect air which does leak, and route it back into the air processing equipment to be recovered. These modules are at the bottom of the site plan, Figure 5, shown with thick walls, also in Figure 5 and at the right of the cut-away, Figure 10. They include the: two-story public space, labs, kitchen, dining area, (Mackenzie, 1987).

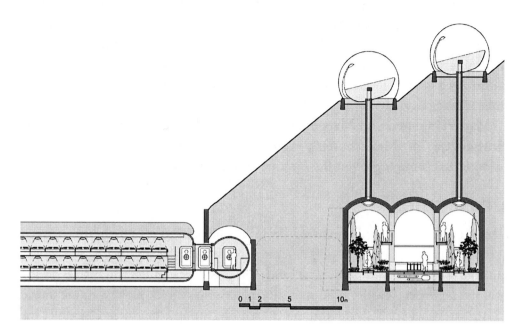

FIGURE 10
Cut-Away View, Greenhouse on Left, Shallow Masonry Covering Inflatable in Center, Deeply Buried Two Story Masonry on Right, Light Collectors in Upper Right

Pressurized Construction - Figure 10 also illustrates a good comparison of different construction techniques to handle the internal air pressure.

Masonry is used over inflatable structures at the edge of the hillside, to transition from the deeply buried masonry vaults to the open area. These are simple inflatable cylinders made from thinner fiberglass or cloth or thin sheet metal. They are used inside masonry vaults for protection. The masonry also holds the hillside back and provides radiation protection. These modules are at the middle of the site plan, Figure 5, shown as thick walls with rounded lines inside them, also at the middle of the cut-away, Figure 10. They include: private suites, waste treatment, greenhouse support, main entry, suit room, rover garage, and some manufacturing spaces.

7. JOINING MARS HOMESTEAD PROJECT

The Mars Homestead Project can use volunteer help as well as material donations and financial assistance to keep projects like this viable. The Mars Foundation manages a number of task forces and sub-projects to continue and further articulate its Earth-based studies for future Martian settlement.

7.1 Prototype Projects - The Foundation plans to sponsor prototype projects, which are small projects suitable for local groups, students, university classes. These projects would design or select equipment to facilitate a self-sustaining settlement on Mars. Other activities include greenhouse experiments, outfitting a single module of the proposed settlement, or developing small robotic projects.

Mars Homestead and Mars Foundation are trademarks of the Mars Institute incorporated in Massachusetts, USA. We can be reached by email at **Info@MarsHome.org** or on the Internet at **www.MarsHome.org.**

8. CONCLUSION

The Mars Foundation's Homestead Project, compared to round-trip missions, is a cost-effective, human-friendly method of settling Mars. With a one-way mission we avoid multiple risks and the costs of return flights. Refining equipment from Earth produces fiberglass, masonry, plastics, and metal construction materials. Semi-automated manufacturing equipment assembles the habitats, laboratories, and greenhouses for permanent use. Thus, we can afford to build a permanent, growing base to both support scientific study, and immediately start settlement. This would cost about the same as three round-trip missions, such as the NASA Design Reference Missions (DRM). The Foundation strongly encourages interested readers of the Journal of Cosmology to join the Mars Homestead project, or suggest complementary projects.

Acknowledgements: Numerous people helped with the technology and other aspects of the Mars Homestead program, and the specific Hillside Settlement design. They including: Adam Burch, April Andreas, Bruce Mackenzie, Damon Ellender, Frank Crossman, Gary Fisher, Georgi Petrov, Ian O'Neill, Inka Hublitz, Jaime Dorsey, James Burk, James Ghoston, K. Manjunatha, Mike Delaney, Randall Severy, Richard Sylvan, Robert Dyck, Susan Martin, William Johns. Special thanks to Georgi Petrov, Phil Smith, Bryan Versteeg, Adam Burch, and NASA for the images.

REFERENCES

Alexander, C. (1977). A Pattern Language. Oxford University Press.

James G., Chamitoff, G., Barker, D. (1998). Resource Utilization and Site Selection for a Self-Sufficient Martian Outpost." NASA/TM-98-206538.

Karentz, D., McEuen S., Land, M., Dunlap, W. (1991). Survey of mycosporine-like amino acid compounds in Antarctic marine organisms: Potential protection from ultraviolet exposure, Marine Biology, ISBN 0025-3162

Mackenzie, B. (1987). Building Mars Habitats Using Local Material, pg 575 in The Case for Mars III: Strategies for Exploration. Stoker, Carol ed., American Astronautical Society: Science and Technology Series v74.

Petrov, G. (2004). A Permanent Settlement on Mars: The First Cut in the Land of a New Frontier. Master of Architecture Thesis, MIT .

Petrov, G. I., Mackenzie B., Homnick M., and Palaia J. (2005). A Permanent Settlement on Mars: The Architecture of the Mars Homestead Project. Society of Automotive Engineers (SAE) International.

Journal of Cosmology, 2010, Vol 12, 4034-4050.
JournalofCosmology.com, October-November, 2010

- 15 -

Sex On Mars: Pregnancy, Fetal Development, and Sex In Outer Space

Rhawn Joseph, Ph.D.
Emeritus, Brain Research Laboratory, Northern California

ABSTRACT

Humans are sexual beings and it can be predicted that male and female astronauts will engage in sexual relations during a mission to Mars, leading to conflicts and pregnancies and the first baby born on the Red Planet. Non-human primate and astronaut sexual behavior is reviewed including romantic conflicts involving astronauts who flew aboard the Space Shuttle and in simulated missions to Mars, and men and women team members in the Antarctic. The possibilities of pregnancy and the effects of gravity and radiation on the testes, ovaries, menstruation, and developing fetus, including a child born on Mars, are discussed. What may lead to and how to prevent sexual conflicts, sexual violence, sexual competition, and pregnancy are detailed. Recommendations include the possibility that male and female astronauts on a mission to Mars, should fly in separate space craft.

Key Words: Sex, Mars, Pregnancy, Fetal Development, Menstruation, Testes, Ovaries, Radiation, Gravity, Sexual Conflicts

1. SEX IN SPACE

PERFORMANCE OF THE SEX ACT DURING a journey to Mars, may require potentially complex sexual gymnastics. On the other hand, any difficulties associated with sexual intercourse in space may turn out to be an easily solved problem of docking and entry as human are notorious for inventing ways of having sex despite all manner of logistical impediments (Joseph 2000a). However, what impact will sexual activity have on team dynamics and morale?

And what if an astronaut became pregnant during the journey? Would the fetus be viable? How would this impact the crew?

NASA has no policy regarding sex in space and its repercussions (Office of Audits, 2010), other than to request, in 2008, that astronauts voluntarily abide by an "Astronaut Code of Professional Responsibility" and maintain "a constant commitment to honourable behaviour." As summed up by the NASA Astronaut Health Care System Review Committee (NASA 2007), "the absence of a code of conduct and its enforcement, and the lack of management action to limit inappropriate activity increases the likelihood of aberrant behaviour occurring and decreases the likelihood of such behaviour being reported." According to NASA's review committee (NASA 2007), and a panel of experts assembled by the National Academy of Science (Longnecker and Molins 2006), this is a serious oversight: if male and female astronauts share a cramped space ship for years, surrounded by stars blazing in the blackness of night, thoughts are bound to turn to sex and romance. Thus, "ignoring the potential consequences of human sexuality is not appropriate when considering extended-duration missions" (Longnecker and Molins 2006), and this includes a human mission to Mars.

2. BIOLOGICAL DESTINY: SEX AND PREGNANCY

Biologically, females serve one purpose: to get pregnant (Joseph 2000a, 2001a,b, 2002). However, the human female is also the only female regardless of species, who is sexually receptive at all times *and* who has evolved secondary sexual characteristics, e.g. the enlarged breasts and derriere, which signal to males and females alike, her sexual availability (Joseph 2000a,b). Almost all non-human primate species that exhibit genital and breast swelling live in multi-male groups and they only develop these secondary sexual characteristic when sexually receptive (Clutton-Brock and Harvey, 1976; Fedigan, 1992; Wallis, 1992). These swellings serve to attract males and to arouse male sexual interest when the female is in estrus (Carpenter, 1942; Chevalier-Skolnikoff, 1974; Fedigan, 1992; Ford & Beach, 1951; Zuckerman, 1932) and the same can be said of the female human derriere and breasts which have evolved and increased in size over the course of human evolution so as to signal continual sexual availability (Joseph 1985, 2000a,b). Moreover, the human female has evolved the cognitive and intellectual capacity to employ cosmetics, perfumes, colorful clothing, push up bras, high heels, and so on, which draw attention to her breasts and derriere, and which emphasize and exaggerate her sexual availability by mimicking the signs of estrus common in other social primates (Joseph 2000a,b). And just as female primates and other female mammals seek sex with high status males whom they most prefer (Allen & Lemmon, 1981; Carpenter, 1942; Chevalier-Skolnikoff,

1974; Fedigan, 1992; Ford & Beach, 1951; Zuckerman, 1932), the same can be said of the human female (Joseph 2000a,b).

3. PREFERENCE FOR HIGH STATUS SEX PARTNERS: ASTRONAUTS

Like other mammals, human females are "choosy" and prefer sex with high status males who can offer prestige and resources (Buss, 1994; Betzig, 1985; Betzig et al., 1988; Symons, 1979; Townsend, 1989). Like their human female counterparts, nonhuman primate females, such as baboons, rhesus, and chimpanzees, prefer and have sex with high ranked males and avoid lower ranking and in particular, the lowest ranked males (Carpenter, 1942, Hausfater, 1975; Lancaster, 1978; Seyfarth, 1977; Smuts, 1987; Tutin, 1975; Zuckerman, 1932).

The Antarctic has long been viewed as an excellent analog for long-duration space missions and the same preference for high ranking males has been observed (Stuster 1996). Female crew members seek sexual intimacies with senior, high status personnel to the exclusion of the other men.

Moreover, female primates will emphasize their sexual availability and may actively pursue desirable males (Fedigan, 1992). Hence, not surprisingly, male astronauts are often targeted by females who have emphasized their sexual availability (Mullane 2007).

As related by former Shuttle Astronaut Mike Mullane (2007), male astronauts are commonly flown all over the U.S. to locations where they often attend coctail receptions and dinners and are actively solicited by young sexy women. "Beside the open bars at our soirees, there were other attractions for the males... young, beautiful women. Lots of them.... a potpourri of pussy.... I had been in enough officer's clubs in my life to know that aviator wings had babe-attracting power... but there was an even more powerful pheromone than jet-jockey wings: The title "astronaut." We males found ourselves surrounded by quivering cupcakes. Some were blatantly on the make, wearing spray-on clothes revealing high-beam nipples and smiles that screamed, "take me." Even the gold bands on the fingers of the married were no deterrent to many of these woman. They were equal opportunity groupies." A common sight in the mornings were the "married colleagues with red-blasted all nighter eyes trailing the odor of alcohol and sex as they exited a motel room with a smiling young woman" (Mullane 2007).

Among social primates, such as the chimpanzee, high status male chimps prefer and preferentially mate with and try to monopolize high status females (Goodall, 1971, 1986; Nishida, 1990; Sade, 1967). Among humans, high status females include movie stars, celebrities, and female astronauts. As detailed by

Shuttle Astronaut Mike Mullane (2007), the male astronauts "warmly welcomed the attention. The women probably welcomed it less so. The major focus was on them.... whenever Judy or Rhea or Anna--the triumvirate of TFNG beauty--walked into the room... There wasn't a man or woman in any public setting who didn't stare... They were particularly dazzling when they were dressed in their dark blue patch-covered flight coveralls. What was it about the women in their flight suits? In them Judy, Rhea, and Anna stole the audience. The flight suits seemed to transform them into fantasy creatures like Barbarella or Cat Woman or Bat Girl."

It can be predicted that being of high status, that astronauts of the opposite sex will be sexually attracted to one another and may act on their sexual desires.

4. ASTRONAUTS HAVE SEX: THE CASE OF LISA NOWAK AND WILLIAM OEFELIEN

The frequency of sex between male and female astronauts is unknown, due in part to NASA's unofficial "mind your own business" policy (Mullane 2007) and its failure to mandate an enforceable code of responsibility which "decreases the likelihood of such behaviour being reported" (NASA 2007; Office of Audits, 2010). Inadvertently, NASA encourages sexual behavior by not restricting it, or training for it; and this could be disastrous for a human mission to Mars.

Michael Collins, whose second spaceflight was as commander and pilot for Apollo 11, is one of only 24 humans to have flown to the Moon. In his book Mission to Mars (Collins, 1990), he points out that the crew of a future mission to Mars will likely be multinational and consist of both males and females from different professional backgrounds. He believed that the presence of women on a long duration mission to Mars, would create tremendous sexual tensions and a "singles bar atmosphere...a charged mixture of sexually unattached competitors, would be a disaster."

Consider the well publicized case of female astronaut Navy Captain Lisa Nowak, and Shuttle pilot Cmdr. William Oefelein who flew together on the space shuttle and engaged in an adulterous relationship which allegedly almost led to the kidnapping and murder of Nowak's rival, Air Force Captain Colleen Shipman.

Nowak, a married mother of three, was a highly regarded professional who had logged more than 1,500 hours in different aircraft, had received a BS in aerospace engineering from the U.S. Naval Academy in 1985; an MS degree in aeronautical engineering and a degree of aeronautical and astronautical engineer from the U.S. Naval Postgraduate School, and who was qualified as Mission Commander and an Electronic Warfare lead in her flight squadron (NASA

2006). She subsequently underwent astronaut training, was qualified as a mission specialist, worked at Mission Control as prime communicator with on-orbit crews, subsequently flew on the Space Shuttle and performed complicated manuevers on a 13-day mission on the International Space Station (ISS) in 2006 (NASA 2006). The Shuttle pilot was Cmdr. William Oefelein (the father of two children and a former "Top Gun" Navy pilot) with whom she was having an adulterous sexual affair.

According to Orlando Police Department investigators (2007a,b), unknown to Nowak, Oefelein was also having sex with Air Force Captain Colleen Shipman who he had met while training for the Shuttle Mission which would carry him and Nowak to the ISS. Subsequently, when Nowak logged onto Oefelein's computer at his home one evening, and discovered emails to and from Oefelein and Shipman which made explicit their sexual activities, Nowak allegedly decided to kidnap and kill her rival. Nowak drove 900-miles from Houston to Orlando, dressed in a trench coat and a wig to disguise her identity, and then attacked Shipman in the parking lot of the Orlando International Airport (Doukopil 2007; Hauser 2007, Kluger, 2007). Detectives found 30 unused diapers in her car, and a rope and a knife which she'd packed for her confrontation with Shipman.

The exact nature of Oefelein's feelings toward Nowak when he turned his attentions to Shipman is unknown. According to Orlando Police Department investigators (2007b), Nowak thought the affair was ongoing. According to Shipman, Oefelein once called Shipman by Nowak's first name when they were in bed (Orlando Police Department investigators 2007b). What we know for a fact is that Nowak and Shipman were willingly having sex with Oefelein while Nowak and Oefelein worked together as astronauts and these affairs were ongoing before and after Oefelein and Nowak flew on a 13 day mission together. We should ask: what would have happened if commander Oefelein, captain Nowak, and captain Shipman, were flying together on a 9 month mission to Mars where they would be together for over 2 years?

One answer comes from studies of female primates. Primate females will compete for access to preferred males, and will fight and threaten one another for the privilege of having sex with these males (Fedigan, 1992). However, primate males also try to monopolize females, and high status males tend to attack and drive off low status males (Goodall, 1971, 1986; Nishida, 1990; Sade, 1967). Male primates will also *rape* high status estrus female primates who resist.

5. ROMANCE AND SIMULATED MISSIONS TO MARS

If a crew of 3 males and 3 females were sent on a journey to Mars, it can be predicted that males would experience sexual interest in the female astronauts, and female astronauts may or may not always reciprocate depending on their hormonal status, and the status and attractiveness of the males vying for their attention. Male astronauts may not always take "no" for an answer.

Consider for example, the case of 32-year-old Dr. Judith Lapierre, a Ph.D. health sciences specialist, sponsored by the Canadian Space Agency, who participated in a 110-day isolation experiment designed to simulate a long duration space mission (Oberg 2000; Warren, 2000). According to Dr. Lapierre, two of her Russian international crew mates became embroiled in a 10 minute violent blood splattering fight, immediately after which she was physically accosted and manhandled by the team commander, a Russian, who began forcibly kissing her and sticking his tongue in her mouth (Oberg, 2000; Warren, 2000).

Dr. Lapierre forcefully protested immediately before, during, and after the assault, and was so frightened of additional sexual attacks that locks were placed on the doors to the passageway linking her test module with the Russian module. She and her crew mates also hid all the knives as they feared more violent outbursts and physical assaults. The experiment, called Sphinx-99, was designed to observe group dynamics under both routine and emergency conditions.

In the case of Dr. Lapierre, it was the commander in charge of the mission who assaulted her and she appealed to outside supervisory personnel for help. On a human mission to Mars, there will be no doors which can lock out sexual predators, and no one from Mission Control who can come to a female astronaut's rescue.

If women accompany men on a human Mission to Mars, are they at risk for rape? Or is the greater risk, falling in love and then pregnancy?

Dr. Judith Lapierre, was the only woman in a crew of eight and who were confined to a replica of the Mir space station. In another simulation, which included more females, romantic relations developed between willing partners. According to Pletser (2010), during the EuroGeoMars project designed to simulate a mission to Mars, in the Utah desert, two crew members engaged in a romantic relationship which involved open displays of affection and physical intimacy such as holding hands, spending exclusive time together in the evenings and while performing chores and various duties. Although privacy prevailed behind closed doors, these romantic activities were not welcome by other team members, and it was felt this behavior was detrimental to the unity of the group and team social activities.

According to an analysis of NASA's (2005) "Bioastronautics Roadmap: a risk reduction strategy for human exploration of space" (Longnecker and Molins 2006), not only are male and female astronauts on extended missions or a mission to Mars likely to have sex, but failing to plan for and ignoring the likelihood of sexual activity could have profound and even life threatening consequences: "Areas of concern for the 30-month Mars mission include the potential psychological and physiological consequences of sexual activity, consequences that could endanger life, crew cohesion, performance, and mission success."

What are some possible consequences? Rape, murder, the monopolization of female astronauts by one or two high ranking males?

6. SEX IN THE ANTARCTIC

The polar environment has been viewed by numerous scientists, and NASA, as an excellent analog for long-duration space missions. The findings are unequivocal: men and women have sex and form temporary romantic and sexual relationships (Leon 2005; Palinkas 2002; Rosnet et al., 2004; Stuster 1996): "One cannot send unmarried men and women" to the Arctic "and not expect them to form bonds" (Leon 2005); bonds which often lead to pregnancies (Ayton 2006; Bowden 1999).

For example, on Australian stations in the Antarctic seven pregnancies were recorded between 1989 and 2006 (Ayton 2006), although "no one wants to become pregnant down there, no one wishes a baby to be born down there" (Bowden 1999). The general belief is that pregnancy in the Antarctic poses increased risk to the mother, to her unborn child, to the team, and to the success of the expedition program (Ayton 2006). Nevertheless, women team members, regardless of nationality, become pregnant.

In fact, just having women as members of the team in the Antarctic can cause significant sexual tension and stress for both men and women (Palinkas 2002; Rosnet et al., 2004; Stuster 1996). Some men harass females for sex, women will behave seductively, sexual frustrations increase, and romantic rivalries develop (Rosnet et al., 2004; Palinkas 2002; Stuster 1996). Moreover, women in these settings are perceived by men and women alike as having tremendous sexual power over men, and thus have greater control over men, as it is the female's choice as to which of the men will receive their sexual favors (Palinkas 2002). Although this "choosiness" can certainly enhance the morale of the male she chooses, group cohesion can be negatively impacted (Suster 1996).

Palinkas (2002) provides an illustrative case, involving "John" who had left a wife and two children back home, was profoundly depressed, and then became embroiled in a sexual relationship with a female crew member, which resulted in an elevation of his mood and increased productivity. Although, temporary sexual relationships between male and female crew members are common in the Antarctic (Palinkas 2000; Stuster 1996), "John's" sexual relationship met with group disapproval.

When a woman chooses to have a sexual relationship during her sojourn in the Antarctic, it is often with senior (rather than junior) personnel, including the station leader who is usually seen as having the most status by the women, as well as an unfair advantage by the other men, thereby creating considerable tension and conflict (Stuster 1996).

On the other hand, most studies report that including women as team members has a very positive effect on morale (Rosnet et al., 2004; Leon 2005), especially if they are married and accompanied by their spouse (Leon 2005; Leon et al., 2003). For example, in a study of three married couples from different countries icelocked on a boat in the High Arctic for a 9 month period, partners provided each other with significant emotional support. Married partners were also instrumental in alleviating interpersonal and group tensions, and contributed to the effective functioning of the team (Leon et al., 2003).

Women also tend to provide emotional support to other team members and try to help solve interpersonal problems, behaviors which are not common in all-male groups (Leon 2005; Leon et al., 2003; Palinkas 2002; Rosnet et al., 2004). Women also serve as "peacemakers" reducing non-sexual competition, tensions, and arguments among the men, and the men often turn to women to share their emotional and personal concerns--though these same men have little interest in listening to these women share their feelings.

7. SEX AND OUTER SPACE

Like other female mammals, the human female is most likely to actively seek sex when she is ovulating and most likely to get pregnant (e.g., Gold & Burt, 1978; Matteo & Rissman, 1984; Udry & Morris, 1968, 1970; Wolfe, 1991). Hence, women become pregnant in the hostile conditions of the Antarctic; and there is no reason to suspect they may not become pregnant on Mars.

According to the Bioastronautics Roadmap (Longnecker and Molins 2006), "the risk of pregnancy might be mitigated by crew selection" and the use of contraceptive medications. Unfortunately, the many "questions about the efficacy, safety, and side effects of contraceptive medications may require that studies to answer these questions be completed prior to crew selection or that other measures be used to mitigate the risk of pregnancy."

But would these pregnancies lead to the first babies born in space, or on Mars?

Reproductive medical issues that pertain to astronauts have generally received scant clinical scientific attention, and there is little information on 1) the effect of long duration spaceflight and gravity on normal menstrual functioning including menstrual efflux and retrograde (intra-abdominal) menstruation, 2), the impact of microgravity on normal hypothalamic and pituitary functioning which in turn impacts and influences gender-specific hormone production and ovarian function, and 3) the effect of space radiation on fertility, reproductive success, and the future childbearing capacity of both men and women.

Of concern is the effect of microgravity (in space) and reduced gravity (once on Mars), on testosterone and estrogen secretions, the menstrual cycle, ovulation, and sperm production and viability. For example, if ovulation ceases the female astronaut may be continuously exposed to high levels of estrogen. By contrast, if the hypothalamus is impacted testosterone and estrogen levels may significantly decrease or increase, all of which would effect sexual and reproductive success. One concern is that "the exercise necessary for long-term cardiovascular and musculoskeletal fitness may be so strenuous enough that it may cause hypothalamic-induced hypogonadism with reduced serum estrogen levels" (Harm et al., 2001).

7.1. MENSTRUATION AND OVULATION If females are to be part of a human mission to Mars, then the long exposure to microgravity may cause menstrual efflux and retrograde menstruation. Many women commonly experience some retrograde intra-abdominal bleeding during menses which remains confined to the pelvis due to the forces of gravity. However, because of lack of gravity these menstrual blood products may induce retrograde menstruation and menstrual efflux during spaceflight.

No symptoms of endometriosis have been as yet detected in female astronauts on the space shuttle (Jennings and Baker, 2008). However, since the longest space shuttle flights are 18 days, and women cycle 28 days, no definite conclusions can be drawn from this finding.

Another concern is the effects of radiation. Radiation exposure has caused endometriosis in various species of primate (Fanton and Golden 1991; Wood et al. 1983). Endometriosis ("inside the womb") is caused by hormonal changes and radiation exposure, and is associated with the flourishing of endometrial-like cells outside the uterine cavity and on the ovaries. A common symptom is severe cramping and pain in the lower back, the legs, and the rectal and vaginal area, thereby making it difficult to walk, sit, or have sex (Ballard et al. 2010; Buyalos and Agarwal, 2000). Moreover, the increased number of endometrial-like cells on the ovaries, can cause anatomical distortions and adhesions which results in infertility (Buyalos and Agarwal, 2000).

According to Jennings and Baker (2008), menstruation during space flight lasting up to 18 days has never proved a problem and no symptoms of retrograde menstruation or pain associated with endometriosis have been reported. Unfortunately, although a few women have flown on the International Space Station for periods longer than 100 days (e.g., Sunita Williams, 194 days, Dr. Peggy Whitson, 350 days) privacy concerns have prevented the collection and reporting of data on female menstrual functioning for long duration space missions.

The hypothalamus, pituitary, gonadal (HPG) axis regulates the ovulatory cycle, and is highly susceptible to environmental factors including stress (Joseph 1998a, 1999a). Stress can exert adverse effects on the ovaries and can lead to ovulation failure (Tau et al., 2002). Although it appears that short term spaceflight does not affect the ovaries, the effects of a long duration mission are unknown. For example, examination of the ovaries of postpartum rats after 9–20 days of gestation showed no effect on ovarian weight or number of preovulatory or atretic follicles (Tau et al., 2002) and female rats ovulated and cycled normally (Serova and Denisova 1982). However, although these females successfully mated with male rats during space flight, no births resulted. Thus, spaceflight might be correlated with negative effects on the capacity to become pregnant and/or on the viability of the fetus while in space.

7.2 Male testosterone, testes and fertility - The ovaries are more radiation resistant than the testes (Suruda 1998), which suggests that males are more likely to become infertile than females following long term radiation exposure, which can be expected in space and on Mars. Perhaps the greater resistance of the female ovum is due to its being approximately 20 times larger than the human sperm cell. The ovum carries the X chromosome, whereas the male sperm may carry either an X or a Y chromosome thus determining if the fetus is male (XY) or female (XX). It is also believed that the male Y chromosome is more fragile than the larger female X chromosome.

In contrast to women, men are at a much greater risk to suffer damage to gametes (Jennings and Baker 2008), i.e. sperm, which then puts the smaller Y chromosome at risk. Damage to sperm can result in infertility or effect the sex ratio of offspring. Likewise, alterations and reductions in testosterone levels can effect the sexual orientation, and the sexual anatomy and sexual differentiation of the developing brain leading to feminization (Joseph et al., 1978, 1982). It is noteworthy that jet pilots and astronauts subjected to high G-force exposure have a higher frequency of daughters than those who experienced low G Forces (Little et al. 1987).

Male rats mated 5 days after space flight to nonflight female rats bred successfully, but their offspring were grossly abnormal. Abnormalities including physical retardation, showed growth retardation, hemorrhages, hydrocephaly, ectopic kidneys, and enlargement of the bladder (Serova et al., 1982). Male progeny also showed reduced epididymis weight at 30 days of age. By contrast, females mated to male rats 2.5–3 months after short duration spaceflight produced healthy, viable offspring (Santy et al. 1982). Unfortunately, these studies tell us nothing about the effects of long-duration flights.

Significant and profound reductions in testosterone levels have been found in both men and male rates following short duration flights (Tau et al., 2002). Reductions in testosterone would effect male sexual functioning and fertility as androgens stimulate sperm production. In most spaceflight studies, reductions in testosterone and the weight of testes have been reported (reviewed by Tau et al., 2002).

7.3 Stress, sex, and the HPG axis - Stress and abnormal environmental infuences exerts profound effects on the immune system, cardiovascular functioning, the central nervous system, especially the hypothalamus, amygdala and hippocampus, and on cognitive and intellectual functions including learning and memory (Joseph 1979, 1982, 1998, 1998a, 1999b, 2003; Joseph and Gallagher 1980). Stress can also significantly reduce testosterone levels. Decreased testosterone levels in male astronauts have been found in association with increased plasma LH, which returned to normal after flight, suggesting that the hypothalamic-pituitary-gonadal (HPG) axis may have been impaired by spaceflight (Tau et al., 2002). The hypothalamus is directly involved in the mediation of sexual behavior and the secretion of sex-specific hormones (Joseph 1982, 1999a).

The HPG axis is influenced by stress (Joseph 1998, 1999a,b) and by radiation and reduced gravity. Chronic stress in spaceflight rats has been indicated by higher adrenal gland weights compared with controls (Tau et al., 2002). In studies of astronauts aboard the MIR Space Station, increased cortisol was reported during spaceflight (Larina et al., 1997) and increased cortisol levels are classically associated with the stress response and injury to limbic system structures which includes the hypothalamus (Joseph 1999a,b). Increased cortisol excretion can also reduce testosterone and other hormones (Joseph 1999a).

Stress will effect fertility in women, and the viability of the fetus (Joseph 2000c). The fetal and infant brain and other physical organs may be severely injured by stress and the secretion of cortisol (Joseph 1999a,b, 2000c, 2003). Thus, if a female astronaut on a long duration mission to Mars becomes pregnant, she may lose the baby, or it may suffer a variety of physical and intellectual abnormalities.

8. PREGNANCY, GRAVITY, EMBRYO AND FETAL DEVELOPMENT

Ultimately, successful reproduction is the production of viable progeny after the female becomes impregnated. It has been demonstrated that rats can successfully mate in hypogravity, although no viable progeny were produced (Serova and Denisova 1982). However, space-faring mammals can become pregnant after they return to Earth. Several studies have reported no detrimental effects of short-duration space flight on pregnancy, reproductive hormones, fetal development, parturition, or lactation in female rats after they return to Earth (reviewed by Tao, et al., 2002). This suggests that once on Mars, males and females may be able to successfully have children. However, the same is not true of pregnancy in space (Jennings and Baker 2008).

There have been many studies done on embryos in microgravity and a variety of species have been used (Crawford-Young 2006; Ma et al., 2008; Ronca 2003; Serova et al., 1982, 1984; Wong and DeSantis 1997). Existing data suggest that space flight is associated with a constellation of changes in reproductive physiology and function. Space flight studies of pregnant mammals have shown a significant reduction in pregnancy weight gain, prolonged parturition, lower birth weights, and increased perinatal mortality (Serova et al., 1982, 1984; Wong and DeSantis 1997). Increased neonate mortality persisted into the F2 generation (Serova et al. 1984).

In a study of the developmental capacity of mouse embryos in the Chinese SJ-8 Satellite it was found that during space-flight, embryos cultured in a sealed culture did not develop whereas the same experiment performed on the ground in the same device showed that the embryos successfully developed (Ma et al., 2008). It was concluded that the space environment, especially the change of gravity harmed the development of the mouse embryo (Ma et al., 2008).

Extended spaceflight missions require prolonged exposure to decreased gravity. Human sexual reproduction and fetal development evolved in relation to living on Earth, standing upright, and the influences of Earth gravity (Joseph 200a,b,c). Therefore, it can be predicted normal fetal development (Joseph 2000c), would be effected by reduced gravity (Ma et al., 2008; Ronca 2003).

The adverse effects of microgravity on embryos, cell structure and function have been demonstrated by experiments performed in space or in altered gravity induced by clinostats. It is now well established that cellular structure, morphology, and genetic expression may be abnormally effected in microgravity and that the cytoskeleton and microtubules are gravity sensitive and may be grossly altered (Crawford-Young 2006; Ma et al., 2008; Ronca 2003). These gravity induced cellular changes exert a variety of deleterious effects on embryogenesis. These include death of the embryo, failure of neural tube closure, and gross abnormalities in the brain, heart, and limbs.

Cells respond to their environment within a three dimensional tissue structure. The number of adhesive structures in a cell and its resulting cell polarity may change the way it will behave in microgravity. For instance cells that are not polarized are more likely to die and undergo apoptosis or develop abnormally as has been demonstrated in microgravity (Crawford-Young 2006).

Microgravity has a significant impact on both cell shape and cytoskeleton (Crawford-Young 2006). Cells show signs of changes in the nucleus and in cell shape, and cells which form layers become disorganized, such that the layers do not develop normally. This would include the brain, the outer coating of which consists of 6 layers, and effect the embroyonic neural tube which is also layered and becomes the brain (Joseph 1982, 1999a, 2000c). Embryonic neural tubes do not close properly due to the changes in cell shape, migration, and adhesion, induced by altered gravity (Crawford-Young 2006).

Embyronic cells of differing types have different surface tensions and will separate into groups and will sort so that the cells with greater surface tension are on the inside of a tissue and those with less surface tension are on the outside. An example of this is the embryonic and fetal brain and heart (Joseph 1982, 2000c). When the brain develops, cells migrate to the surface area to form six layers consisting of different cell types. The heart also develops as a process of cell migration. If cell migration movements are altered in a developing animal in response to microgravity, the heart and other structures formed by the somites will not develop, causing early embryonic death (Crawford-Young 2006). Thus, any situation where cell mobility or cell lamellipodiae are involved could be affected, such that the fetus and its brain, heart, and internal organs, do not develop normally Hence, there have been no births of mammals in space (Tou et al., 2002; Ronca, 2003).

Early embryonic death could also be a direct result of microgravity induced fluid shifts, alterations in cardiovascular functioning, muscle wastage, skeletal demineralization, and decreased red bood cells, all of which would affect the ability to sustain a pregnancy.

And what if women already pregnant were to embark on a long duration space flight to Mars? As summed up by Sekulić et al., (2005): "During space flight it is impossible to apply the existing countermeasures against microgravity deconditioning of the muscular and cardiovascular systems to the fetus. Absence of gravitational loading during the last trimester of gestation would cause hypotrophy of the spinal extensors and lower extremities muscles, reduction in the amount of myosin heavy chain type I in the extensor muscles of the trunk and legs, hypoplasy and osteopeny of the vertebras and lower extremities long bones, and hypotrophy of the left ventricle of the heart muscle. Because of decreased capacity of postural and locomotor stability, acquisition of the gross developmental milestones such as sitting, standing and walking could be delayed.

NASA Medical Standards for space flight (JSC 11570) specifically disqualify any woman who is pregnant from space flight (NASA 1977) because of concerns regarding the adverse effect of toxins, microgravity and radiation on embryogenesis and fetal development, and to prevent on-orbit pregnancy accidents such as preterm labor, ectopic gestations, or spontaneous abortion. Thus all female astronauts are tested for pregnancy repeatedly beginning 20 days before launch (Jennings and Baker 2008).

Female astronauts use essentially every form of contraception to prevent pregnancy and are encouraged to continue using contraceptives during flight. On long duration missions continuous menstrual suppression is often used (Jennings and Baker 2008).

9. MOTHERS AND BABIES IN SPACE

There is now considerable evidence that biologically meaningful interactions between mothers and offspring are changed in the weightlessness of space (Crawford-Young 2006; Ronca 2003). Studies of young rat litters launched at 9 days of postnatal age or earlier, have been characterized by compromised maternal–offspring interactions and behavioral abnormalities (Ronca 2003). Although alterations in gravity would have a profound impact on the maternal-infant relationship, the stress of space flight would also be a factor. Prolonged and chronic stress would effect the mother, fetus, infant and child and disturbances in the mother-infant relationship would have severe effects on the brain and mind of the child (Joseph 1982, 1998, 1999a,b, 2000a,c). In addition, prenatal stress is a direct cause of fetal mortality, abnormal brain functioning, abnormal nursing behavior, and increased postnatal mortality.

However, a major concern must be the effects of pregnancy on the crew of a craft journeying to Mars. It can be surmised that all aspects of the mission would be put in jeopardy. Crew mates would unlikely to be supportive as their ability to perform their duties or to live comfortably would be impacted. Stress levels would rise, as would irritability, resulting in considerable hostility and anger directed toward the mother and father unless, perhaps, she had sex with multiple astronauts and the identity of the father was unknown.

10. RADIATION ON MARS: FERTILITY AND EMBRYONIC
   DEVELOPMENT

Another concern is the effects of radiation on pregnancy during a long duration mission and after arrival on Mars. Several studies have reported no detrimental effects of short-duration space flight on pregnancy, reproductive hormones, fetal development, parturition, or lactation in female rats after they return to Earth (Reviewed by Tao, et al., 2002). But would female astronauts be able to become pregnant and produce viable offspring on Mars?

The guidelines of The National Council on Radiation Protection and Measurements guidelines limit radiation exposure to 500 mrem for an entire pregnancy and to only 50 mrem per month. On the International Space Station, radiation exposure may approach 35,000 mrem or more (Harm et al., 2001), and these levels would significantly impact the viability of the fetus, producing a range of severe to profound abnormalities (Dekaban 1968; Diamond et al., 1973; Otake et al., 1996; Reyners et al., 1992) and would likely kill a developing human fetus.

If a female astronaut became pregnant during a mission to Mars, she would be subjected to massive doses of galactic cosmic radiation (GCR). GCR consists of heavier nuclei from He to Fe and high energy deep penetrating protons; and which would be very difficult to shield against (ICRP 1991; Straume et al., 2010). The catastrophic effects of radiation not only result from radiation-induced damage in the irradiated cells but in adjacent cells thereby causing widespread tissue damage and triggering genomic instability (NCRP 2000, 2006, NRC 1990, 2006; Straume et al., 2010 and references therein). Radiation of this sort would causes a wide variety of abnormalities in a fetus and the developing central nervous system which is especially sensitive to radiation.

Infants and children exposed to high doses of radiation prenatally, suffer significant intellectual deficits, including mental retardation, the severity of which increases with radiation dose, but which also varies depending on the time and length of exposure (Straume et al., 2010). According to Straume et al., (2010) intellectual deficits are greatest in those children exposed to radiation during 8th to 15th week of gestation with sensitivity continuing until the 25th week. However, for the first 8 weeks and after 25th weeks the fetal brain is more radiation resistant.

Unfortunately, even if the female astronaut is not pregnant when exposed to radiation, genetic alterations may be passed on to progeny such that genomic instability and abnormalities become transgenerational (NRC 1990, 2006; Straume et al., 2010 and references therein). However, even if she may become pregnant depends on if she and her ovaries can be shielded from radiation.

Exposure to chronic gamma rays will damage the ovaries and can induce temporary or permanent sterility (ICRP 1984, 1991; NRC 1990).

However, it is not just the human female on Mars, but the fertility of the men which may be effected by radiation. Sperm production can be profoundly reduced and effected by radiation, thereby causing radiation-induced fertility which may be permanent or temporary depending on dose and length of exposure (NRC 1990). Straume et al., (2010), argue, however, that nominally shielding in interplanetary space or on the surface of Mars would protect sperm production.

According to Straume et al. (2010), fortunately, "permanent sterility in men requires at least 5 Sv chronic low LET radiation, which is not anticipated for Mars missions... and... we would not expect decreased fertility from exposures to the radiation received during transit to/from Mars or living in a Martian base... though the possibility that human females may also have a sensitive prenatal stage should be given serious consideration."

Thus, it would appear that if male and female astronauts are properly shielded during the journey to and after arriving on Mars, that they will remain fertile and should be able to successfully reproduce.

On Earth, almost all female astronauts have delayed child bearing until after they have completed one or two space flights. Fertility and the ability to bear children has been negatively impacted in almost half of these women. As summed up by Jennings and Baker (2008), "the average maternal age at the time of delivery for the 15 children born to 13 U.S. female astronauts after flight is 41 years. The mean maternal age of the 12 postflight pregnancies that ended in spontaneous abortion is also 41 years.... because of the relatively advanced maternal age of female astronauts, there has been considerable need for infertility services and assisted reproductive technology... the success rates have been low."

11. SEX ON MARS: THE FIRST MARTIANS

Humans have sex and pregnancies result. If a baby is born on Mars, what predictions can be made?

From the moment of conception on Mars, the developing embryo-fetus-neonate-child, and its genome, will be subjected to a Martian environment markedly different from Earth, the most obvious distinctions being gravity, sunlight, radiation, temperature, and the biosphere. It is now well established that environmental influences early in life profoundly effects intelligence, learning, memory, vision, language, social-emotional functioning and the development, size, functioning, and interconnections of neurons and the brain (Casagrande & Joseph, 1978, 1980; Joseph 1979, 1982, 1998a,b, 2003;

1999a,b; Joseph, & Casagrande, 1978,1980; Joseph & Gallagher 1980); a function of the environment acting on gene activation vs suppression.

It has also been demonstrated "that populations contain a surprising amount of unexpressed genetic variation that is capable of affecting certain typically invariant traits" (Rutherford & Lindquist, 1998). There are thousands of "silent" genes within the human genome and the genome of other species, which code for or express functions which are as yet unknown (IHGSC 2001). However, fluctuations in temperature, oxygen levels, and diet can activate these "silent genes" (e.g., de Jong & Scharloo, 1976; Dykhuizen & Hart, 1980; Gibson & Hogness, 1996; Polaczyk et al., 1998; Rutherford 2003; Wade et al., 1997) which may express novel traits (Joseph 2009; Rutherford, 2003).

Environmental influences on gene expression are mediated through regulatory proteins such as Hsp90 (Feder and Hofmann 1999; Rutherford 2003; Sangster et al., 2004). Hsp90, for example, is a highly conserved multifunctional protein which targets multiple signal transducers and act as "molecular switches" which control gene expression in eukaryotes ranging from yeast to humans (Feder and Hofmann 1999; Rutherford 2003; Sangster et al., 2004). Hsp90 does not act alone but is part of a network that includes other proteins such as Hsp70, and p23 (Pratt and Toft 2003). These proteins "normally suppress the expression of genetic variation affecting many developmental pathways" (Rutherford & Lindquist, 1998). These proteins prevent DNA expression by acting as a buffer between silent genes and the environment. However, changes in the environment can directly impact regulatory genes and change the configuration of these proteins thereby removing their buffering influences, such that silent genes are then activated which triggers the expression of silent characteristics (Joseph 2009). As demonstrated by Rutherford and Lindquist (1998, p. 341) Hsp90 acts as an "explicit molecular mechanism that assists the process of evolutionary change in response to the environment" and it accomplishes this through the "conditional release of stores of hidden morphological variation.... perhaps allowing for the rapid morphological radiations that are found in the fossil record."

As a variety of genes will be effected by the environment of Mars, then the genome and the development of a fetus conceived on Mars will be differentially effected as compared to the genome and the development of a fetus on Earth.

It must also be recognized that the testes, ovaries, and genome of the parents will have been subjected to markedly adverse environmental conditions as they traversed space to reach Mars (cf Crawford-Young 2006; Ma et al., 2008; Ronca 2003; Straume et al., 2010), and again once on Mars. It is also highly likely subsequent genetic alterations will be passed on to offspring born on Mars (cf

NRC 1990, 2006b; Straume et al., 2010). As the environment acts on gene selection in parents and offspring, and as alterations in the environment and the genome effect evolutionary innovation and extinction (Joseph 1993, 2000b; 2009), then not just development but the evolution of humans on Mars will also be differentially effected as compared to humans of Earth.

Hence, if a child is conceived and born on Mars, we can predict at least three possible outcomes:

1) The child may suffer from mild to gross genetic, physical, and intellectual abnormalities (cf Serova et al., 1982, 1984; Ma et al., 2008; Otake et al., 1996; Reyners et al., 1992; Ronca 2003; Wong and DeSantis 1997).

2) The child will be completely healthy (cf Santy et al. 1982; Tao, et al., 2002).

3) Due to environmental influences on gene activation and differential and adaptive gene selection, although the child is completely healthy and *normal* it will be so different from children born on Earth, that it may appear*abnormal* simply because it is adapted, to varying degrees, to the Martian environment. The child will not be defective. Nor will it represent a new "race". Rather, it may represent a new species of humanity: The first Martian.

Naturally, all prospective parents will have to be counseled, before embarking to Mars, on these possible eventualities.

12. CONCLUSIONS AND RECOMMENDATIONS

Women have been an integral part of United States space crews since 1983, and world wide over 50 women have been selected as astronauts thereby providing considerable data on menstruation control and hygiene, contraception, and urination. As summed up by Jennings and Baker (2000) "there are no operational gynecological or reproductive constraints for women that would preclude their successful participation in the exploration of our nearby solar system." Perhaps the age of the female astronaut is the only major limiting factor (women of child bearing age only), if the goal is the colonization of Mars.

Women and men are sexual beings. The likelihood is male and female astronauts, traveling in the same space craft, will have sex during the long duration space-flight to Mars and after they arrive on Mars, even if substantial rules and steps are taken to prevent it. Therefore, training and preparations must be taken to anticipate all possible consequences and to regulate, guide, and modulate sexual activity; not for the purpose of preventing sex, but to prevent catastrophe.

Consider, for example, the possibility that the commander of the space craft may monopolize the female astronauts who prefer to mate with him because of his higher status relative to those he commands. Estrus female chimps may copulate up to 50 times in a day but only with a few of the available males who generally tend to be of high status (Goodall, 1986; Tutin, 1975). In one colony of free ranging rhesus monkeys, consisting of 150 adult females and 52 males, most of the estrus females actively sought out and mated with the same three dominant males (Carpenter, 1942). In some colonies, only about 20% of the males were responsible for 80% of the matings (Freedman, 1979). In one study of 25 male and 25 female captive baboons, five of the males possessed all the females (Zuckerman, 1932). In the military, one case came to light where a major general was having sexual relations with the wives of four of his subordinates (Burns, 1999). In studies of humans in long duration Antarctic analogue (Mars-like) conditions, females also prefer high ranking senior personnel, including station chiefs, often to the exclusion of those with a junior status (Stuster 1996).

Female primates will also attack and fight among themselves for the opportunity to have sex with a high ranking male (Fedigan, 1992). In 2007, captain Lisa Nowak allegedly planned to kidnapped and murder captain Colleen Shipman, her rival for the affections of space shuttle commander Oefelein.

Naturally, if a few males monopolize the available females, the other male astronauts will respond negatively and this may lead to violence. This can be avoided by a rule which relieves the monopolizer of command in cases of sexual monopolization, thereby stripping any male of the high status which made female astronauts prefer him to the other male astronauts.

Astronaut Michael Collins (1990), advocates sending only married couples on these voyages to better be able to survive the hardships of space. "An element of stability, of old-shoe comfort, would be introduced by having one's husband or wife to fall back on." Studies of married couples in the Antarctic support this view. Married couples are a source of stability and married females have a very positive effect on morale not just on their spouse but on the emotional stability of the group (Rosnetet al., 2004; Leon 2005; Leon et al., 2003). Married partners have been reported to have a calming effect and help to reduce interpersonal and group tensions thereby contributing to the effective functioning of the team (Leon et al., 2003).

However, the reality is: divorces are common and relationships often come to an end, and if either or both the male and female astronaut decided to change partners during the course of a voyage to Mars there may be conflict. The changing of partners could be a frequent occurrence during the mission to Mars and once on the Mars surface. Naturally feelings of jealousy, betrayal, and emotional upset would be expected. Therefore, if married couples are to be selected for a mission to Mars, the focus should be on those with very strong emotional bonds.

If the long term goal is the colonization of Mars, then it must be recognized that many older female astronauts are no longer capable of becoming pregnant or having children (Jennings and Baker 2008). This may be due to age or factors associated with exposure to the conditions of space travel. Therefore it is important to send younger versus older female astronauts; i.e. women who are in their early child-bearing years.

Although male and female astronauts could be trained to "share and share alike" so that sexual favors are provided equally to one and all, perhaps a better solution might be to send two space craft, one with an all male crew and another with an all female crew. Not only does this solve the problems which may arise from male-female sexual couplings in space, but it would prevent any possibility of pregnancy.

Naturally, humans are going to have sex and women are going to get pregnant. Pregnancies during a mission to Mars must be avoided. Pregnancies may lead to the death or abnormal development of the fetus. Pregnancy coupled with muscle atrophy, bone mineral loss, radiation, and cardiovascular activity may put the mother's life and the entire mission to Mars at risk.

Therefore, if issues of human sexuality are not addressed, and if necessary precautions are not taken, sex in space could lead to pregnancies, conflict, violence, and catastrophe. By contrast, once safely on the Red Planet, sex on Mars and the subsequent birth of the first Martian, would truly make humans a two planet species, and would be the first step to human colonization of the cosmos.

REFERENCES

Ayton, J. (2006). Women's health in Antarctica, O & G, 8, 22-44.

Ballard, K.; Lane, H.; Hudelist, G.; Banerjee, S.; Wright, J. (2010). Can specific pain symptoms help in the diagnosis of endometriosis? A cohort study of women with chronic pelvic pain. Fertility and Sterility 94, 20.

Bao-Hua Ma, et al. (2008). Microgravity, Real-Time Micrography of Mouse Preimplantation Embryos in an Orbit Module on SJ-8 Satellite, Science and Technology, 20, 127-136.

Betzig, L. L. (1985). Despotism and Differential Reproduction: A Darwinian View of Hisotyr, New York, Aldine de Gruyter.

Betzig, L., Borgerhoff Mulder, M., & Turke, P. (1988). Human Reproductive Behavior: A Darwinian Perspective. Cambridge, Cambridge University Press.

Bowden, T. (1999). Silence Calling, Allen & Unwin.

Burden H, Zary J, Lawrence I, Jonnalagadda P, Davis M, Hodson C. (1997). Effect of spaceflight on ovarian-hypophyseal function in postpartum rats. J Reprod Fertil. 109, 193–197.

Burns, R. (1999). Retired General Demoted. The Washington Post, September 3, 1999.

Buss, D. M. (1994). The Evolution of Desire: Strategies of Human Mating. New York. Basic Books.

Buyalos RP, Agarwal SK (2000). Endometriosis-associated infertility. Current Opinion in Obstetrics & Gynecology 12, 377–381.

Carpenter, C. R. (1942). Sexual behavior of the free ranging rhesus monkey. J. Comp. Psychol. 33:113-162.

Casagrande, V. A. & Joseph, R. (1978). Effects of monocular deprivation on geniculostriate connections in prosimian primates. Anatomical Record, 190, 359.

Casagrande, V. A. & Joseph, R. (1980). Morphological effects of monocular deprivation and recovery on the dorsal lateral geniculate nucleus in prosimian primates. Journal of Comparative Neurology, 194, 413-426.

Chevalier-Skolnikoff, S. (1974) Male-female, female-female, and male-male sexual behavior in stumptial monkey. Archives of Sexual Behavior. 3, 95-116.

Clutton-Brock, T. H, & Havey, P. H. (1976). Evolutionary rules and primate societies. In P.P.G. Bateson & R.A. Hinde (Eds.), Growing Points in Ethology (pp. 123-144). Cambridge, Cambridge University Press.

Crawford-Young, S. J. (2006). Effects of microgravity on cell cytoskeleton and embryogenesis Int. J. Dev. Biol. 50: 183-191.

Collins, M. (1990). Mission to Mars, Grove Press.

Dekaban, AS. (1968). Abnormalities in children exposed to x-radiation during various stages of gestation: tentative timetable of radiation injury to the human fetus. Int J Nucl Med 9: 471-477.

De Jong, G., & Scharloo, W. (1976). Environmental determination of selective significance of neutrality of amylase variants in Drosophilia. Genetics 84, 77-94.

Diamond, EL, Schmerler H, and Lilienfeld AM. (1973). The relationship of intra-uterine radiation to subsequent mortality and development of leukemia in children. A prospective study. Am J Epidemiol 97: 283-313.

Doukopil, T. (2007). Lisa Nowak's Strange Spacewalk, Newsweek, July 10, 2007.

Dykhuizen, D., & Hart, D. L. (1980) Selective neutrality of 6PDG alozymes in E. coli and the effects of genetic background. Genetics 96, 801-817.

Feder, M.E., and Hofmann, G. E. (1999). Heat-shock proteins, molecular chaperones, and the stress response: evolutionary and ecological physiology. Annu Rev Physiol 61:243–282.

Ford, C. S. & Beach, F. A. (1951). Patterns of Sexual Behavior. New York, Harper.

Fedigan, L. (1992). Primates and Paradigms: Sex Roles and Social Bonds. Elden Press, Montreal.

Fanton, JW, and Golden JG. (1991). Radiation-induced endometriosis in Macaca mulatta. Radiat Res 126: 141-146.

Freedman, D. (1979). Human Sociobiology, New York, Free Press.

Gibson, G., & Hogness, D. S. (1996). Effects of polymorphism in the Drosophilia regulatory gene Ultrabithorax on homoetic stability. Science 271, 200-203.

Gold, A. R., & Burt, A. D. (1978). Rise in female-initiated sexual activity at ovulation and its suppression by oral contraceptives. New England Journal of Medicine, 299, 1145-1150.

Goodall, J. (1971). In the Shadow of Man. Houghton Mifflin, Boston.

Goodall, J. (1986). The Chimpanzees of the Gombe: Patterns of Behavior. Cambridge U. Press, Cambridge.

Goodall, J. (1990). Through a Window. Houghton Mifflin Co, Boston.

Harrison AA. (2001). Spacefaring: The Human Dimension. Los Angeles: University of California Press.

Hauser, C. (2007). Astronaut Charged with Attempted Kidnapping. New York Times, June 2, 2007.

Hausfater, G. (1975). Dominance and reproduction in baboons. Contribution to Primatology, Basel, S. Karger.

ICRP (1984). Non-stochastic effects of ionizing radiation. ICRP Publication 41. Pergamon Press.

ICRP (1991). Recommendations of the International Commission on Radiological Protection, ICRP Publication 60, Pergamon Press.

IHGSC (2001). International Human Genome Sequencing Consortium. 2001. Initial sequencing and analysis of the human genome. Nature 409:860–921.

Jennings RT, Baker ES. (2000). Gynecological and reproductive issues for women in space: a review. Obstet Gynecol Surv, 55, 109–116.

Jennings, R. T. and Baker, E. S. (2008). Gynecologic and Reproductive Concerns. In Barratt, M. R. and Pool, S. L. (eds). Principles of Clinical Medicine for Space Flight. Springer.

Joseph, R. (1979). Effects of rearing environment and sex on learning, memory, and competitive exploration. Journal of Psychology, 101, 37-43.

Joseph, R. (1982). The Neuropsychology of Development. Hemispheric Laterality, Limbic Language, the Origin of Thought. Journal of Clinical Psychology, 44, 4-33.

Joseph, R. (1985). Competition between women. Psychology, 22, 1-11.

Joseph, R. (1993). The Naked Neuron: Evolution and the Languages of the Body and Brain. New York, Plenum Press.

Joseph, R. (1998a). The limbic system. In H.S. Friedman (ed.), Encyclopedia of Human health, Academic Press. San Diego.

Joseph, R. (1998b). Traumatic amnesia, repression, and hippocampal injury due to corticosteroid and enkephalin secretion. Child Psychiatry and Human Development. 29, 169-186.

Joseph, R. (1999a). Environmental influences on neural plasticity, the limbic system, and emotional development and attachment, Child Psychiatry and Human Development. 29, 187-203.

Joseph, R. (1999b). The neurology of traumatic "dissociative" amnesia. Commentary and literature review. Child Abuse & Neglect. 23, 715-727.

Joseph, R. (2000a). Female Sexuality: The Naked Truth. University Press.

Joseph, R. (2000b). The evolution of sex differences in language, sexuality, and visual spatial skills. Archives of Sexual Behavior, 29, 35-66.

Joseph, R. (2000c). Fetal brain behavioral cognitive development. Developmental Review, 20, 81-98.

Joseph, R. (2001a). Biological Substances to Induce Sexual Arousal and as a Treatment for Sexual Dysfunction. United States Department of Commerce: Patent & Trademark Office, January 12, 2001 #60/260,910.

Joseph, R. (2001b). Biological Substances to Induce Sexual Arousal, Sexual Behavior, Ovulation, Pregnancy, and Treatment for Sexual Dysfunction. United States Department of Commerce: Patent & Trademark Office, February # 60.

Joseph, R. (2002). Biological Substances to Induce Sexual Arousal and as a Treatment for Sexual Dysfunction. Patent Pending: United States Department of Commerce: Patent & Trademark Office, February, 2002 #10/047, 906.

Joseph, R. (2003). Emotional Trauma and Childhood Amnesia. journal of Consciousness & Emotion, 4, 151-178.

Joseph, R. (2009). Extinction, Metamorphosis, Evolutionary Apoptosis, and Genetically Programmed Species Mass Death. Journal of Cosmology, 2, 235-255.

Joseph, R., & Casagrande, V. A. (1978). Visual field defects and morphological changes resulting from monocular deprivation in primates. Proceedings of the Society for Neuroscience, 4, 1978, 2021.

Joseph, R. & Casagrande, V. A. (1980). Visual field defects and recovery following lid closure in a prosimian primate. Behavioral Brain Research, 1, 150-178.

Joseph, R., Hess, S., & Birecree, E. (1978). Effects of sex hormone manipulations on exploration and sex differences in maze learning. Behavioral Biology, 24, 364-377.

Joseph, R., and Gallagher, R. E. (1980). Gender and early environmental influences on learning, memory, activity, overresponsiveness, and exploration. journal of Developmental Psychobiology, 13, 527-544.

Kluger, J. (2007). "Houston, She's Got Some Problems". TIME, August 2, 2007.

Lancaster, J. B. (1978). Sex and gender in evolutionary perspective. In H. Katchadourain (Ed.), Human Sexuality (pp. 51-80). Berkeley, University of California Press.

Larina IM, Bystrikzkaya AF, Smironva TM. (1997). Psycho-physiological monitoring in real and simulated spaceflight conditions. J Gravit Physiol, 4, 8113–8116.

Leon, G. R. (2005). Men and women in space. Aviation, Space, and Environmental Medicine, 76, Supplement, B84-8.

Leon, G. R., et al. (2003). Women and couples in isolated extreme environments: Applications for long-duration missions Acta Astronautica. 53, 259-267.

Little, B.B., Rigsby, C.H. and Little, L.R. (1987). Pilot and astronaut offspring: possible G-force effects on human sex ratio. Aviation Space Environ. Med. 58: 707-709.

Longnecker, D., and Molins, R. (2006). Bioastronautics Roadmap: a risk reduction strategy for human exploration of space' David Longnecker and Ricardo Molins, Editors, Committee on Review of NASA's Bioastronautics Roadmap, National Research Council.

Matteo, S., & Rissman, E. F. (1984). Increased sexual activity during the midcycle portion of the human menstrual cycle. Hormones & Behavior, 18, 249-255.

McMichael, W. H. (1997). The Mother of All Hooks: The Story of the U.S. Navy's Tailhook Scandal, Transaction Publishers.

McMorris, M. (2000). Orgies, Adultery and Don't Ask Don't Tell. The Cornell Daily Sun. March 14, 2007.

NASA (1977). Medical Evaluation and Standards for Astronaut Selection: NASA Class II - Mission Specialist (JSC-11570). Prepared by Space and Life Sciences Directorate, NASA, Lyndon B. Johnson Space Center, Houston, Texas.

NASA. (2005). Bioastronautics Roadmap: A risk reduction strategy for human space exploration. NASA/SP-2004-6113. Hanover MD: NASA Center for Aerospace information.

NASA (2006). "Astronaut Bio: Lisa M. Nowak". NASA. October 2006. http://www.jsc.nasa.gov/Bios/htmlbios/nowak.html.

NASA (2010). NASA Astronaut Health Care System Review Committee, February-June, 2007: Report to the Administrator http://www.nasa.gov/pdf/183113main_NASAhealthcareReport_0725FINAL.pdf.

NCRP (2000) National Council on Radiation Protection and Measurements, Radiation Protection Guidance for Activities in Low-Earth Orbit. NCRP Report 132, Bethesda, MD.

NCRP (2006) Information Needed to Make Radiation Protection Recommendations for Space Missions Beyond Low-Earth Orbit. NCRP Report 153. Bethesda, MD.

Nishida, T. (1990). The Chimpanzees of the Mahale Mountains. Sexual and Life History Strategies. Tokyo, Tokyo University Press.

NRC (2006) Health Risks from Exposure to Low Levels of Ionizing Radiation: BEIR VII, National Research Council, National Academy Press, Washington, D.C.

Oberg, J. (2000). Violence and Sexual Assault in "Space" Galaxyonline.com.

Office of Audits. (2010). NASA's Astronaut Corps: Status of corrective actions related to health care activities. Report No. IG-10-110-016 (Assignment No. A-09-005-01). NASA: Office of Inspector General. pp.43.

Orlando Police Department (2007). Second charging affidavit, Orlando Police. Orlando Sentinel/Orlando Police, June 2, 2006.

Orlando Police Department (2007). Orlando Police Department forensic examination report. Orlando Sentinel/Orlando Police, December 3, 2007.

Otake, M, Schull WJ, and Lee S. (1996). Threshold for radiation-related severe mental retardation in prenatally exposed A-bomb survivors: a re-analysis. Int J Radiat Biol 70: 755-763.

Palinkas, L. A., (2002). On the ice. Individual and group adaptatin in Antarctica. Online Articals, 2002 - bec.ucla.edu.

Pletser, V. (2010). A Mars Human Habitat: Recommendations on Crew Time Utilization, and Habitat Interfaces. Journal of Cosmology, 12, 3928-3945.

Polaczyk, P. J., Gasperini, R., & Gibson, G. (1998). Naturally occurring genetic variation affects Drosophilia photoreceptor determination. Developl. Genes Evol. 207, 462-470.

Pratt, WB & Toft, D.O (2003). Regulation of signaling protein function and trafficking by the hsp90/hsp70-based chaperone machinery. Exp Biol Med 228:111–133.

Reyners, H, Gianfelici de Reyners E, Poortmans F, Crametz A, Coffigny H, and Maisin JR. (1992). Brain atrophy after foetal exposure to very low doses of ionizing radiation. Int J Radiat Biol 62: 619-26.

Ronca, A. E. (2003). Mammalian Development in Space. Advances in Space Biology and Medicine, 9, 217-251.

Ronca A, Alberts J. (2000). Physiology of a microgravity environment selected contribution: effects of spaceflight during pregnancy on labor and birth at 1G. J Appl Physiol, 89, 849–854.

Rosnet, E. et al., (2004). Mixed-Gender Groups: Coping Strategies and Factors of Psychological Adaptation in a Polar Environment Aviation, Space, and Environmental Medicine, Volume 75, Supplement 1, July 2004 , pp. C10-C13.

Rutherford, S.L. (2003). Between genotype and phenotype: protein chaperones and evolvability. Nat Rev Genet 4:263–274.

Rutherford, S. L., & Lindquist, S. (1998). Hsp90 as a capacitor for morphological evolution. Nature 396, 336-342.

Sade, D. S. (1967). Determinants of dominance in a group of free-ranging rehsus monkeys. I: S. Altmann, (Ed). Social Communication among Primates. New York. Wiley.

Sangster, T.A, et al., (2004). Under cover: causes, effects and implications of Hsp90-mediated genetic capacitance. BioEssays 26:348–62.

Santy P, Jennings R, Craigie D. (1989). Reproduction in the space environment: part I. Animal reproductive studies. Obstet Gynecol Surv, 45, 1–6.

Seyfarth, R. M. (1977). A model of social grooming among adult female monkeys. Journal of Theoretical Biology, 65, 671-698.

Serova L, Denisova L. (1982). The effect of weightlessness on the reproductive function of mammals. Physiologist, 25, S9–S12.

Serova L, Denisova L, Apanasenko Z, Kuznetsova M, Meizerov E. (1982). Reproductive function of the male rat after a flight on the Kosmos- 1129 biosatellite. Kosm Biol Aviakosm Med, 16, 62–65.

Serova L, Denisova L, Makeev V, Chelnaya N, Pustynnikova A. (1984). The effect of microgravity on prenatal development of mammals. Physiologist, 27, S107–S110.

Slobodan R. Sekulić et al., (2005). The fetus cannot exercise like an astronaut: gravity loading is necessary for the physiological development during second half of pregnancy. Medical Hypotheses, 64, 221-228.

Smuts, B.B. (1987). Sexual competition and mate choice. In: B.b. Smuts (Ed), Primate Societies. New York.

Stuster, J. (1996). Bold Endeavors: Lessons from Polar and Space Exploration. Annapolis, MD: Naval Institute Press.

Suruda, A. (1998). Reproductive Hazards of the Workplace, edited by Frazier LM, and Hage ML.. New York: Van Nostrand Reinhold.

Symons, D. (1979). The evolution of human sexuality. New York, Oxford University Press.

Thompson, M. (1998). Sex, The Army And A Double Standard A general allegedly coerced an officer's wife into an affair. Why did he get off so easy? Time Magazine, May 4, 1998.

Straume, T., Blattnig, S., an Zeitlin, C. (2010). Toward Colonizing Mars. Perspectives on Radiation Hazards: Brain, Body, Pregnancy, In-Utero Development, Cardio, Cancer, Degeneration. Journal of Cosmology, 12, 3992-4033.

Townsend, J. M. (1989). Mate selection criteria: A pilot study. Ethology and Sociobiology, 10, 241-253.

Tutin, C.E.G. (1979). Mating patterns and reproductive strategies in a community of wild chimpanzees. Behavioral Ecology and Sociobiology, 6, 29-38.

Udry, J. R., & Morris, N. (1968). Distribution of coitus in the menstrual cycle. Nature, 220, 559-596.

Udry, J. R., & Morris, N. (1970). Effects of contraceptive pills on the distribution of sexual activity in the menstrual cycle. Nature, 227, 502-503.

Wade, M., Johnson, N.A., Jones, R., Siguel, V., & McNaughton, M. (1997). Genetic variation segregating in natural populations of Tribolium castaneum affecting traits observed in hybrids with T. fremani. Genetics, 147, 1235-1247.

Wallis, J. (1992). Chimpanzee genital swelling and its role in the pattern of sociosexual behavior. American Journal of Primatology, 28, 101-113.

Warren, M. (2000). A Mir kiss? No, it was sex assault, says astronaut. UK Telegraph, March 28, 2000.

Wichman HA. (2005). Behavioral and health implications of civilian spaceflight. Aviat Space Environ Med. 76:B164-B171.

Wong A, DeSantis M. (1997). Rat gestation during spaceflight: outcomes for dams and their offspring born upon return to Earth. Integr Physiol Behav Sci., 32, 322–342.

Wolfe, L. D. (1991). human evolution and the sexual behavior of female primates. In J. D. Loy and C. B. Peters (Eds.), Understanding Behavior. New York, Oxford University Press.

Wood, DH, Yochmowitz MG, Salmon YL, Eason RL, and Boster RA. (1983). Proton irradiation and endometriosis. Aviat Space Environ Med 54: 718-724.

Zuckerman, S. (1932). The Social Life of Monkeys and Apes. London, Routlege.

Journal of Cosmology, 2010, Vol 12, In press
JournalofCosmology.com, October-November, 2010

- 16 -

Terraforming Mars: Generating Greenhouse Gases to Increase Martian Surface Temperatures.

N. N. Ridder[1,2], D.C. Maan, Ph.D.[1], L. Summerer, Ph.D.[1],

[1]ESA Advanced Concepts Team, Keplerlaan 1, Postbus 299,
2200 AG Noordwijk, The Netherlands,
[2]Climate Change Research Centre, University of New South Wales, Sydney,
2052 NSW, Australia

ABSTRACT

Increasing the temperature and atmospheric pressures on Mars are considered to be the two main requirements for making Mars habitable for human life. Among the several methods reported to achieve such a change, the release of artificial greenhouse gases to increase temperatures by about 20 degrees at the Mars poles in order to trigger the evaporation of CO_2 ice is considered as being one of the more feasible approaches. This study assesses the warming potential of four fluorine based greenhouse gases (GHGs), namely CF_4, C_2F_6, C_3F_8 and SF_6, in the Martian atmosphere using for the first time a state-of-the-art three dimensional Martian global circulation model. The temperature increase due to these GHGs will be assessed using the most effective mixture of these gases as determined by published laboratory experiments. The paper discusses the details of the methodology and the preliminary results.

Key Words: Mars, planetary engineering, terraforming, GCM, greenhouse gases, Martian atmosphere, habitability

1. INTRODUCTION

1.1 PLANET FORMATION AND ATMOSPHERIC EVOLUTION - It is widely believed that all planets in our solar system were formed similarly involving the same processes within the proto-planetary disk around the sun (Yung & DeMore, 1999). However, due to the masses of the planets and their individual distance to the sun their development varied strongly leading to eight very different planets, two of which, Earth and Mars, having possibly shared a similar early history.

Earth and Mars possess rocky surfaces with a clear separation to the atmosphere. Their atmospheres have lost most of the originally contained light gases like hydrogen and helium through escape processes and are believed to have formed by outgassing from the planet's interior. Therefore it is assumed that Earth and Mars had quite similar atmospheres in the early stages of their development. However, when compared now, the atmospheres of these two neighbour planets show severe differences in composition and density. While Mars has a thin, carbon dioxide (CO_2) dominated atmosphere with only traces of molecular nitrogen (N_2) and molecular oxygen (O_2), Earth developed a thick N_2 dominated atmosphere with a substantial O_2 content (Jacob, 1999).

The reasons for these differences are manifold. The most important ones are the mass difference between the planets, which leaves the atmosphere of the lighter Mars more affected by escape processes (Lammer et al., 2008), and the evolution of life on Earth (Kasting & Siefert, 2002).

1.2 The Evolution of Earth's Atmosphere - The first life forms on Earth were anoxic microorganisms whose metabolism substantially altered the Earth's climate, oceans, and land masses and atmosphere by liberating various minerals, metals, and gasses including oxygen (Joseph, 2010). Methane (CH_4) was likely one of the major metabolic products in this process. CH_4 has a 10^3 times longer lifetime in the absence of atmospheric oxygen compared to its present residence time of 10 years in the current atmospheric composition of Earth. Due to this longer lifetime the atmospheric content of CH_4 in the early Earth's development could have risen to over 1000 ppm assuming a biogenic release similar to today's 535 Tg (CH_4)/year (Kasting & Siefert, 2002).

Following this first transition of the Earth's atmosphere from a CO_2 dominated to a presumably thick, high CH_4 atmosphere like on Titan, the transition to an atmosphere with a high O_2 content took place (Joseph 2010). This oxygenation of the atmosphere was made possible by the release of O_2 produced by photosynthesis of cyanobacteria, and and increased substantially around 2.45 Ga ago (Farquhar, Bao, & Thiemens, 2000; Holland & Bengtson, 1994); and which made it possible for multi-cellular eukaryotes, equipped with mitochondria, to evolve (Joseph, 2010). Evidence for this socalled great oxidation event can be found in geologic records, which show that older sedimentary rocks lost their iron content via weathering rather than via iron oxidation that can be found in rocks after about 2.0 Ga. This lack indicates a lower partial pressure of oxygen in the ancient Earth's atmosphere (Kasting, 1993). Another way of determining the timing of the great oxidation is the analysis of signature of the degree of mass-independent fractionation (MIF) of sulphur isotopes (Farquhar et al., 2000). Under present atmospheric conditions MIF of sulphur photochemistry is uncommon in the lower atmosphere and is homogenised quickly. In an anaerobic environment however the MIF signature is preserved and can be detected in the geologic record (Kump, 2008). This method indicates that the increase in atmospheric oxygen took place about 2.45Ga ago (Zahnle, Claire, & Catling, 2006). The precise cause of the sudden rise of atmospheric oxygen content is still not fully understood and remains subject to ongoing research (Kump, 2008), though accumulating evidence points to biological activity as a major contributor (Joseph, 2010).

1.3 The Evolution of the Martian Atmosphere - It is believed that long ago Mars was a warmer and wetter planet; conditions conducive to the origin and evolution of complex life (Yung et al., 2010). But because of changing geochemical forces which effected energy flow it became impossible for complex life to dwell upon the surface or to continue to evolve. In contrast to Earth, Mars presumably never harboured complex life. Although there is substantial debate about the possibility of microbial life (Houtkooper and Schulze-Makuch, 2010; Levin, 2010; Sephton, 2010; Yung et al., 2010), if extant, they are not in sufficient numbers to significantly alter or to maintain the atmosphere. Whatever atmosphere the planet had was lost after the initial outgassing from its crust. The high atmospheric loss was most likely caused by the comparably smaller mass of the planet and the missing magnetic field that would have protected the Martian atmosphere from powerful solar winds and the impact of loss processes through, for instance, sputtering and ion pickup at the top of the atmosphere (Lammer et al., 2008). The remaining thin atmosphere, the resulting lack of significant greenhouse warming on Mars (which is only 5% of Earth) and the absence of shielding against high energetic ionizing radiation, are conditions that would most likely inhibit the evolution of life on the planet's surface. Therefore, it is highly unlikely that, unlike Earth, Mars has ever developed a biologically engineered atmosphere. Martian geological history remains however subject to substantial speculation and large uncertainties due to the limited amount of data (Levine et al., 2010a,b).

1.4 The Concept of "Terraforming" - In 1971, a publication by Sagan triggered a still ongoing discussion about the deliberate alteration of the Martian climate with the aim to make the planet habitable. Sagan published the theory of the Long Winter Model (LWM), suggesting that the Martian climate alters between an icy state in which most atmospheric components are stored frozen on the surface and a state where the atmosphere is thick and sufficient to allow a warm and wet surface (Sagan, 1971). Even though today this hypothesis is contested, it was the beginning of the discussion about what is usually referred to as "terraforming", a term first coined in science fiction literature, and inspired a variety of subsequent implementation proposals.

Amongst the first ones, Burns and Harwit (1973) proposed to modify the planet's precession cycle in order to provoke the transition between the cold state of the Martian climate to the warm summer state. First proposed implementation options include changing the orbit of the Martian moon Phobos and introducing material from the asteroid belt into the Martian system (Burns & Harwit, 1973). With further insight, such a Burns-Harwit manoeuvre is widely considered as ineffective, especially since the Long Winter Model no longer seems to be an accurate description of the Martian climate system.

Sagan (1973) suggested a less drastic modification of the Martian climate: changing the polar cap albedo to warm Mars' surface and thus triggering the release of CO_2 into the atmosphere leading to a runaway greenhouse effect (Sagan, 1973). Even though Mars would not be fully habitable after this process the planet would possess a denser CO_2 atmosphere, which could provide enough greenhouse warming to heat the Martian surface to the point where liquid water can exist, fulfilling the prerequisites for ecopoiesis, i.e. the fabrication of an uncontained, anaerobic biosphere on the surface of a sterile planet (Fogg, 1995).

However, modern observational data show that it is rather unlikely that the amount of CO_2 stored in the Martian polar caps will be sufficient to trigger the necessary runaway greenhouse effect. Even if the polar caps did contain a sufficient amount of CO_2, a sufficient change of the polar caps' albedo seems currently out of reach: calculations show that the amount of material with the albedo of black carbon necessary to change the albedo sufficiently would be approx. 10^8 tonnes. The transport of such masses from Earth to Mars does not seem feasible in any foreseeable future. Alternative suggestions are therefore to use Martian soil or plants or material from asteroids. Even if such processes were feasible, ensuring a long-enough change in the albedo might prove difficult due to weathering and the harsh Martian climatic conditions (Fogg, 1995; Levine et al., 2010c).

Following these first proposals a variety of other partially even more ambitious concepts to terraform Mars were published. These include (1) changing the orbital eccentricity of Mars' orbit around the Sun, (2) changing the obliquity of Mars's spin, (3) channelling of volatile-rich cometary nuclei into the Martian atmosphere, (4) seeding of Martian atmosphere with heat-absorbing, cloud-forming particles, (5) heating the polar caps using large spaceborne mirrors, (6) devolatising of the carbon within the Martian crust, (7) inducing large-scale drainages of potential Martian aquifers, (8) the introduction of microbes, bioengineered to survive the harsh environment on the Martian surface, (9) the addition of bioengineered plants to lower the surface albedo and (10) the introduction of super-greenhouse gases (GHGs). An overview of many of these methods has recently been published by Beech (Beech, 2009).

The entire concept of active interference in the climate of another planet raises also a series of ethical, philosophical, economic and legal questions (Arnould, 2005; Pompidou & Audouze, 2000; Reiman, 2010). These issues are not addressed in the present paper. The authors are fully aware that concepts related to the deliberate introduction of terrestrial life forms would raise currently insurmountable difficulties related to planetary protection requirements (Rummel et al., 2010). The reader is referred to the Planetary Protection Policy of the Committee on Space Research (COSPAR) as well as recent publications in the field (Arnould, 2010; Conley & Rummel, 2010; COSPAR, 2005; Guarnieri, Lobascio, Saverino, Amerio, & Giuliani, 2009; Kminek et al., 2010; C. P McKay, 2009; Nicholson, Schuerger, & Race, 2009; Rummel et al., 2010).

For the purpose of this paper, the term Martian "climate engineering" is preferred to "terraforming".

1.5 Feasibility of Martian Climate Engineering Concepts - While some of the afore mentioned proposals seem completely out of reach for foreseeable futures considering contemporary technological capabilities, their assessment and especially the potential effects on Mars and its atmosphere are often scientifically interesting, such as e.g. if the alteration of the Martian orbit as proposed by Burns and Harwit would introduce a substantial perturbation of the equilibrium of the solar system (Beech, 2009). The deliberate introduction of bioengineered plants and microbes as well as space borne mirrors to melt the ice caps seem to be also out of reach at this stage of the technological development. Proposals dealing with the devolatization of the Martian crust or the drainage of aquifers are problematic as they require a sufficient abundance of the respective material in order to introduce enough greenhouse warming to allow ecopoiesis. Considering the lack of evidence about their existence from recent measurements by Mars missions such as ESA's Mars Express and NASA's Mars Odyssey, it is not clear if these proposals could ever be realised (Picardi et al., 2005).

One of the proposals considered as potentially feasible in a far but foreseeable future is the introduction of GHGs into the Martian atmosphere as suggested by Lovelock and Allaby (1984). Lovelock and Allaby originally suggested the injection of chlorofluorocarbons (CFCs) into the Martian atmosphere to increase the planet's greenhouse effect. However, the use of CFCs as GHGs to warm Mars would be limited based on the lack of a shield to protect the lower Martian atmosphere from high energetic radiation. The lower atmosphere of Earth, i.e. the troposphere, is protected through the ozone layer, which reaches from approx. 15 km to 40 km. On Mars such a layer is missing hence CFCs are photolysed at a high rate, which would make it necessary to produce them continuously. Furthermore, the dissolving of CFCs produces ozone-depleting species, e.g. highly reactive chlorine, which prevent the production of an effective shield against ionizing radiation similar to the one on Earth (Fogg, 1995).

A possible way to prevent the release of O3 depleting gases was presented in the study by Marinova et al. in 2005 who analysed a set of four fluorine based GHGs in regards to their warming potential in the Martian atmosphere by introducing these into a one-dimensional radiative convective model of the Martian atmosphere (Marinova, McKay, & Hashimoto, 2005). This assessment is of special interest as it shows the ability of these gases to warm the Martian surface significantly without the release of ozone depleting products that would prevent the development of an UV-radiation shield (Table 1).

Even though some bacteria and higher life forms are known to be able to withstand high radiation levels, the shielding from excessive natural radiation would likely be one key question to be to support the introduction of life forms like bacteria that can pursue the transformation of the Martian climate.

TABLE 1

Temperature increase due to greenhouse gases on present Mars (P(CO₂) = 6 mbar. Cases which resulted in a surface temperature over 260K were discarded. Table adapted from (Marinova et al., 2005)

	10^{-6} mbar	10^{-5} mbar	10^{-4} mbar	10^{-3} mbar	10^{-2} mbar	10^{-1} mbar
CF_4	0.019 K	0.143 K	0.497 K	1.817 K	5.160 K	10.100 K
C_2F_6	0.052 K	0.348 K	1.530 K	5.410 K	13.600 K	31.000 K
C_3F_8	0.065 K	0.562 K	2.910 K	10.100 K	33.500 K	-
SF_8	0.112 K	0.506 K	1.920 K	5.010 K	9.800 K	19.700 K
Best combination	0.112 K	0.677 K	3.330 K	12.300 K	37.500 K	-

1.6 The Assessment of Martian Climate Engineering Proposals using Numerical Climate Models - The assessment of all Martian climate engineering proposals in regards to their warming potential relies on the numerical description of the Martian climate system. Over the last decades the quality and accuracy of these numerical climate models has increased significantly. The high variety of models with different dimensions and resolutions increased accordingly

and has allowed new insights into the climate of Mars in a way that has not been possible before (Wordsworth et al., 2010). Especially highly sophisticated global circulation models (GCM), which are also used for simulating Earth's climate support this new view on our neighbour planet. These continuously improved GCMs allow the assessment of past, present and future climates on Mars in more detail than ever before. Despite this development, scientific assessment of Martian climate engineering proposals has been scarce up until recent times. However the tools to perform such assessments in a scientifically rigorous way matured sufficiently to allow such research questions to be addressed in a scientific sound way. This paper reports on the assessment of the effects of the four fluorine-based gases used by Marinova et al. (2005) on the Martian climate, by using for the first time a three-dimensional GCM. The method and first results of this study are presented.

The study used the model developed at the *Laboratoire Météorologie Dynamique* (LMD) (Forget et al., 1999, 2007; Spiga & Forget, 2009). The results aim to provide new insights into the reaction of the Martian climate system to artificially introduced gases in the atmosphere. In contrast to the study of Marinova et al. (2005), the present study not only gives information about the global temperature in the new balanced climate state, but also monitors the progress and typical timescales of these climate adjustments. The main aim of this paper is to give an overview of the applied method, the approach and discuss the preliminary results obtained during the continuing work on this project.

Section 2 summarizes the model that is used. Section 3 gives a detailed description of the changes to the radiative scheme of the model to account for the greenhouse warming of the injected gases; section 4 describes the first simulations that were performed in this study and the preliminary results. An outlook to future work on this ongoing project and other work in this area is given in section 5.

2. THE MODEL

The present study uses the LMD GCM described by Spiga and Forget (2009). This model is continuously improved and regularly updated. Its horizontal resolution can vary over a broad range, including a few hundreds of kilometres to only a few meters, which makes it applicable to global and locally targeted simulations.

The model has been validated using measured data (e.g. from Viking and Pathfinder missions, from the Miniature Thermal Emission Spectrometer (Mini-TES)) and is reported to represent well a number of observed phenomena in the Martian atmosphere, such as wind patterns, regional / daily temperature patterns, dynamic phenomena such as convective motions, overlying gravity waves, and dust devil–like vortices (Spiga & Forget, 2009).

2.1. Model Structure - The model calculates the temporal evolution of atmospheric and surface temperature, surface pressure, wind and tracer concentrations, i.e. variables that control or describe the Martian meteorology and climate, on a 3D grid (Forget et al., 2007). The dependencies and interactions between the different variables are parameterized based on physical phenomena and integrated over time. The LMD GCM consists of two parts, a fully compressible non-hydrostatic dynamical core and the physical part to incorporate parameterizations for the Martian dust, CO_2, water, and photochemistry cycles. The operations of the two parts are as follows:

- a "dynamical part" contains the numerical solution of the general equations for atmospheric circulation. This part is very similar to GCMs that model Earth's climate.

- a "physical part" that is specific for Mars and calculates the tendencies due to radiative transfer, condensation and sub-grid dynamics.

The calculations for the dynamical part are made on a 3D grid with horizontal exchanges between the grid boxes, whereas the physical part can be seen as a juxtaposition of atmosphere "columns" that do not interact with each other (Forget et al., 2007).

2.2. Radiative Transfer - The radiative scheme is embedded in the physical part of the model. The radiative scheme of the LMD GCM is a classical molecular band model in which the infrared spectrum is divided into four spectral intervals. The strong CO_2 15 μm band, extending from 11.5 to 20 μm, is divided into two wide bands representing the central strongly absorbing part (14.2-15.7 μm) and the wings. The rest of the spectrum is divided into the CO_2 9 μm band (5-11.5 μm) and a far infrared band (20-200 μm). Upward and downward heat fluxes are calculated and the radiative cooling rates of the vertical layers are derived from the flux divergence. For each spectral interval, the upward and downward fluxes (both oriented upward) are computed by

$$F_{\Delta v}^{\uparrow}(z) = F_{\Delta v}^{\uparrow}(s)\,\tau_{\Delta v}(0,z) + \pi \int_0^z B_{\Delta v}\left(T_{z'}\right)\frac{\partial \tau_{\Delta v}(z',z)}{\partial z}dz' \qquad (1)$$

and

$$F_{\Delta v}^{\downarrow}(z) = \pi \int_0^{\infty} B_{\Delta v}\left(T_z\right) \frac{\partial \tau_{\Delta v}\left(z', z\right)}{\partial z} dz, \qquad (2)$$

where $B_{\Delta v}$ is the spectral-band-averaged Planck function, $\tau_{\Delta v}$ is the spectral-band-averaged transmission function, T_z is the temperature at level z, and $F_{\Delta v}(s)$ is the flux emitted or reflected by the surface. The transmission $\tau_{\Delta v}$ is a function of the absorber amount, evaluated with the Padé approximant (Dufresne, et al. 1974).

The absorber amount is computed at each time step and point on the grid by integrating the density over the depth of the considered layer. In the original model, pressure and Doppler broadening of spectral lines are taken into account for CO_2., following Rodgers and Walshaw (Berman et al., 1976; Dufresne et al., 2005).

3. ADJUSTING THE RADIATIVE TRANSFER

In the version of the LMD GCM described in Forget et al. (1999) the radiative transfer through the Martian atmosphere can be affected by the presence of CO_2 gas, mineral dust, water vapour, water ice particles and CO_2 ice particles (Forget et al., 1999, 2007). In addition to the absorption coefficients of the species already present in the model, the absorption coefficients of the four fluorine based GHGs CF_4, C_2F_6, C_3F_8 and SF_6, are introduced into the radiative scheme of the model. The absorption of these gases is a function of wavelength (see e.g. Fig. 1 for the wavelength-dependent absorption of the most potent of these gases, C_3F_8).

To implement the absorption of the four artificial GHGs, the transmission function of the atmosphere is adjusted. The new atmospheric transmission value is computed by following the approach of Marinova et al. who identified different absorption bands for each gas and fitted the band-averaged transmission data (based on laboratory measurements) to a sum of exponential functions of the column density N of the absorbing gas (in molecules m^{-2}) (Dufresne et al., 2005)

$$\tau_{mb} = \sum_{i=1}^{n} a_i e^{-k_i N} \qquad (3)$$

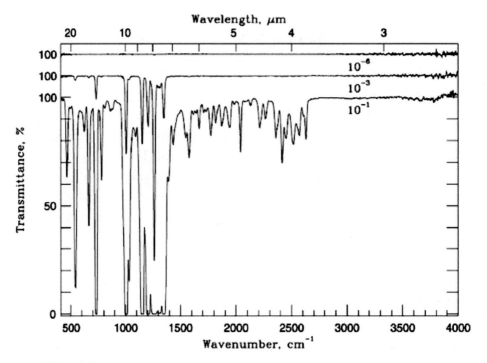

FIGURE 1

Transmission spectra for C₃F₈ at concentrations of 10^{-6}, 10^{-3} and 10^{-1}
(Marinova et al., 2005)

where T is the transmission averaged over the spectral interval, a_i is a weighing factor, k_i is the absorption coefficient (in m^2 per molecule) and n varies between 1 and 3, depending on what order exponential fit is needed to produce a good fit. N is a measure for the absorber amounts of the absorbing gas, expressed in column density (molecules m^{-2}). In the present study these fits for 68 different absorption bands are used to compute the new opacity of the atmosphere. In the model N is derived from the partial pressure, which is computed by taking a specified percentage of the total atmospheric pressure.

After computing the transmission of the 68 narrow bands, these numbers are translated into the new widebandaveraged transmission ($\tau_{\Delta v}$ in Eq. (1) and Eq. (2)). $\tau_{\Delta v}$ is derived from the spectrally averaged results of the narrow band transmission (nb in Eq. (4)) weighted by the Planck function and the width of the narrow spectral bands (Dufresne et al., 2005).

$$\tau_{\Delta v} = \frac{\sum_{i=1}^{n} \left(B_{\Delta v}(T)\right)_i \left(\tau_{nb}\right)_i \left(\Delta v\right)_i}{\sum_{i=1}^{n} \left(B_{\Delta v}(T)\right)_i \left(\Delta v\right)_i} \qquad (4)$$

where $(B_{\Delta v})_i$ is the Planck function averaged over narrow spectral band i and n is the number of narrow bands that make up the considered wide spectral band. In this approach, the Doppler and Lorentz broadenings are not taken into account.

TABLE 2

Proportional contributions of four GHGs for producing the best combination of various total gas amounts (adapted from Marinova et al., 2005)

	10^{-6} mbar	10^{-5} mbar	10^{-4} mbar	10^{-3} mbar	10^{-2} mbar
CF_4	0.0%	0.0%	0.0%	0.0%	0.0%
C_2F_6	0.0%	5.0%	10.0%	15.0%	7.5%
C_3F_8	0.0%	60.0%	67.5%	62.5%	82.5%
SF_8	100.0%	35.0%	22.5%	22.5%	10.0%

4. SIMULATIONS

In order to test the adjustments in the radiative scheme and of the adapted model, a preliminary short-duration simulation is performed in which the climate evolution is modelled over four Martian years. In this simulation the greenhouse gasses are introduced at concentrations of 0.2% (CF_4), 0.2% (C_2F_6), 0.4% (C_3F_8) and 0.2% (SF_4) of the total atmospheric pressure (so that in total these gases sum up to 1% of the atmospheric pressure). These partial pressures are taken constant in time.

4.1 Spatial and temporal resolution - The tendencies in the radiation were calculated at a temporal resolution of 12 times a day, which is relatively coarse compared with the default setting of twice per hour. The atmospheric dynamical tendencies were calculated at the default frequency of 480 times a day. The applied horizontal spatial resolution is 32 x 24 and the Martian atmosphere is subdivided into 25 layers in the vertical.

4.2 Settings of the model parameters - The model has been run with the "default" settings as suggested in the User Manual for the LMD GCM (Forget et al, 2007), except for the following parameters:

- The temporal resolution for the physical part of the model is set to 12 times a day instead of 48 times per day (the parameter iphysiq is set to 40)

- The NLTE radiative scheme is not used (callnlte = F)

4.3 Computation Details - The simulation was performed using a single processor on an Intel Xeon x5355 2.6 GHz 8 processor computer with 8 GB RAM. The simulation of the evolution of one Martian year takes about 10 hours of calculation time. The code has not yet been optimised for speed and further improvements in the simulation speed can therefore be expected.

5. DISCUSSION AND CONCLUSIONS

5.1 Discussion of Results - The Martian surface temperature profile as used by the model is shown in Fig. 2. This profile is typical for the Northern Hemisphere spring, at zero solar longitude. Temperatures vary between 150K at the poles and 280K at some regions at the equator.

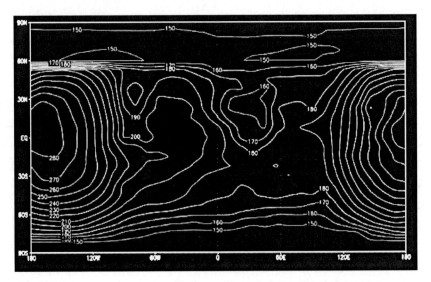

FIGURE 2

Typical Martian surface temperatures profile. (values in K, isothermal line intervals in 10K).

Modelled changes in the Martian surface temperature due to the implemented radiative effects of the GHGs after 5, 10 and 15 Martian years are plotted in Figure 3, 4 and 5 respectively. The figures show temperature changes averaged over the 5th, 10th and 15th simulated Martian year respectively. Positive values indicate regional warming in the simulation with the implemented greenhouse gases compared with the default run, whereas negative values show a cooling. The figures show that regional changes in the order of 0.01 K to 0.2 K start to evolve after only a few years of simulation. The alterations appear regionally grouped and are concentrated around the poles. The effect of the implemented greenhouse gasses does not clearly increase in time.

The evolution of the global surface temperatures in a climate with and without greenhouse gases are plotted in figure 6. The effects of the GHGs are small relative to the absolute value of the temperature and are not visible in this figure. Figure 7 shows the evolution of the temperature differences. The figure shows that the surface temperature changes are fluctuating in time at a scale of -0.05 K to 0.2 K around a value of about 0.05 K. There is no trend observed in the evolution of the climate system under the influence of the implemented GHGs. It might well be the considered time scale is too short to distinguish a clear trend. At this timescale the expected trend is small and could be masked by the observed fluctuations in the effect of the greenhouse gases.

FIGURE 3
Difference in the Martian surface temperatures between the model simulations with and without the introduction of the four artificial greenhouse gases averaged over the 5th simulated Martian year.

5.2. Ongoing Project and Next Steps - The subsequent step is to perform long-term simulations in which the Martian climate is able to converge towards a new equilibrium. When a new balance is reached, the warming potentials of the different GHGs can be determined and compared with and validated against the results obtained with the one-dimensional model by Marinova et al. In order to be able to perform longer simulations of at least at a temporal scale of about 100 years, the model's computational time needs to be optimised, which likely involves

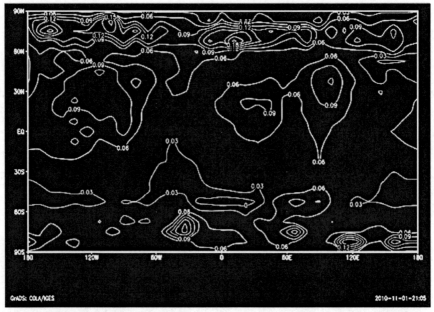

FIGURE 4

Difference in the Martian surface temperatures between the model simulations with and without the introduction of the four artificial greenhouse gases averaged over the 10th simulated Martian year. (values in K, isothermal line intervals: 0.03 K)

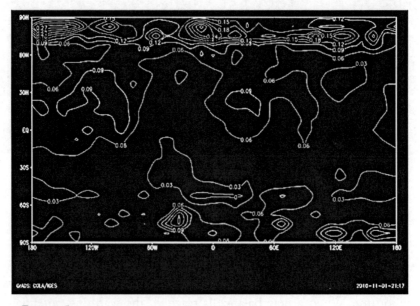

FIGURE 5

Differences in the Martian surface temperature between the model simulations with and without the introduction of the four artificial greenhouse gases averaged over the 15th simulated Martian year (values in K, isothermal line intervals: 0.03 K)

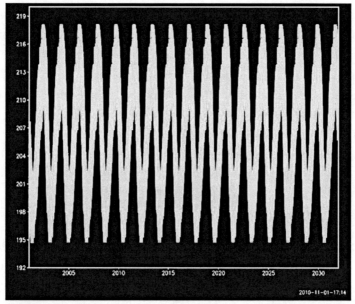

FIGURE 6

The evolution of the global surface temperatures in a climate with and without greenhouse gases. Because the effect of the greenhouse gases is small, differences between the different runs are not visible (Temperature in K on y-axis).

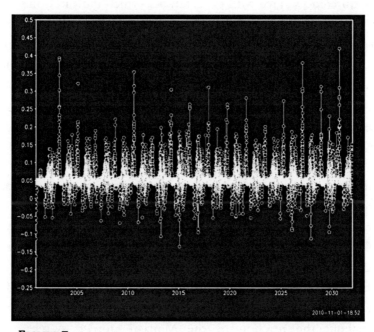

FIGURE 7

Evolution of the temperature differences over the study period and averaged over the entire Martian surface. (Temperature values in K on y axis)

also a parallelisation of part of the code. Following a full three-dimensional simulation of the atmospheric concentrations of the injected GHGs used in the

297

assessment by Marinova et al. (2005) (Table 1), these will then be used as open optimisation parameters to determine their relative concentrations for the optimal warming potential. The adopted approach is to start with the assessment of the temperature increase caused by one GHG at a time. Both, the global temperature increase and the spatial pattern of the changes need to be considered. Following this, an assessment of mixtures of GHGs is performed including, but not limited to, those given in Marinova et al. (2005). The final step is the inclusion of atmospheric chemistry and the lifetimes of the chemicals in the model simulations.

6. SUMMARY AND CONCLUDING REMARKS

The state of the art LMD GCM model has been modified to allow for the assessment of the effects of the introduction of four different strong artificial greenhouse gases, CF_4, C_2F_6, C_3F_8 and SF_6, in the Martian atmosphere. Short-term simulations have been performed to validate the modifications in the adapted model. These simulations show small temperature changes (in the order of 0.01 to 0.6 K) after a few Martian years following their introduction into the Martian atmosphere. The magnitudes and the spatial distribution of the temperature differences agree with expectations, which indicates that the adjusted model is a good representation of the chosen scenario. First simulations in the order of decades allowed obtaining insight into short to medium term trends and evolution of the Martian climate under such conditions. The changes are as expected relatively small within the studied timeframe and centered around the poles. The introduction of the four species at a total of 1% of the atmospheric pressure results in localized surface temperature changes in an order of 0.05 K within only a few Martian days. The respective concentrations of the four gases have been kept constant at 0.2% (CF_4), 0.2% (C_2F_6), 0.4% (C_3F_8) and 0.2% (SF_4) of the total atmospheric pressure and assumed homogenously distributed. The different model runs were started at solar longitude $L_s=0$ (Northern Hemisphere spring) and lasted for 15 Martian years.

Despite of the rather basic implementation of the radiative effects of the GHGs and assumptions such as a fixed lifetimes for all species, a uniform mixture of the GHGs within the atmosphere and the neglect of chemical reactions with other atmospheric species, the results of this study give first indications about the reaction of the Martian climate system to such an interference. Though conclusions about the long term changes of the climatic state of the Martian atmosphere cannot be deduced from the limited runtime of the model, this study contributes to the understanding of the sensitivity of the

Martian climate to external interferences and intends to build a foundation for more detailed and increasingly realistic studies.

REFERENCES

Arnould, J. (2005). The emergence of the ethics of space: the case of the French space agency. Futures, 37(2-3), 245- 254.

Arnould, J. (2010). Purposeful Panspermia: The Other Conquest of Space? Ethical Considerations. Journal of Cosmology, 7, 1726–1730.

Beech, M. (2009). Terraforming: The Creating of Habitable Worlds (1. ed.). Springer.

Berman, S., Kuhn, W. R., Langhoff, P. W., Rogers, S. R., Thomas, J. W., MacElroy, R. D., & Averner, M. M. (1976). On the habitability of Mars. An approach to planetary ecosynthesis. Springer.

Berman, S., et al., (1973). Towards a more habitable Mars - or - the coming Martian spring. Icarus, 19(1), 126-130.

Conley, C. A., & Rummel, J. D. (2010). Planetary protection for human exploration of Mars. Acta Astronautica, 66(5-6), 792–797.

COSPAR. (2005, March 24). COSPAR Planetary Protection Policy. Retrieved from http://cosparhq.cnes.fr/Scistr/Pppolicy.htm

Dufresne, J. L., Fournier, R., Hourdin, C., & Hourdin, F. (2005). Net Exchange Reformulation of Radiative Transfer in the CO_2 15 micrometer Band on Mars. Journal of the Atmospheric Sciences, 62(9), 3303–3319.

Farquhar, J., Bao, H., & Thiemens, M. (2000). Atmospheric Influence of Earth's Earliest Sulfur Cycle. Science, 289(5480), 756-758. doi:10.1126/science.289.5480.756.

Fogg, M. J. (1995). Terraforming: Engineering Planetray Environments. Society of Automotive Engineering. 12, 4.

Forget, F., Hourdin, F., Fournier, R., Hourdin, C., Talagrand, O., Collins, M., Lewis, S. R., et al. (1999). Improved general circulation models of the Martian atmosphere from the surface to above 80 km. Journal of Geophysical Research, 104, 24155-24176.

Forget, F., Millour, E., Dassas, K., Hourdin, C., Hourdin, F., & Wanherdrick, Y. (2007). User Manual for the LMD Martian Atmosphere General Circulation Model. May 2007.

Fouquart, Y. (1974). Utilisation des approximants de pade pour l'étude des largeurs équivalentes des raies formées en atmosphère diffusante. Journal of Quantitative Spectroscopy and Radiative Transfer, 14(6), 497–508.

Guarnieri, V., Lobascio, C., Saverino, A., Amerio, E., & Giuliani, M. (2009). Search for Life on Mars and ExoMars Planetary Protection Approach.

Holland, H. D., & Bengtson, S. (1994). In Early Life on Earth. In Nobel Symposium, 4, 237–244.

Jacob, D. (1999). Introduction to Atmospheric Chemistry. Princeton University Press.

Joseph, R. (2010). Climate change: The first four billion years. The biological cosmology of global warming and global freezing. Journal of Cosmology, 8, 2000-2020.

Houtkooper, J. M., and Schulze-Makuch, D., (2010). The Possible Role of Perchlorates in Martian Life. Journal of Cosmology, 5, 930-939.

Kasting, J. F. (1993). Earth's early atmosphere. Science, 259(5097), 920-926.

Kasting, J. F., & Siefert, J. L. (2002). Life and the Evolution of Earth's Atmosphere. Science, 296(5570), 1066-1068.

Kminek, G., Rummel, J. D., Cockell, C. S., Atlas, R., Barlow, N., Beaty, D., Boynton, W., et al. (2010). Report of the COSPAR Mars Special Regions Colloquium. Advances in Space Research. 12, 43.

Kump, L. R. (2008). The rise of atmospheric oxygen. Nature, 451(7176), 277-278.

Lammer, H., Kasting, J. F., Chassefière, E., Johnson, R. E., Kulikov, Y. N., & Tian, F. (2008). Atmospheric Escape and Evolution of Terrestrial Planets and Satellites. Space Science Reviews, 139(1-4), 399-436.

Levin G.V.(2010). Extant Life on Mars. Resolving the Issues. Journal of Cosmology 5. 87-91.

Levine, J.S., Garvin, J.B. and Head III, J.W. (2010a). Martian Geology Investigations. Journal of Cosmology, 12, 3636-3646.

Levine, J.S., Garvin, J.B. and Elphic, R.C. (2010b). Martian Geophysics Investigations. Journal of Cosmology, 12, 3647-3657.

Levine, J.S., Garvin, J.B. and Hipkin, V. (2010c). Martian Atmosphere and Climate Investigations. Journal of Cosmology, 12, 3658-3670.

Lovelock, J. E., & Allaby, M. (1984). The Greening of Mars. Warner Brothers Inc.

Marinova, M. M., McKay, C. P., & Hashimoto, H. (2005). Radiative-convective model of warming Mars with artificial greenhouse gases. Journal of Geophysical Research, 110, 15.

McKay, C. P. (2009). Biologically Reversible Exploration. Science, 323(5915), 718.

Nicholson, W. L., Schuerger, A. C., & Race, M. S. (2009). Migrating microbes and planetary protection. Trends in microbiology, 17(9), 389–392.

Picardi, G., Plaut, J. J., Biccari, D., Bombaci, O., Calabrese, D., Cartacci, M., Cicchetti, A., et al. (2005). Radar Soundings of the Subsurface of Mars. Science, 310(5756), 1925-1928.

Pompidou, A., & Audouze, J. (2000). The ethics of space policy. (UNESCO, Ed.). Unesco.

Reiman, R. (2010). On Sustainable Exploration of Space and Extraterrestrial Life. Journal of Cosmology, 12, 3894-3903.

Rummel, J. D., Race, M. S. Conley, C. A. Liskowsky, D. R. (2010). The integration of planetary protection requirements and medical support on a mission to Mars, Journal of Cosmology, 12. 3834-3841.

Sagan, C. (1971). The long winter model of Martian biology: A speculation. Icarus, 15(3), 511-514.

Sagan, C. (1973). Planetary engineering on Mars. Icarus, 20(4), 513-514. d

Sephton, M. A. (2010). Organic Geochemistry and the Exploration of Mars. Journal of Cosmology, 5, 1141-1149.

Spiga, A., & Forget, F. (2009). A new model to simulate the Martian mesoscale and microscale atmospheric circulation: Validation and first results. Journal of Geophysical Research, 114(E2), E02009.

Wordsworth, R., Forget, F., Millour, E., Madeleine, J., Eymet, V., & Haberle, R. (2010). Three-Dimensional Modelling of the Early Martian Climate and

Water Cycle. In Lunar and Planetary Science Conference. Proceedings, 41, 1–5.

Yung, Y. L., et al., (2010). The Search for Life on Mars. Journal of Cosmology, 5, 1121-1130.

Yung, Y. L., & DeMore, W. B. (1999). Photochemistry of planetary atmospheres. Oxford University Press. Zahnle, K., Claire, M., & Catling, D. (2006). The loss of mass-independent fractionation in sulfur due to a Palaeoproterozoic collapse of atmospheric methane. Geobiology, 4(4), 271-283.

VII

MARKETING MARS: PAYING FOR THE ADVENTURE

Journal of Cosmology, 2010, Vol 12, 4068-4080.
JournalofCosmology.com, August, 2010

- 17 -

Marketing Mars: Financing the Human Mission to Mars and the Colonization of the Red Planet

Rhawn Joseph, Ph.D.
Emeritus, Brain Research Laboratory, Northern California

ABSTRACT

The conquest of Mars and the establishment of a colony on the surface of the Red Planet could cost up to $150 billion dollars over 10 years. These funds can be easily raised through a massive advertising campaign, and if the U.S. Congress and the governments of other participating nations, grant to an independent corporation (The Human Mission to Mars Corporation, a hypothetical entity), sole legal authority to initiate, administer, and supervise the marketing, merchandizing, sponsorship, broadcasting, and licensing initiatives detailed in this article. It is estimated that $10 billion a year can be raised by clever marketing and advertising thereby generating public awareness and enthusiasm, and through the sale of Mars' merchandise ranging from toys to clothing. With clever marketing and advertising and the subsequent increase in public interest, between $30 billion to $90 billion can be raised through corporate sponsorships, and an additional $1 billion a year through individual sponsorships. The sale of "naming rights" to Mars landing craft, the Mars Colony, etc., would yield an estimated $30 billion. Television broadcasting rights would bring in an estimated $30 billion. This comes to a total of up to $160 billion, and does not include the sale of Mars' real estate and mineral rights and other commercial ventures.

Key Words: Financing the Mission to Mars, Colonization, Red Planet, International Effort, Marketing, Branding, Licensing, Merchandising, Real Estate, Mineral Rights, Broadcasting.

1. ONWARD TO MARS

IT IS ESTIMATED THAT THE CONQUEST of Mars and the establishment of a colony on the surface of the Red Planet could cost up to $150 billion dollars over 10 years (Day 2004; Zubrin 1996). The benefits of making humans a two-planet species and the technological innovations and revolution a human Mission to Mars would engender, would be unparalleled, with humanity the ultimate beneficiary.

Many in the scientific and corporate community believe a Human Mission to Mars and the establishment of a permanent Mars' base, will be feasible only if led by a public enterprise independent of the U.S. government. As detailed in this proposal, the $150 billion can be raised by "The Human Mission to Mars Corporation," (a hypothetical entity) if given an exclusive mandate and exclusive licensing rights by the U.S. Congress and other participating nations.

Our objective: The Greatest Adventure in the History of Humanity." Our goal: The Conquest of Space. Our battle cry: "Onward to Mars."

Of course, battle cries will not get us to Mars. It will take money. Those funds can be easily raised through advertising and clever marketing, the selling of exclusive broadcast and all media rights, the licensing and selling of Mission to Mars-related merchandise, paid commercial endorsements by astronauts, paid corporate sponsorships (The Human Mission to Mars is sponsored by...), individual sponsorships, the selling of Mars real estate and mineral rights, and the auctioning of naming rights to corporations who will bid against one another to have the Mars Landing Crafts and Mars' Colonies and Base Camps named after their companies (e.g. the Google Mars Express, the Microsoft Mars Lander) with bidding starting at $10 billion dollars.

2. ESTIMATED COSTS

Most estimates envision a Mars' mission with expenditures of less than $25 billion. For example, in 2002, the European Space Agency (ESA) proposed a joint mission with Russia which would cost $20 billion. This was a two spacecraft proposal, one carrying a six-person crew and the other the supplies. The mission would take about 440 days to complete with three astronauts visiting the surface of the planet for two months. Russia originally envisioned a manned Mars mission by 2015 (New Scientist, July, 2002).

In 2007, NASA chief administrator, Michael Griffin suggested a human mission to Mars could cost as little as $11 billion. However, NASA's vague goal would be to put humans on Mars after the year 2035 (AFP Sep 24, 2007).

NASA's current five-year budget is around $86 billion and the $11 billion estimate for a Human Mission to the Red Planet may be unrealistic.

Thus, it is possible that a two year round trip journey to Mars could be accomplished with expenditures of around $20 billion whereas a more ambitious mission involving the establishment of a permanent Mars' base would cost considerably more.

According to NASA, a single space shuttle cost around 1.6 billion dollars. Estimates are that the entire space shuttle program, since the program became operational in 1981, has cost $145 billion, with much of those costs having accrued in the first 10 years. Therefore, it could be estimated that a Mission to Mars and the establishment and maintenance of a permanent colony, with space craft journeying to and from the Red Planet, could cost around $145 billion over a 10 year period.

3. RELATIVE COSTS: CONQUEST OF SPACE VS WAR ON EARTH

Other than paying for one of the greatest achievements of all time and the technological revolution that would result, is it worth $145 billion in expenditures, over a 10 year period, to conquer an entire planet and to lay claim to the vast wealth which may lay beneath the surface? To put this into perspective, consider the costs and benefits of the U.S. war against Iraq which commenced in 2003. In 7 years, and as of September 2010, the U.S. has spent nearly 1 trillion dollars on the war in Iraq.

According to an analysis and statistics provided by The Brookings Institution's Iraq Index and the U.S. Congressional Research Service, as of September 2010, the United States has spent and approved the spending of over $900 billion to fight the war in Iraq. Whereas NASA has estimated it could cost $11 billion to fund a human mission to Mars, the U.S. government has lost and cannot account for nearly $10 billion allocated for the Iraq War and has wasted and mismanaged another $10 billion according to Congressional hearings held in February of 2007. In addition, the U.S. government paid KBR, a former subdivision of Halliburton, over $20 billion to supply the U.S. military in Iraq with food, fuel, and housing which is the same amount the ESA and Russia estimate could pay for a mission to Mars.

The contrasts are stark: $145 billion to conquer an entire planet, vs a trillion dollars to fight a war which many believe was unnecessary and accomplished nothing of substance.

With U.S. Congressional approval, the mandate of The Human Mission to Mars Corporation (THMMC), would raise approximately $150 billion to make the conquest and colonization of Mars a reality by the end of the next decade. How this can be accomplished will now be explained.

4. MARKETING MARS

Advertising increases public awareness and enthusiasm, not just to buy products (Brigs and Stuart 2006), but to attend and watch movies and sporting events (Gerbrandt 2010). To generate public demand for a mission to Mars requires that the message be repeated in a variety of mediums, TV commercial, print ad, radio ad, and online (Brigs and Stuart 2006). Over 30 major corporations spend over $1 billion each year in advertising all of which significantly impacts public awareness and increases sales (Brigs and Stuart 2006). Hollywood movie studios effectively use advertising in a variety of mediums (particularly TV and online) to successfully generate public interest in very short time periods (Gerbrandt 2010). According to Brigs and Stuart (2006), the numbers prove that the "surround-sound" approach is a big winner. They also note it is best to display the product name and logo for the duration of an ad. Likewise, the human mission to Mars must be advertised and marketed as a product and as entertainment, and must use a product name and logo to generate brand identity. Further, the marketing campaign must be targeted and tailored to those who might be the most interested in what a human mission to Mars might offer, i.e. adventure, drama, and life and death competition with clear winners and losers.

World wide, sports is a $185 billion dollar a year industry which generates much of its income from television and radio broadcasting, merchandizing, sponsorships, advertising, online and mobile media, magazines and periodicals, and athlete endorsements (Miller 2009).

The Human Mission to Mars, can be marketed and sold as the ultimate sports and reality TV extravaganza with the conquest of an entire planet as the ultimate prize. Astronauts from around the world, each with their compelling life stories, would compete against one another to be selected for the Mars' teams; Mars' teams would compete against one another to be the first to land on the Red Planet, and all astronauts would be competing against the possibility of death. Astronauts would be marketed for what they are: heroes and athletes in superb physical and mental condition.

Merchandise, from toys to clothing, featuring anything and everything associated with the Human Mission to Mars, can be marketed and sold, including official astronaut jerseys, with the names of favorite astronauts emblazoned on front and back. Then there are product endorsements by the most popular astronauts, with all income going to support and pay for the Human Mission to Mars.

It is important to recognize sex differences in behavior and thinking; a field of study this author pioneered many years ago (Joseph 1979, 1985, 2000, 2001a,b, 2002; Joseph and Gallagher 1980; Joseph et al., 1978). From a marketing perspective it is important to target traditional female interests: love, romance, and the prospect of a Martian marriage and the first baby to be born on Mars. Mars' fashions, Mars' styles, magazines and books featuring the lives and loves of the astronauts with all income going to support and pay for the Human Mission to Mars.

5. MARKETING REALITY

What could be more "real" than a Human Mission to Mars, where Astronaut heroes must overcome a grueling, competitive ordeal, with the "survivors" winning the right to face death while taking part in the ultimate adventure of all time? "Reality TV" is the number 1 money maker for broadcast and cable TV (Rose 2009). When the TV show "Survivor" was first aired in 2000, it spawned a revolution in "reality" TV broadcasting, with 51.69 million viewers watching the summer phenomenon's final episode, more than any other program that season (Kissell 2000). "Survivor" knockoffs have since been produced in every major TV market and country throughout the world.

Is the world interested in astronauts? The 1983 Hollywood film, "The Right Stuff" which featured the competitive life-threatening ordeals of the first astronauts, was among the top 12 films, variably ranked 6 to 12, among all feature films released, from October 21, 1983 through March 4, 1984 (BoxOfficeMojo.com).

Then there is the reality of the landing on Mars; the establishment of a Mars' colony; and the ultimate adventure: exploration of the Red Planet. How many might "tune in" world-wide to watch the first humans land on Mars, and what might advertisers, sponsors, merchandisers, and television broadcasters pay for the privilege of association or to broadcast all that transpires?

Lessons learned from the marketing of professional sports may provide the answers.

6. MARKETING AND MERCHANDIZING: LESSONS FROM PROFESSIONAL SPORTS

The following international marketing model, is in part based on the marketing strategy employed by the UEFA Champions League, which has become one of the world's premier sports competitions with television stations in all continents bidding for television rights (Clegg, 2010; Thompson and Magnus 2003).

Specifically, the UEFA Champions League, consists of soccer (football) teams from throughout Europe, and is broadcast to over 200 countries with a worldwide cumulative audience tuning in over 4 billion times per season. The Champions League final between Barcelona and Manchester United in May of 2009, drew an audience of 109 million, which is three million more than the 2010 Super Bowl between the Pittsburgh Steelers and the Arizona Cardinals (BBC 2010).

The UEFA marketing model has demonstrated that the key to successful income generation begins with successful marketing and a centralized marketing approach directed by a single organization (in this case, The Human Mission to Mars Corporation). The centralized organization must have complete responsibility for negotiating all media rights and for creating "brand" and "name" identity. However, financial success also require the cooperation of three key partners: Television, Merchandizers, and Sponsors (Thompson and Magnus 2003).

6.1. Branding - If the Human Mission to Mars begins as an international effort, with astronauts from countries throughout the world vying and competing to become members of the Mars' team, this would guarantee world wide interest and enhance all income possibilities. Astronauts from various countries could be marketed as national heroes and celebrities, who wear uniforms designating their home nations, thus creating national Brand identities and promoting the purchasing of Mission to Mars merchandise world wide.

6.2. Rights To Astronauts - Exclusive media rights to all astronauts and their families must be acquired by THMMC such that they can only be interviewed, appear as guests on TV shows, or provide product endorsements with paid THMMC approval. This would include exclusive right to their images and pictures, life stories, family stories

6.3. Astronauts as Athletes and Heroes - Astronauts would be accurately marketed as athletes and heroes and would be required to compete against each other, in a variety of Olympic style events, reality TV events, and game-show events, to win a place on one of the Mars teams. Mars teams would also compete against one another to win the right to journey to Mars. These competitions would be aired on TV, as sporting and reality TV events. As astronauts from nations throughout the world would participate initially, this would ensure world wide interest.

6.4. Merchanizing and Logo - This would include an official "Logo" featuring the concept of a Human Mission To Mars and a musical anthem to generate excitement, passion and feelings of prestige. The official "Logo" would appear on all products and merchandise so that buyers would know that the merchant was an active supporter, and that part of the proceeds were going to support the Mission to Mars.

According to the Sports Licensing Report (2010), licensing and merchandizing income from Major League Baseball (MLB), the National Football League (NHL), and National Basketball League (NBA), exceeds $10 billion a year. Apparel are the most popular merchandise, particularly team jerseys with the names of the most popular stars (e.g. Brett Favre, Vikings; Drew Brees, Saints; Peyton Manning, Colts) emblazoned on the front and back.

Merchandizing The Human Mission to Mars (THMM) requires that astronauts be rightly characterized as heroic and glamorized as stars and celebrities.

Merchandizing would also include THMM-related toys, equipment, vehicles, action figures, posters, magazines, books, comics, video games, puzzles, DVDs, and movie-TV-Book adaptations. Consider for example, "Star Wars" movie merchandise which encompasses everything from lunch pails to bed sheets. Annual income from Star Wars merchandise and video-games is approximately $ 1.5 billion a year (Greenberg, 2007). In total Star Wars merchandise has grossed over $14 billion in retail sales (Greenberg, 2007; Sydell 2007).

If THMMM were paid 25% of the gross sales of merchandise this could come to at least $2.5 billion a year and $10 billion in five years. If THMMM manufactured and distributed as well as sold this merchandise (thus cutting out the "middle men"), the profits could be more than $10 billion a year.

6.5. Corporate Sponsors - Each sponsor would receive exclusivity in its product area on TV, with commercial airtime spots and program sponsorship. Laws would also be passed to make it impossible for non-sponsors to associate with the Mars Mission. Further the number of sponsors would be limited domestically and internationally thereby increasing the advertising and prestige value of sponsorship. Sponsor packages would include identification on TV interview backdrops and in the VIP and press areas. This would generate a "multiplying media effect" offering high levels of recognition to the sponsors for which they would pay dearly.

Bidding for sponsorships for the Human Mission to Mars, would begin at $1 billion and could easily range as high as $10 billion per sponsor. General Motors, for example, paid $1 billion to sponsor the U.S. American Olympic team for the 2008 Summer Olympics in Beijing (MSNBC 2010). However, the price of sponsorship varies depending on if one is sponsoring an entire national team, or the Olympics in general. For example, the top 12 official sponsors for the 2008 games paid over $866 million. Sponsorships for the 2012 games in London is expected to gross nearly $3 billion of which over $1 billion will come from companies headquartered in England (Rossingh 2010).

Therefore, world wide, and given that the Human Mission to Mars would generate world wide attention for years, it can be estimated that THMMC could easily raise $10 billion if limited to 10 sponsors paying $1 billion each.

Sponsorships can also be licensed to limited time periods, such as sponsoring the initial Part 1 of the Human Mission to Mars (from inception to entering Mars' orbit), or Part 2 (the landing on the surface of Mars) or Part 3 (exploring the Red Planet). Again, even if limited to 10 sponsors for each Part of the mission, this would come to $30 billion at a minimum. Increase the sponsors to 30 for each Part, and we have $90 billion.

6.6. Individual Sponsors - Children, teenagers, and adults, from throughout the world, will want to feel they are part of this effort and they should be made to feel they are not just part, but have contributed. These individuals would influence and recruit their peers, might organize fund raising efforts and clubs, consume Mars merchandize, and watch Human Mission to Mars programming.

Therefore, individual sponsorship will also be offered scaled in terms of age-range and income. For example, children (or their parents) who contribute $10 to $100.00 will receive official certificates naming them "Mars Cadets". Wealthy adults who contribute $100,000.00 (or who raise $100,000 in contributions) would receive an official Plaque naming them a "Mars Pioneer" and would be invited to events featuring and allowing them to meet the astronauts. Those contributing or raising $1 million would be designated "Mars Ranger" and would be given opportunities to attend special events and meet with individual astronauts. If 1,000 wealthy adults from around the world contribute (or raise) at least $1 million that comes to $1 billion. It is estimated that at least $1 billion could be raised from individual sponsors.

6.7. Protecting Exclusive Sponsoring and Brand Name Rights - Using the Olympics as a illustrative example, the overall Olympics brand consists of many different elements such as official names, phrases, designs, trademarks, and so on; and the same would be true of the Human Mission to Mars. As is the case with the Olympics (and the NFL, MLB, NBA, etc), these brands would require legal protection to combat "ambush marketing" (Soldner 2010). For example, the Olympic Symbol Protection Act of 1995 provides the legal framework to protect the exclusive rights of official sponsors and to control advertising methods of competitors who seek indirect means of associating their products with the Olympic Games (Soldner 2010).

Therefore, the U.S. Congress, and all other nations who wish to participate in the Human Mission to Mars, would need to pass similar laws. Thus, all merchants and corporations who wish to be associated with the Mars mission must pay for the privilege. These laws would ensure that sponsors pay top dollar for the privilege and that adequate funds will be available to make the Mars endeavor a success.

6.8. Naming Rights - Corporations and other entities or persons, would be offered the right to bid for "naming rights" for terms lasting for Part 1, or Part 2, or Part 3 of the Human Mission to Mars. These would include "naming rights" for A) each "Mars Team" of astronauts, B) the Mars crafts ferrying the astronauts to Mars, C) the Mars landers, D) the Mars base camp, E) Mars Colony, F) Mars Vehicles, G) and so on.

It is estimated that corporations would pay up to 10 billion dollars for "naming rights". Consider, for example, the cost of "naming rights" for basketball, baseball and football stadiums. Staples paid $100 million dollars to rename the L.A. Arena, home to the Los Angeles Lakers basketball team, the "Staples Center." Minute Maid paid $170 million to name a ballpark in Houston, the "Minute Maid Park". Reliant Energy paid over $300 million to rename a football stadium, in Houston. Numerous corporation have done the same for massive exposure in a very limited geographical area. These include Citigroup, Bank of America, Wachovia, Citizens Bank, and so on, with Citigroup paying $400 million to name the New York Mets' baseball stadium "Citi field." "Named" stadiums can be found in Australia, Japan, China, Finland, Canada, Israel, the United Kingdom etc. (Badenhausen 2006).

Since corporations are willing to pay up to $400 million to have their name on a single stadium located in a single city and with limited national and world wide exposure, and on a structure which will quickly become forgotten and insignificant in the annals of history, it can be predicted they would be willing to pay billions for "naming rights" to Mars landing craft or the Mars base camp or colony as their names will go down in history and become famous world wide. Therefore, naming rights will be auctioned to corporations who will bid against one another to have the Mars Landing Crafts , Colonies, Base Camps and vehicles named after their companies (e.g. the Google Mars Express, the Microsoft Mars Lander) with bidding starting at $10 billion dollars. It is estimated that at least $30 billion can be raised through the sale of "naming rights."

6.9. Television and Broadcasting - All aspects of training, competition, selection, the journey to and landing and exploration of Mars, would be filmed 24/7, and then sold or licensed to broadcasters, documentary film-makers, and online and print media outlets. THMMC could also work with producers to cut (or direct) this footage to create 30 minute programs in the style of TV shows such as "Survivor", "Dancing with the Stars", "American Idol", as well as game and other reality-type shows.

A single U.S. American realty TV show, American Idol, brings in on average $625000 for a single 30 second ad, earning over 7 million every 30 minutes (Forbes Reports, April 2009). The Human Mission to Mars and the actual landing and exploration of the planet will be the greatest reality show in the history of humanity and it too could be packaged as a 30 minute program aired several times a week with a superbowl like extravaganza planned for the actual landing on the red planet.

Exclusive broadcast rights could also be sold to national broadcasters who would assume responsibility for distribution and advertising. Or conversely, The Human Mission to Mars Corporation (THMMC) could establish The THMMC Channel which would license and distribute live and canned footage to every television broadcaster that provides national territorial coverage. Since the astronauts would initially represent different nations world wide, this would insure world wide interest and a willingness of national broadcasters to give air time to the Mission to Mars.

The UEFA Champions League has the exclusive right to sell all TV rights on an exclusive basis to a single broadcaster per territory for a period lasting several years (Thompson and Magnus 2003). Estimates are that total broadcast income, world wide, is over $2 billion a year (2008/09 UEFA Financial Report). However, these funds to not include income from sponsorships, videos, DVDs, advertising, merchandise, or product endorsements. Further, only a fraction of the billions in annual income are derived from the U.S.A.

In the U.S.A. there is relatively little interest in UEFA Champions League as compared to baseball, basketball, and football. The United States of America is the richest nation on Earth, America will be leading the way to Mars, and America is therefore the number one market for the Human Mission to Mars. Thus, to fully grasp the income potential the U.S. market must be considered separately.

7. USA BROADCAST RIGHTS: LESSONS FROM BASEBALL, BASKETBALL, FOOTBALL

Sports is about conquest: invading and conquering territory and overcoming a determined foe. The Human Mission to Mars would be the greatest adventure in the history of humanity. Mars' astronauts would be competing against death for the greatest prize of all time: The conquest of an entire planet. U.S.A.-based broadcasters would therefore compete and bid against one another to broadcast this great sporting adventure. It is estimated that American broadcasters might pay up to $30 billion dollars for exclusive rights to broadcast this footage. Consider the following:

7.1. Baseball - A single sports franchise, the Chicago Cubs earns approximately $42 million a year just for local broadcasting rights sold to Comcast SportsNet Chicago and WGN TV and which comes to $420 Million over 10 years (Peters 2010). In 2003, Fox Sports paid baseball owner Tom Hicks, $550 million in a fifteen year deal, for rights to broadcast the Dallas Stars and Texas Ranger baseball games (Andrews Dallas Star, 2003). Collectively, over 15 professional baseball teams in the U.S. earn $256 Million a year for local broadcast rights, with the New York Yankees Yes TV network grossing over $340 million in 2006 (Jacobson 2008). In addition Major League Baseball Advanced Media earns over $400 million a year by streaming videos of live baseball games (Jacobson 2008). In 2007, DirectTV paid $700 million to Major League Baseball, for a seven year deal which allows subscribers to see up to 80 out-of-market baseball games a year. Collectively Major League Baseball grosses over $1 billion a year in broadcast fees.

7.2. Basketball - The National Basketball Association (NBA) earns $930 million a year from broadcasting rights, i.e. approximately 10 billion over 10 years (Cohen 2007; Gradizadio Business Report 2010; Schnietz et al., 2005). In 2010, CBS and Time Warner Cable Network paid the National College Athletic Association $10.8 billion dollars for a 14 year deal giving them the TV, Internet, and wireless rights to air the annual men's basketball tournament held each spring and featuring 68 college basketball teams (Szalai 2010).

7.3. Football - Between 2004 and 2006, the National Football League (NFL) received over 20 billion dollars from CBS, NBC, FOX, and ESPN to broadcast games locally and nationally, in deals spanning from 4 to 7 years (Sports Business Daily 2007).

8. BROADCAST RIGHTS TO PART 1, 2, AND 3 OF THE HUMAN MISSION TO MARS

Sports as entertainment grosses over $185 billion dollar a year. Major sports, such as American baseball, basketball and football, gross over $10 billion a year in broadcasting, licensing, merchandizing, and media rights.

A Human Mission to Mars followed by the exploration and colonization of the red planet, would be a multi-year endeavor, and broadcasting and related sponsorship rights could be sold separately for each phase of the endeavor. Part 1: Competition between astronauts, Picking the crews, the launch and the journey to Mars. Part 2: Arriving in orbit and then landing on Mars. Part 3: Exploration and colonization.

What might be the potential income of a multi-hour program featuring Part 2, the landing and first human on Mars? The American Superbowl may provide an answer. According to TNS Media Intelligence 30-second commercials during the 2010 Super Bowl sold for around $2.8 million each, totaling around $62 million every 60 minutes (with 11 commercials every 30 minutes on average). However, whereas 100 million watch the Superbowl, 2 billion might watch humans land on Mars and then emerge from the space craft to wonder the Martian surface over the following days. Part 2, the landing and first wonderings on the surface of the Red Planet, could average out to $10 billion dollars in advertising revenue over 24 hours. Then there are the sales of DVDs of this footage which could be packaged and sold within a few hours delay in real time.

Therefore, using American and world wide sports as a guide, including viewership for world wide premier sporting events such as World Cup, Championship League, and the Super Bowl, coupled with the examples from American reality TV, it could be estimated that broadcasters would pay a minimum of $10 billion for exclusive rights to Part 1, 2 and 3 for a total of $ 30 billion dollars minimum.

9. MINERAL RIGHTS AND MARTIAN REAL ESTATE

Article II of the 1967 Outer Space Treaty, which was ratified by the United States and 61 other countries explicitly states that "Outer Space, including the moon and other celestial bodies, is not subject to national appropriation by claim of sovereignty, by means of use or occupation, or by any other means." This treaty, however, says nothing about personal or corporate claims of private ownership or individual or corporate rights to extract and mine minerals and ores. Nor is the planet Mars explicitly mentioned in the 1967 Treaty.

Although the Space Treaty does not bar private ownership of "celestial bodies", this does not mean that someone can simply say: "I own Mars". Legal precedent requires possession. Consider, for example, maritime salvage law (also known as Admiralty and Maritime Law, and the Law of Salvage), which explicitly states that to claim ownership, the party making the claim must first make contact with and secure the property which must be beyond or outside a nation's national territory (Norris, 1991; Shoenbaum, 1994). In terms of "salvage" the original owner is entitled to a percentage of whatever is recovered. In the case of Mars, there are no original owners (and if there were, they are long dead and gone).

Therefore, although some may argue that the 1967 treaty bars national ownership of Mars, the treaty does not apply to private ownership. This means that those who first arrive on Mars, may claim Mars (or all areas of Mars explored by humans) as private property. They may also sell portions of this property to other private parties or corporations. What might humans of Earth pay to own an inch or acre of Mars?

Traditionally, mineral resources within national territory, belong to the government ruling that territory. Corporations and individuals must license the right to extract and sell those resources. Therefore, if those who first take possession of Mars form a government, they may claim ownership of all mineral and other resources (e.g. minerals, metals, gemstones, ores, salt, water). However, in the early history of the United States, private owners owned both "surface rights" and "mineral rights" and they had the right to sell, lease, or give away these rights.

According to the Mars Mineral Spectoscropy Database of Mount Holyoke College, a wide variety of over 50 minerals may exist on Mars. Gold, silver, platinum, and other precious metals are likely to exist in abundance above and below the Martian surface; spewed out by volcanoes, and produced by ancient hydrothermal activity and circulating goundwater which acted as a concentrater.

Therefore, once humans land on Mars, Martian mineral rights can be sold to the highest bidders, and Martian real estate can be sold by the inch or acre, with all these funds going to support the Human Mission to Mars and the colonization of the Red Planet.

10. CONCLUSIONS: ONWARD TO MARS

A succession of Presidents and NASA administrators have voiced interest in a human mission to Mars. However, they have also proposed vague, fanciful dates so many decades into the future. Even if a serious 20-year or 30-year plan were to emerge, it would have to survive for decades through multiple NASA and U.S. government administrations to ultimately succeed. Success is not just unlikely, but will be too late, as the ESA, China, Russia, Japan, and other nations are already planning on making it to Mars in the next two decades. The United States of America, the American people, and American business will be the big losers.

The Human Mission to Mars must commence now, and it must be an international effort. The conquest of Mars and the establishment of a colony on the surface of the Red Planet could cost 150 billion dollars over 10 years. These funds can be easily raised if the U.S. Congress and other participating nations, grants and enacts legislation to give sole marketing, licensing, and fund-raising authority to an independent corporation (such as the hypothetical Human Mission to Mars Corporation) which initiates and supervises the marketing, merchandizing, sponsorship, broadcasting, and licensing initiatives detailed in this article. The United States Congress and all participating nations must also enact legislation and pass laws to protect these fund-raising efforts and those who sponsor, donate to, and partner with THMMC to make a Human Mission to Mars a reality.

The sole mission of The Human Mission to Mars Corporation should be to raise $150 billion to fund a Human Mission to Mars and the colonization of the Red Planet, and this can be accomplished by initiating and following the detailed plans discussed in this article.

It is estimated that $10 billion a year can be raised through clever advertising and marketing and the sale of merchandise. Following a massive advertising campaign which increases public interest, between $30 billion to $90 billion can be raised through corporate sponsorships, and an additional $1 billion a year through individual sponsorships. The sale of naming rights would yield an estimated $30 billion. Television broadcasting rights would bring in an estimated $30 billion. This comes to a total of between $100 billion to $160 billion, and does not include other commercial ventures and the sale of real estate and mineral rights.

NASA can't do it. The United States government can't do it. An International effort can.

Our battle cry: "Onward to Mars."

REFERENCES

Andrews Dallas Star (2003). CBA: The TV Issue. August 10, 2003 Andrews Dallas Star.

Badenhausen, K. (2006). The Ten Biggest Stadium Naming Rights Deals. Forbes November 14, 2006.

BBC (2010). Champions League final tops Super Bowl for TV market. BBC Sport, January 31, 2010.

Briggs, R. and Stuart, G. (2006). What Sticks. Kaplan Business.

Clegg, J. (2010). Creating a Stage for Champions, The Wall Street Journal, May 19.

Cohen, R. (2007). NBA extends TV deals with ESPN/ABC, TNT. USA TODAY June 27.

Day. D.A. (2004). Whispers in the Echo Chamber. The Space Review, 22, 404-478.

Gerbrandt, L. (2010). Does movie marketing matter? Hollywood Reporter, June 10, 2010.

Gradizadio Business Report (2005). 5-Forces Industry Analysis. Volume 13, No 4.

Greenberg, A. (2007). Star Wars' galactic dollars. Forbes, May 24.

Jacobson, D. (2008). The revenue model: Why baseball is booming. July 11, 2008 CBS Interactive Business Network.

Joseph, R., Hess, S., & Birecree, E. (1978). Effects of sex hormone manipulations on exploration and sex differences in maze learning. Behavioral Biology, 24, 364-377.

Joseph, R. (1979). Effects of rearing environment and sex on learning, memory, and competitive exploration. Journal of Psychology, 101, 37-43.

Joseph, R., and Gallagher, R. E. (1980). Gender and early environmental influences on learning, memory, activity, overresponsiveness, and exploration. journal of Developmental Psychobiology, 13, 527-544.

Joseph, R. (1985). Competition between women. Psychology, 22, 1-11.

Joseph, R. (2000). The evolution of sex differences in language, sexuality, and visual spatial skills. Archives of Sexual Behavior, 29, 35-66.

Joseph, R. (2001a). Biological Substances to Induce Sexual Arousal and as a Treatment for Sexual Dysfunction. United States Department of Commerce: Patent & Trademark Office, January 12, 2001 #60/260,910.

Joseph, R. (2001b). Biological Substances to Induce Sexual Arousal, Sexual Behavior, Ovulation, Pregnancy, and Treatment for Sexual Dysfunction. United States Department of Commerce: Patent & Trademark Office, February # 60.

Joseph, R. (2002). Biological Substances to Induce Sexual Arousal and as a Treatment for Sexual Dysfunction. Patent Pending: United States Department of Commerce: Patent & Trademark Office, February, 2002 #10/047, 906.

Kissell, R. (2000). Survivor finale racks up phenomenal ratings. Variety, August 25, 2000.

MSNBC (2007). GM's $1 billion Olympic sponsorship. MSNBC, New York.

Miller, R. K., (2009). Sports Marketing 2009. Richard K. Miller & Associates Research Reports. MarketResearch.com

Norris, M. J. (1991). Benedict on Admiralty: The Law of Savage. § 2, at 1-4 (7th ed. 1991).

Peters, J. (2010). Buying Cubs could be a steal for Ricketts. Yahoo! Sports Sep 16.

Rose, L. (2009). TV's biggest moneymakers. Forbes, April 8.

Rossingh, D. (2010). London Olympics close to reaching $1.1 billion sponsor target, Bloomberg News, July 27.

Schnietz, K. E., et al., (2005). A look at Porter's 5-Forces Industry Analysis as a tool to assess the key success factors in the National Basketball Association. Gradizadio Business Report, Volume 8, No 4.

Shoenbaum, T. J. (1994). Admiralty and Maritime Law. § 14-1, at 782.

Soldner, A. (2010). Ambush marketing vs. sponsorship values at the London Olympic Games 2012. IMR management report. Sports Sponsorship & the Law.

Sports Business Daily (2007). NFL Media Rights Deals For '07 Season. Sports Business Daily, September 6.

Sports Licensing Report (2010). The Licensing Letter's Sports Licensing Report EPM Communications, Inc. NY.

Sydell. L. (2007). Star Wars' Merchandise Still Sells After 30 Years. NPR, Morning Edition, WDC.

Szalai, G. (2010). CBS, TW to share March Madness rights. Hollywood Reporter, April 22.

Thompson, C., and Magnus, E. (2003). The UEFA Champions League Marketing. Team Marketing, AG. Assist Magazine, February 2, 49-50.

Zubrin, R. (1996). The Case for Mars. New York: The Free Press.

Journal of Cosmology, 2010, Vol 12, 4081-4089.
JournalofCosmology.com, October-November, 2010

- 18 -
The Mars Prize and Private Missions to the Red Planet

C.A. Carberry[1] Artemis Westenberg[2] and Blake Ortner[3]

[1]Executive Director, Explore Mars, Inc., 576 Cabot Street, Beverly, Massachusetts [2]President, Explore Mars, Inc., Amiranten 12, 2904 VB Capelle aan den Ijssel, The Netherlands, [3]Project Leader, ISRU Challenge, Explore Mars, Inc.; 12 Tavern Road, Stafford, VA.

ABSTRACT

Since its beginnings in the mid-twentieth century, space exploration has been the exclusive domain of a small group of national governments. The primary reason for this was the fact that space exploration is extremely expensive and highly complicated to achieve successfully. This monopoly may be weakening and within the next one to two decades, it may be possible that the private sector will be launching private robotic and human missions to Mars. While NASA should play a substantial role in space exploration in the next few decades, finding ways to empower the private sector to play a major role in exploration should be considered a vital goal of United States space policy. This may be able to be accomplished through a series of prizes, tax incentives, and other strategies that can stimulate a major private sector commitment to human Mars exploration.

Key Words: Mars exploration, Prizes, commercial, Mars, X-Prize, private, exploration, incentives

1. INTRODUCTION

A NEW CORPORATE PARTNERSHIP WAS announced in 2008 called Virgle. Virgle was heralded as a partnership between Google and Virgin for an historic goal. As the Virgle website stated, "Earth has issues, and it's time humanity got started on a Plan B. So, starting in 2014, Virgin founder Richard Branson and Google co-founders Larry Page and Sergey Brin will be leading hundreds of users on one of the grandest adventures in human history: Project Virgle, the first permanent human colony on Mars" (Virgle.com 2008).

The Virgle website boasted videos of Virgin founder Sir Richard Branson as well as Sergey Brin and Larry Page, the founders of Google. All of them enthusiastically endorsed the Virgle concept. Google's Brin stated, "We're going to select the very first settlers of the planet Mars." Virgin founder, Sir Richard Branson, added that, "We are looking for volunteers. We believe there are people out there who would like to participate in this grand adventure..." (Virgle.com 2008).

This concept seemed perfectly plausible based on advancements in commercial space flight and the combined resources of both Google and Virgin. It would have been one of the most significant announcements in history if it were not for the fact that Virgle was announced on April 1, 2008 (April Fool"s Day).

Despite the fact that Virgle was just an extremely well executed hoax, it stimulated some very intriguing questions – most notably – would a corporate partnership or consortium like Virgle really be able to launch a private mission to Mars? There are many people who believe that a private mission to Mars is not only possible, but perhaps the only way that the United States will be able to get there (Joseph 2010). They feel that NASA has become too bureaucratic to develop an affordable human Mars mission; that a human mission would fall victim to a lack of long-term political will in Congress and cannot be carried through multiple Administrations.

However, how would such a mission transpire? There have been numerous proposals over the last fifteen years aimed at stimulating the private sector to engage in Mars exploration. These include creating a massive Mars Prize; the creation of a corporate consortium (like Virgle); tax incentives to stimulate potential "Virgles"; and/or individual billionaires who want to personally make history. All these concepts have advantages and disadvantages, but all are much more plausible today than they were just ten years ago.

Despite the advances in the private sector space industry, the main reason that has prevented government agencies and private entities from mounting such a mission is cost. Until less expensive and more efficient methods are developed, it is unlikely a private entity will be able to launch a successful mission.

2. THE MARS PRIZE

The Mars Prize concept is an extremely ambitious version of a concept that was successful in the early days of aviation. The most commonly cited example was the $25,000 Orteig Prize. This prize was offered in 1919 to the first person to fly solo non-stop from New York to Paris. Several years later in 1927, this prize was won when Charles Lindbergh made his historic flight (Randolph 1990).

Of course, this was not the first prize for exploration projects. In 1714, the British government offered the Longitude Prize to the first person who could accurately determine longitude, which led to major advances in navigation (Sobel 1996).

The key question is: Can this same concept be applied to space exploration and particularly, exploration of Mars? Over the past couple of decades, estimates for a NASA-run human mission to Mars have ranged anywhere from $150 billion to $1 trillion (Flatow 2009; Zubrin 1996; Day 2004). If this is true, it is highly unlikely that a private mission of any kind will be achievable in the near future. There are many individuals, however, who believe that a human mission to Mars can be accomplished at a dramatically lower cost if a market model is utilized.

In 1994, Robert Zubrin and United States Representative Newt Gingrich came up with the Mars Prize bill that would offer a $20 billion prize to the "first private organization to successfully land a crew on Mars and return them to Earth..." (Zubrin 1996).

At the time, this was quite an innovative concept. The Mars Prize bill predated the X-Prize by two years and few people were taking this type of program seriously. Although Zubrin's 1996 estimate for his Mars Direct plan for sending humans to Mars was $30 billion, he hypothesized that a privately developed mission would be substantially less expensive. Using a market model, it could cost $4 to $6 billion. This estimate was based on using Titan, Atlas, Delta, or Russian Energia launch vehicles. Zubrin's model also predated any of the current commercial launch vehicles that are now in development (Zubrin 1996).

Gingrich did not actively promote the Mars Prize concept for over a decade, but he also did not abandon a prize based Mars exploration program altogether. In an April 2007 speech, Gingrich proposed a $20 billion prize again which would be tax free. He noted that being tax free is extremely important because Americans do not like paying taxes. He claimed that a tax free $20 billion prize would be psychologically more attractive than a $40 billion prize with taxes. As with the Gingrich-Zubrin concept of 1994, the first team to get to Mars and return safely would win the prize. (Gingrich 2008) It is not surprising that former Speaker Gingrich revived the Mars Prize concept. Two years after the Mars Prize bill was proposed (and essentially died), Peter Diamandis and a group of other visionaries founded the X-Prize which offered a $10 million prize to the first non-government team to successfully launch a human occupied spacecraft into space twice within a two week period. Eight years later this prize was won by Burt Rutan's SpaceShipOne, which had been financed by Microsoft co-founder,

Paul Allen. In addition, over $100 million was invested in this contest by the various competing teams; $25 million was invested by Paul Allen alone (Brekke 2004).

While this achievement represented only a tiny fraction of the complexity and cost of what a Mars mission would entail, it represented a paradigm shift in what was possible and what individuals and corporations may be willing to invest in. At that moment, a Mars Prize did not appear to be nearly as farfetched. It also inspired the next step for the X-Prize Foundation with the announcement in 2007 of the $30 million Google Lunar X-Prize (Diamandis 2008).

When asked if the Google Lunar X-Prize could lead to Mars related prizes, Tiffany Montague, Director of Google's space initiatives stated, "I don't think there is any reason that it wouldn't. I do think that we need to walk before we can run. The preamble to that is demonstrating that we can send rovers successfully to the Moon" (Montague 2010).

Could this concept be applicable to sending humans to Mars? In 2008, X-Prize founder, Peter Diamandis proposed Mega X-prizes including a human mission to Mars. However, in a recent interview, Diamandis stated that it was unlikely there would be a Mega X-Prize geared to a human mission to Mars. "I don't see a Mars Mega-X Prize... An incentive prize works when there's a long-term business model and the prize can drive numerous teams to spend the money to play. A private Mars mission is likely a $5B - $10B endeavor and you won't see multiple teams each raising this level...If we ever re-invented launch technology to reduce the cost by 100-fold, then I think a "humans to Mars prize" would make a lot of sense" (Diamandis 2010).

While not likely to reduce launch costs by 100-fold, if SpaceX can deliver on its goal of dramatically reducing launch costs, it may bring a Mars mission down to the level where an X-Prize may be viable. In a September 7, 2010 email interview with the first author (Carberry), Elon Musk said he thought a privately financed mission would only cost $2 billion and that a prize would only have to be $1 billion. He added that it would take "ten years from starting fundraising to landing back on Earth." If Musk is correct, not only would hundreds of individuals in the United States alone have the means to fund such a prize, but hundreds would also be able to fund teams to compete.

3. STEPPING STONE APPROACH

Even if a government agency, wealthy individual(s), or corporate consortium does not present itself in the next few years, the Mars Prize concept is still viable. There are many intermediate technologies, science projects, and processes needed for a Mars mission that can be achieved for far less money than the Mars Prize proposed by Newt Gingrich or even a $1 billion prize that Elon Musk thinks would be sufficient. Although much attention is focused on launch systems, there are other technologies that have the potential of reducing overall launch costs. Technologies like In Situ Resource Utilization (ISRU) have the potential of dramatically reducing mission mass and costs (Landis 2007). It will also make Mars (and other missions) far more sustainable. Some of these enabling technologies include:

1) $10,000-100,000: There is real work that the private sector can engage in to advance robotic and human Mars exploration in this cost range. Example: Explore Mars, Inc. is launching an ISRU competition starting with a $50,000 prize. Other technologies that could be advanced in this cost range including agriculture research, dust mitigation techniques, rover design, and many others.

2) $100,000-$1 million: A private organization or universities can send a payload into low Earth orbit. The Mars Society built two analog Mars bases in this dollar range that continue to provide valuable information on Mars exploration (Pletser 2010). In addition, flight-ready ISRU hardware, at least for a sample return mission, can be built. This technology can potentially reduce the cost of missions and can lessen the number of metric tonnes needed to be launched into space (Landis 2007).

3) $1 million-$100 million: Numerous classes of robotic Mars missions are achievable. Prior to the creation of SpaceX, Elon Musk had estimated that he could land the Oasis lander on Mars for under $20 million (personal communication). Assuming the launchers of SpaceX and other companies continue to advance, that class of mission will probably be substantially less expensive. Mission-ready hardware for a human mission can also be developed well within this cost range.

4) $100 million-$1 billion: Major pieces of hardware can be developed and flown into space in this range. If cost estimates from SpaceX and other private sector advocates are correct, a heavy lift vehicle could be developed for under $1 billion. Solar Electric Propulsion (SEP) cargo tugs are also possible in this cost range. In addition to providing cargo service, SEP tugs could also provide on-site power for ISRU at Phobos or Near Earth Objects (NEOs). There is also the potential that a sample return mission could be achieved in this range. Despite the fact that most traditional estimates for a Mars Sample Return mission are

between $2 billion and $8 billion, utilizing commercial models and using ISRU may present some low cost options. Elon Musk believes that such a mission could be accomplished for less than $500 million in the private sector (Landis 2007; Mars 1995).

5) $1billion-$10 billion: It may be possible to send a lean mission to Mars. Numerous people believe that a private mission could be achieved within this range and that a prize could also be in this category of funding (Elon Musk believes a prize may only need to be $1 billion.) This is also the range where a more novel one way mission to Mars is quite viable from a mission expense perspective.

6) $10 billion +: This is the class of prize that former Speaker Newt Gingrich has been promoting. Assuming cost estimates from Elon Musk, Robert Zubrin, and others are correct, this level of prize should almost guarantee that commercial competitors stand a reasonable chance of making a profit. However, this level of prize would probably require government sponsorship.

4. HAS A PRIVATE MISSION ALREADY BEGUN?

One could argue that the private sector effort has already begun. SpaceX was founded partially to advance a private Mars mission. In 2001 and 2002, Elon Musk had plans to send a Mars greenhouse called Oasis to Mars. As Musk investigated the launch options that existed at the time, he was not satisfied with the existing American and Russian options. As a result, SpaceX was created. When asked during an interview if SpaceX was already running a private Mars mission, Musk said, "No, but we are slowly building the elements needed for that mission from internal cash flow. The greenhouse idea is what drove me to understand that rockets are the fundamental thing holding back humanity" (Foorahar 2008).

Despite this statement, Mars is a primary objective of Elon Musk and SpaceX. The mission has changed since 2002, as the SpaceX long-term goal is not just focused on Mars – but it seems quite apparent that Elon Musk and SpaceX is focused on getting *humans* to Mars. In *The Observer*, Musk was quite clear about the Mars goal. "One of the long-term goals of SpaceX is, ultimately, to get the price of transporting people and products to Mars to be low enough and with a high enough reliability that if somebody wanted to sell all their belongings and move to a new planet and forge a new civilization, they could do so" (Harris 2010).

Their long-term intentions were made even clearer with the unveiling of drawings of the Falcon XX heavy lift vehicle that would be larger than the Saturn V rocket. According to Tom Markusic, Facility Director of the SpaceX McGregor Rocket Development Center, if the United States decided to direct heavy lift vehicle funding to SpaceX, they could land humans on the surface of Mars between 2020 and 2025 (Harris 2010).

Regardless of whether Space X is "officially" running a private humans to Mars program, if they continue to successfully launch their Falcon 9 rocket and demonstrate that they will be able to safely launch humans into orbit at dramatically reduced cost, this could significantly advance the possibility of a Virgle-like consortium. When asked if such a consortium may be possible in the next one to two decades, Director of Google Space Initiatives Tiffany Montague said, "…there is no reason to assume that it couldn't be a reality. I think each company has to make wise decisions on what the investments are for short and long term, and I can't really speculate on what companies might end up joining the consortium, but I don't think it is improbable that the solution to long-term space exploration should come from the commercial sector" (Montague 2010).

Despite his doubts concerning a mega Mars X-Prize, Peter Diamandis is a strong advocate of a private mission to Mars. "I think privately funded missions are the only way to go to Mars with humans because I think the best way to go is on "one-way" colonization flights and no government will likely sanction such a risk. The timing for this could well be within the next 20 years. It will fall within the hands of a small group of tech billionaires who view such missions as the way to leave their mark on humanity" (Diamandis 2010).

5. TAX INCENTIVES, ADVERTISING, SELLING SCIENCE AND SAMPLES

In a proposed piece of tax legislation drafted by Joseph Webster (2001) for The Mars Society, called the "Mars Exploration Tax Policy Act of 2001," he proposed:

1) A limited exception from all federal income taxes with a carefully tailored cap for the first corporation to successfully conduct a human mission to Mars.

2) An exemption from all federal income taxes for all mission related revenues earned by corporations competing to conduct a human mission to Mars.

This bill proposed that in addition to the tax incentives, competing companies could earn revenue by selling media rights, sponsorships, excess payload and crew space (Webster 2001).

There would also be a wide range of other potential ways to raise funds through a Mars mission. Joseph (2010) proposes what he calls "marketing Mars" and the creation of an independent corporation with a mandate from Congress to sell the exclusive rights to name the Mars landing crafts and to broadcast the landing on Mars. He also proposes selling Mars mineral rights and property rights. However, would this be enough to make the mission profitable (or at least break even)?

Even if such a mission did succeed in profitably landing humans on Mars, what would the sustainability of the profits be? Unless new profit variables present themselves, it would be difficult for a private entity to maintain a profit for more than a few years, unless NASA or other government agencies began to pay the winning consortium for access to launch services, facilities on Mars, and other resources.

6. ONE WAY TO MARS

Dramatic reductions of launch costs could go a long way to drive down overall costs. However, another option discussed in the last few years is the "One-Way" option (Schulze-Makuch and Davies, 2010). Traditional mission architectures all assumed that the astronauts would return to Earth. There are a growing number of people who feel that the first mission may very well be a one-way trip (Boston, 2010; Schulze-Makuch and Davies, 2010).

According to Schulze-Makuch and Davies (2010) A human mission to Mars is so hugely expensive that is makes more sense to have one-way human mission to Mars as this would cut the costs several fold and ensure a continuous commitment to the exploration of Mars. They suggest the sending of four astronauts initially, two on each of two space craft, each with a lander and sufficient supplies, to stake a single outpost on Mars. The astronauts would be re-supplied on a periodic basis from Earth until they became increasingly proficient at harvesting and utilizing resources available on Mars. Eventually the outpost would reach self-sufficiency.

This viewpoint is also argued by X-Prize founder Peter Diamandis. "The cost, complexity and risk of round trip missions is too high...Government space mission budgets always get cut and compromised." Diamandis added, "and the science and meaningful long-term infrastructure is what gets cut out. With a one-way mission, you have to make sure you have the long-term infrastructure in place" (Diamandis 2010).

In an NPR interview, Arizona State University physicist Lawrence Krauss echoes this viewpoint, arguing that a one-way mission should be attempted, particularly since it does not require a return vehicle (Flatow 2009). Krauss stated, "From a monetary point of view, we'd save a lot because really, whenever you want to bring people back, you have to send the fuel for the voyage along on the way out... And it's not just a factor of two. It's a huge factor in terms of just the cost of the fuel and the mass of the rocket ship." Krauss states this is not a new idea. "There's a long tradition of that in human history. The explorers didn't necessarily expect to come back. Certainly, colonists and pilgrims never expected to go home" (Flatow 2009).

It should be noted, however, that explorers to the American continent knew that in most cases, there would be a plentiful supply of air, food, and usable space that did not require the cutting edge use of technology. Even with the use of ISRU technology, these basic necessities will be far more tenuous than they were for the early European settlers in America. Krauss also fails to point out that ISRU propellant production can remove or drastically reduce the need for fuel for the return voyage.

Despite the fact that the first people on Mars would face the large possibility of not surviving beyond a few years, there are still apparently many people who would still willingly take the risk. In a *Mars Exploration Magazine* article, Google's Tiffany Montague was asked if she would go to Mars if she had the chance. She replied, "Oh, hell yes! Not only would I go, I would elbow everyone else out of the way – even if you told me it was a one-way trip. Hopefully you send me with the means to make my own greenhouse, but even if it was going to be a very short abrupt one-way trip, I'd still go and I'd still elbow everyone else out of the way" (Montague 2010). Although the tone of Montague's statement is intended for comedic effect, there is no question that she would go one-way to Mars if given the chance.

The question is not can we mount a private one-way mission to Mars relatively cheaply? The question is whether it is politically and ethically feasible? The United States government would never send a one-way mission. It has become far too risk averse to even consider such a concept.

The question therefore is, would the private sector? Companies and private individuals are very much susceptible to political pressure and public outcry. Once it became known that a consortium or individual was going to send someone on what could be a suicide mission to Mars (even if the person funding the mission was the person making the trip), there would be a massive national and international debate on the topic. While this could have positive aspects, it could also present some very negative consequences, particularly if Congress and other bodies create legal and regulatory roadblocks – this could also hinder other space exploration efforts.

Another problem with a one-way mission to Mars is that there will be the limited *direct return* from the Mars travelers. There would be nothing so inspiring as Mars explorers returning from their voyages and sharing their first hand accounts with the world. This could be just as inspiring to the population of Earth as any electronic updates they may provide. As with the returning tales of Antarctic, African, and other explorers of the late 19th and early 20th centuries, returning Mars explorers will certainly inflame our passion for exploration.

It will probably be more productive to reduce overall costs. In addition to reducing launch costs, the development of technologies such as SEP and ISRU could effectively reduce the cost as much as eliminating the return vehicle – without the perception of a "suicide mission" (Cassady 2010).

7. CURRENT GOVERNMENT CASH INCENTIVES AND PRIZES

The United States has been cognizant of developments in commercial space and the success of the X-Prize. As a result, NASA created the Commercial Orbital Transportation Services (COTS) program. COTS provides seed money to commercial space companies to develop cargo services to the International Space Station. Thus far, SpaceX has been the main recipient of these funds, while Rocketplane Kistler and Orbital Sciences have received lesser amounts. (Bergin 2008; Berger 2006) NASA has also created the Centennial Challenges and the Obama administration has proposed major increases in the level of funding for new commercial options.

8. CONCLUSION

If the United States is going to lead a mission to Mars in the upcoming decades, it will almost certainly be achieved through the use of at least some elements of the new commercial model. However, we have now reached an era where the major question is not whether the private sector has the capacity to get a human mission done, but whether a traditional government program will be able to build enough political momentum to maintain a strong and steady program over more than a decade.

This is not to say that an entirely private program is better than the traditional approach or a public-private hybrid version. On the contrary, the hybrid method is probably the path that stands the best chance of mission success, but it is also subject to far more political turbulence concerning funding and the overall balance and focus of the program. In order to alleviate some of this turbulence, there must be more unity between the traditional and the "new space" companies. NASA and the established aerospace community should not fear or dismiss these new approaches to space exploration. The new space companies, and their advocates, need to recognize that there is strong value in how the traditional space community approaches mission design. Both need to think about new and efficient methods of designing missions, whether by reducing launch costs or embracing technologies like in situ resource utilization.

Even if the United States government does decide to embrace a true hybrid version or aim for Mars in a more traditional fashion, government should still create an environment that could stimulate a major private effort. If a Virgle-like consortium or a group of billionaires start seriously considering the feasibility of a private mission, that would be a good time to create major tax incentives or a tax-free prize as suggested. While NASA should play a substantial role in space exploration in the next few decades, finding ways to empower the private sector to also play a substantial role in exploration should be considered a vital goal of United States space policy.

REFERENCES

Berger, B. (2006). SpaceX, Rocketplane Kistler Win NASA COTS Competition. Space.com August 18, 2006.

Bergin, C. (2008). SpaceX and Orbital win huge CRS contract from NASA" NASA Spaceflight.com, December 23, 2008.

Boston, P. (2010). Location, location, location. Lava caves on Mars for habitat, resources, and the search for life. Journal of Cosmology, 12, 3957-3979.

Brekke, D. (2004). SpaceShipOne Wins the X-Prize. Wired. October 4, 2004.

Cassady, J. (2010). One Way to Mars? Mars Exploration Magazine, Fall, 2010.

Day. D.A. (2004). Whispers in the Echo Chamber. The Space Review, March 22, 2004.

Diamandis, P. (2010) An Interview with Peter Diamandis. Mars Exploration Magazine. October 2010.

Diamandis, P. (2008). Long-Term X-Prizes." Lecture for The Long Now, September 12, 2008.

Flatow, I. .(2009). Traveling To Mars On A One-Way Ticket. (Interview with Lawrence Krauss, a physicist at Arizona State University). NPR September 11, 2009.

Foroohar, K. (2008). Musk Makes Rockets for Stars as Tesla Taunts Ferrari on Earth. Bloomberg. May 2, 2008.

Gingrich, N. (2007). Bureaucratic Space. Lecture at the Young America's Foundation. April 2007.

Harris, P. (2010). Elon Musk: I'm planning to retire to Mars. The Observer, August 1, 2010.

Joseph, R. (2010). Marketing Mars. Journal of Cosmology, 12, 4068-4080.

Landis, G. et al. (2007). Mars Sample Return with ISRU. Seventh International Conference on Mars. 2007.

Pletser, V. (2010). A Mars Human Habitat: Recommendations on Crew Time Utilization, and Habitat Interfaces. Journal of Cosmology, 12, 3928-3945.

Mars (1995). Mars Sample Return Mission Utilizing In Situ Propellant Production. NASA Advanced Design Program, University of Washington, Department of Aeronautics and Astronautics.

Montague, T. (2010). An Interview with Tiffany Montague. Mars Exploration Magazine October 2010.

Randolph, B. (1990). Charles Lindburgh: New York : F. Watts.

Schulze-Makuch, D., and Davies, P. (2010). To boldly go: A one-way human mission to Mars. Journal of Cosmology, 12, 3619-3626.

Sobel, D (1995). Longitude: The True Story of a Lone Genius Who Solved the Greatest Scientific Problem of His Time. New York: Walker.

Virgle.com (2008).

Webster, J. H. (2001). Mars Exploration Tax Policy Act of 2001. unpublished white paper, 2001.

Zubrin, R. (1996). The Case for Mars. New York: The Free Press.

VIII

VOLUNTEERS FOR A ONE WAY JOURNEY TO MARS

- 19 -
To Boldly Go to Mars:
Volunteers for a One Way Journey, From Around the World

Over 1000 Volunteers for a One Way Mission to Mars

IN OCTOBER OF 2010, the Journal of Cosmology received an unexpected and surprising email. A 70-year old man was volunteering for a one-way mission to Mars. Emails from over a dozen hopeful volunteers were received the next day from men of all ages, and the pattern continued until November when a flood of emails were received, over 80 in a single day. Teenage boys and old men, firemen, police, military personnel, engineers, and college students from around the world were all eagerly volunteering for a one way mission to Mars, and detailing their qualifications. Emails were even received from women volunteering their husbands.

But why?

> "I didn't know what to make of it at first," says Dr. Lana Tao, executive Director of the online Journal of Cosmology. "True, we had just published a special edition, The Human Mission to Mars, Colonizing the Red Planet, but we had not solicited volunteers and there was nothing in any of the 55 chapters we published as part of this special edition which would make anyone think we were accepting applications."

> "And then it dawned on me. They were responding to chapter 12, To Boldly Go: A One Way Mission to Mars. Sure enough, Dr. Davies and Dr. Schulze-Makuch, the authors of this chapter, were also receiving emails from hopeful volunteers."

> "By December we had received over 400 emails from men volunteering to go to Mars, and then, on January 10, the flood

began again. At least one an hour from men, women, and boys, from around the world. Yes, even women were volunteering."

But Why?

"A journey to Mars would be the greatest adventure in the history of humanity" says Dr. Joel Levine, Senior Scientist Science Directorate NASA, and editor of the "Human Mission to Mars" text.

As summed up by Dr. Edgar Mitchell (the 6th man to walk on the Moon) "As a species we have always had an incredible curiosity and because of it the thought of exploration and exploitation of new frontiers has always excited our imagination and motivated our efforts. We now stand on the threshold of becoming a space faring civilization."

The following are just a sample of the over 1,000 letters from volunteers the world over.

****

MY NAME IS MARK. I am currently completing high school in an international scholarship school in Italy, the United World College of the Adriatic. I came across the invitation for the 'A One Way Mission to Mars' publication and felt inspired. I hoped to write this piece in the perspective of a somewhat younger generation, particularly as an aspiring scientist. This may not apply for your publication, but I nevertheless enjoyed writing it, and look forward to the final version.

In 1990, the spacecraft "Voyager 1 turned its camera back towards earth and took a picture. In that picture, there is a tiny blue speck, filling less than an eighth of a pixel. That is Earth. It is a perspective of Earth, and of our existence as a species, that we do not often get to view. Carl Sagan reflected on this image and appealed to human humility, that we might take a moment and realize that we are truly on an isolated rock in cold universe and make the effort to act accordingly. Yet while this image is humbling, it also demonstrates why we matter, and why this voyage matters.

To the extent of our knowledge, this blue rock of ours is the only one that contains sentient life. Within the vastness of space, amongst every growing and dying galaxy, we have a unique position. We are a particular phenomenon with a particular opportunity. We can voyage outside of our little moment in space and time.

We have done it before. A few brave souls have crossed the expanse and walked upon the Moon. How many people has that mere act inspired? How many children looked up at the night sky and wondered which twinkling light was next?

Now, more than 40 years later, we have gone no farther. We plucked the low-lying fruit and stopped there. That wonder has dwindled in my generation. We are told that anything worth seeing is too far away, any trip to expensive, that the risks of any voyage are too great. We are content to stay put on our pale blue dot.

But the first step on Mars' red rocks will kindle the wonder that has driven our species to accomplish so much. We will learn a lot from our time there, and not necessarily about the planet. Those moments on an alien planet will be a reminder and a resolution. A reminder that we are not bound to this ground. A resolution that we will always push forward, always make the steady advance to realize our particular place in the Universe.

Those few who undertake that mission may make the ultimate sacrifice. They will be remembered as the best, brightest, and bravest among us. That sacrifice will not be made for mere data, but for the rest of us. They will show us it can be done, that the oft-dreamed-of voyages are not as far away as we had thought.

From then on, we will look towards the sky and not see distant points of light, but a map. It will not be one traveled by my generation, or the next, but we may pave the way for one many years hence. They will be able to look back at our generations and those first, bold steps. They will know the risks taken, and the costs paid.

That generation can look back at this pale blue dot and remember this humble beginning of ours. It will no longer be an image of our isolation, but the affirmation that it is humanity's determination, curiosity, and ingenuity that will carry us into the stars.

****

I AM A 35 OLD GUY. I think it's realy important finding some new places to live in, because the earth has a limited area wich are possible to live comfortable on an the population of humanity grows even faster every year. Personally i think that, even if it's possibly will be tecnically realistic for mankind continue living on the earth, we still wery much need more space to make it possible live comfortable, for example feel the freedom in walking alone in a pice of untoced nature (even if it's once was created in an artificial way or is a land of stone and dust like the original surface on march of today, but it's still an extremly amazing lanscape created by nature it self and i think there will be no problems joining it outdoor as much as those we see here on earth) and i think when people feel good and comfortable with their lives there will be much less of war and violence amnong os. So this is not only a question for me as a single person, it's an important question for the next generations and we hawe to begin somewhere so thay will hawe a good starting point. I dont see the one way traveling as a sacrifice, but a way that might lead to opportynitys for the next generation as well as a more challening and interesting live for my self during being a part of the mission to build up av new world there, i think at least i can do something that possible matters.

Further and perhaps most important is that I realy hope that if we can make it on March, we can go further and possibly live on other planets in other solar systems in the future.

****

BEATRIZ, 28, FEMALE, EARTHLING. Spanish by birth. Alien by US Department of State.
To live an unexpected, endless adventure.
Not to return,
no expectations,
no fame no glory.
To see never-ever seen images.
To satiate inner curiosity.
To escape an incomprehensible society.
To die having seeing beyond the stars.
To enjoy absolute solitude.
To have time to reflect, to introspect.
To have time to enjoy memories of the past.
To have time to relive the first kiss, the first embrace, the first love.
To walk, sing and dance.
To write about feelings and passions.
To scream of joy and fear.
To cry and laugh out loud.

Child-like creative minds with curious naive eyes, child adults, passionate artists and lovers, sensitive souls living with intensity should be sent there. For only these would truly appreciate the beauty of the unknown.

****

I BELIEVE THAT HUMANS should explore the universe as a way to ensure our survival as a species. Our world has limited resources and natural disasters are always a risk. A mission to Mars is the beginning of a New Era for humankind. We are not meant to be confined in this planet. We must care for it dearly, but the universe is also our home, and we need to explore it. Space exploration will, not only, guarantee our survival but allow us to learn even more about the universe and our existence.

I would love to be a part of this mission and be one of the first men and women who will begin the colonization of Mars, and prepare the way for future generations of explorers. This will be the biggest adventure of humankind. A unique event in history.

The fact that it will be a one way trip makes it even more exciting because i'll be able to spend the rest of my life preparing a new colony for the world and improving the knowledge of our universe.

Being away from family and friends is always difficult. But I believe they'll understand what this voyage implies and the meaning and honor of it. The opportunity of meeting and working with new people, with this common goal, is also very exciting. If I could, I would love to take my family with me, I wouldn't mind raising a family in this new world. This is clearly the next step for our species and our evolution.

M.K.D. 27 years old, male, from Brasil.

****

AT AN EARLY AGE I HELD a deep fascination of the cosmos. Inspired by popular media titles such as *Star Wars, Star Trek, Dune,* and Macross, I began to ask questions. What is life in space really like? Are there other planets out there which could support life? Where can we find life if it does not exist in our Solar System? Is the fate of our species to remain isolated on Earth or are we destined to spread our existence to the far reaches of known space? The answer to this last question is obvious.

Humans are a naturally curious and innovative species. Tens of thousands of years ago, our ancestors required a way to keep warm in the cold northern hemisphere. The creation of fire and clothing granted us this. When we desired a method to identify the perfect time to pick and gather our crops, stone calendars marked the position of the sun and stars to aid us. As we stared out at the vast ocean, we wondered what was on the other side. We crafted ships and sailed to the ends of the Earth. Not only did we find the Americas and amazing new cultures, we also discovered our world was round, changing our perspective of the place we call home. Our desire to master the secrets of the unseen led us to develop a method of splitting the atom. The list of our accomplishments goes on and on.

In 1969, the human race went one step further and set foot on the moon. Neil Armstrong was right in saying it was "one small step for man, one giant leap for mankind". Our race has entered the Space Age. The time has come for us to begin stretching our existence beyond our tiny planet. To do this, we *must?* colonize Mars.

The red planet has served as a focal point for nations in the last sixty years. The knowledge that the first nation to set foot on a foreign planet would be forever remembered in history is indeed very attractive. However, I want you to think about another possible outcome from the exploration and colonization of Mars. The Human race would no longer be a species bound to a single planet. We would emerge as an inter-planetary species. In time, Mars could be used to mine minerals rare to Earth, develop new fuel sources, grow large quantities of food, create new medicines, and serve as a platform for the exploration of the outer-planets and neighboring star systems.

In conclusion, the colonization of Mars is not critical to balance of power in the world, but to the survival and advancement of the Human race. If ever offered a ticket to become one of the first colonists in this grand expedition, I would not hesitate to say yes. To do anything less would be to deny the human spirit of its never ending quest to explore and understand the universe.

J.P.P. Age 24

****

UNLIKE MANY APPLICANTS PERHAPS, I like this planet. It's been a nice home. I've been to all of the continents, even Antarctica. I only mention that to help you better understand my position. It's just that at this stage of my life I feel I can do more. I'm sixty-six. What better way to wind down my life than to do something extraordinarily beneficial for mankind by helping him to understand another world? And in doing so to perhaps better understand his own.

When I was younger, I wouldn't have dreamed of volunteering for a one-way mission to Mars. Being on the cusp of what most people would consider old age I see things differently. My wife passed away years ago. I have no close family. I am, however, in possession of excellent health. I run regularly, and my favorite participatory sport is Orienteering. The friends I've told about volunteering are at best supportive, and at worst, understanding.

Other than my science background, all I have to offer is a somewhat dubious talent for writing and art. I don't know if that would be important in a decision to include someone such as myself, but I hope so. The men and women who are chosen should be exceptional only in the ability to endure, and to care for one another. They should otherwise be representative of all of humanity.

The suggested topics we are to write about the importance of Mars, who you would want to bring, what do you hope to accomplish are either self-evident or dependent upon developing technology. Mars is the closest planet. It is the most like Earth. It's natural that we should go there first, and, if possible, establish the first colony. What could be accomplished is a matter of the tools we are given. The main goal of the one-way trip is to establish a viable colony, little else. But I assume that there will be other work done: astronomy, geology, biology and psychology, to name a few. It would make little sense to found a colony, and have the inhabitants just sit there for the rest of their lives. I'm assuming there would be frequent contact with Earth for the colonists, if desired. But I see that becoming less of an issue as time goes by and the colonists begin to identify more with Mars, and less with Earth. It will never cease completely, though.

I am under no illusions about going. I am too old. The mission is at best twenty years in the future. I do regret that I was born twenty years too soon, but only to a point. I'm happy to have lived when I could go in my back yard and see the first Sputnik overhead, or to have been at Cape Canaveral for the launch of Apollo 11, and among the crowd early in the morning at Edwards Air Force Base to see the landing of the first space shuttle. If I'm still around when this mission goes, and still have control of my faculties both mental and physical, I hope you'll save me a seat.

T.S. Male, 66, U.S.A.

****

341

MARS. IT IS MUCH MORE than a barren red planet hosting a plentiful bounty of challenge. It is a planet with tons of precious metals and ore lying within its boundaries waiting for life, extraterrestrial life, to seize it. But metals and ore alone do not make it important for us extraterrestrials to imprint our civilization upon Mars. What makes it important for us to go to Mars is that it's been an icon uttered in ancient times now revived over the past century making us question, making us draw out our minds from the intellectual slumber and awaken new ways of thinking. Now that we are awake those questions may start to be answered. By traveling to Mars and solving the challenges she provides not only will make us a greater, more mature civilization, we will grow into yet other ways of understanding everything around us and our place in an infinite future. That, above all else, is what makes Mars important to all.

Some may weigh the social impact of a journey to Mars as being too great. Leaving loved ones and close friends behind would be tough (not to mention the inherent excitement found with hiking, games and so on). But I believe it is worth the risk when you imagine yourself on the other side of the looking glass. Undoubtedly technology would continue to evolve making it easier to interact with everyone. Besides, how cool would it be to have a Facebook page while on Mars?

But a journey to Mars requires pioneers willing to risk their lives to surmount unknown challenges that will allow civilization to follow. I would go, not because I consider myself a pioneer, but because I like challenges that stimulate creativity and problem solving. Sure, a journey to Mars could be a one-way ticket or life could be extinguished before even making it to Mars, but striking out into the unknown, adapting, and flourishing is what makes life worth living. And, assuming the journey to Mars was successful, I would work to establish an outpost and expand it for others who would follow. Then I would explore the region, analyze it and put it to good use. What kind of use? I have no idea but it would be something cool…maybe constructing a big sign "Eat at JM's" with an arrow pointing at the first restaurant on Mars, and a special deal on Mars dusty drinks. Now that sounds tasty!

Let us rise to the challenge, gain understanding, and make our civilization better. Let's go to Mars!

Joe, 40, USA

****

I AM CURRENTLY A SOPHOMORE in college, but when the time comes, I intend to be at the forefront of this mission. I wholeheartedly believe that this proposed mission to Mars is the beginning of the future of our race. My stepfather pointed me in the direction of your article after I had just explained extensively that this is what I desire for the future of our kind. I believe this to be an amazing opportunity just waiting for us to take hold. The possibility of not only establishing a colony, but having the space and distance from Earth to conduct research. The technology that the masses have been taught to fear from science fiction novels, could be developed and worked on without endangering the whole of our species. The possibility for expansion, the isolation providing focus, this is what I intend for my future. I wish to travel with my husband, and start a family, possibly a next step in the evolution of our species. I wholeheartedly volunteer and pledge my passion and life to this cause.

****

I KNOW THAT I AM ONE OF THE HUNDREDS if not thousands who wants to take part in it, but I strongly believe that I may be helpful for the rest of the team.

First of all I am from Europe - nice touch to the colonists' demography. Whats more - I am a member of the country who gave Copernicus to the world of astro science. I am a young (22 years) Video Production graduate. I would love to produce a series of films and photographic reports after we will land on the surface of Mars. I am a very social person, with great communication and group-work skills. Also I am an IT fanatic and a quite nice off-road driver.

If in 20 years, you will really be ready for this trip- you can count on me.

****

MY NAME IS T. W. I am 28 years old and have spent the last 8 years in the space world. I have worked with satellite systems as well as Missile Warning. I also spent last 4 years as a launch control officer for the shuttle and all other launches from NORAD. I would love to try to prove myself to have a chance to be the lucky few to become part of the Mars journey. If you think I might have a remote chance please get back with me and I can get into further details about my background.

Oh ya I am also a jack of all trades from making toast to ice to even a PB and J. Sorry this is my personality. Life is all about risks and laughs and being serious at times.

****

I'M NINETEEN BOY WHO IS INTERESTED in travel to Mars. I'm a loner, if I lived by the rules, which I confess - I lived as a hermit in the woods. I live in Poland, in a small town. Stupidity overwhelms me today's society, which invests in the economy, poor production of material goods. Do not focus on the technological development of the world, which stands in place by the stupid system of values of all humanity.

****

HI, I KNOW YOU'VE PROBABLY received a very large amount of applications from people volunteering themselves for the one-way colonization mission to mars but I would like to put my name in the pot so to say. There are many reasons why I believe I am qualified for this missions including, but not limited to, the fact that I am extremely independent, driven, focused in my work, technically savvy with both fixing machines and thinking outside the box to bring solutions to the table, and a deep interest in being more than status quo. But like any story, there are always two sides. My lack of real world experience and mild asthma would likely disqualify me for any sort of exo-atmospheric flight, which I believe to be a load of bollocks considering what I know I am physically capable of. In space it is more important to have a cool head than the ability to lift 2x your body weight on earth, but I digress. Thank you for your time and I look forward to any response you might find the time to respond with.

****

I WOULD GO TO COLONIZE THE PLANET Mars. I am 45 and nearing the appropriate age and also I feel a registered nurse would be a great assest to the project. I would like to be considered for the project when the time comes.

****

WHAT AN INSPIRATIONAL ARTICLE I have read on your pages! Not only do I agree with the article about a one way mission to Mars being highly feasible, I am willing to volunteer to do it (I'm 43 years old, in good health, and have already had 3 children). I believe strongly that the only way to improve life for all species is to get off Earth, and to learn to live in space (artificial habitats) and other planets. Whilst the technology is still a bit out of reach, I think it is viable using electromagnetic and electrostatic shielding to prevent too much radiation harm occurring to astronauts. I note that it is not uncommon to have gold shielding satellites, so I wonder if it is possible to use plastic insulated gold wiring (finely woven) instead of sheet gold, with a current of electricity applied to create an electromagnetic and electrostatic field around the crew areas? From my understanding, electric currents have an electromagnetic field at 90 degrees to the direction of flow, which may be used to deflect cosmic radiation in a similar manner to Earth's field. If gold is not the best for this, I submit that there are other materials which could be used to attain the desired effect. It is the purpose of life to spread, and we are potentially the best vehicle for that to occur.

****

I WOULD GO ON A ONE WAY trip to mars. i am 45,run daily,have a reasonably high i.q.,and would love to "boldly go where no man has gone before" it seems soothing the way the ocean is, space i mean. i would miss plumbing and all, but people are very adaptable... as long as earth could ship some books occasionally......

****

HELLO, MY NAME IS A. R. I'm a 21 year old, currently a junior in college, and I'd also like to volunteer for the one way mission to Mars. I recently read an article about people doing so, and I would hate to sit idly by and not offer myself up for such an opportunity. I'm quick with computers, get along with all manner of people, and though I have no specialties inherently beneficial to the mission I would be willing to drop everything to learn. Lagged communication wouldn't be a problem as I rarely utilize it, other than texting of course. Being cooped up in a ship and on a biodome also wouldn't bother me as I am a night owl and typical gamer. Earth would be missed. But not crippling so. I'm also certified in SCUBA, which i believe is a requirement of NASA for such endeavors.

Where do I sign up?

****

I AM A 20 YEAR OLD MALE social sciences student of Ottawa University, Ottawa, Ontario, Canada. My ethnicity Vietnamese. 5'11", mesomorph, non-smoker, occasional drinker. My IQ rating is 135, whatever that may mean to you. Interests include video games, music, and fiction. I want to be volunteer because of a couple of reasons. Should the expedition be either a success of a failure, my name should go down in human history as a member of the team that left Earth. My own share of fame you could say. My more adventuress side tells me that this would be an amazing experience. To settle on a new planet which would be my new home is definitely a scary leap for me, but I've decided to take it. But who knows, maybe Aliens will come and tell us, it isn't humanity's time to leave our planet yet, and zap us all, and that would not be cool.

****

I AM A POLICE OFFICER from NYC and would volunteer for a one way mission to mars. If I could convince my wife to come of course. Sign me up and I'll work on her.

****

I WANT TO VOLUNTEER MY HUSBAND, Brad. We were high school sweet hearts and he was a super athlete, played varsity baseball, football, and soccer. Then he got fat and just sits in front of the TV, angry all the time, drinking beer. Mars would be perfect for him, and would get him out of the house. Take him, please!

****

I'M 13 YEARS OLD, and I'm a South Korean. I was selected at the school talented class, and I am a good computer user. My dream is a astronomer, and I'm usually thought of the new world, the big world in the universe. How tiny we are, how many misconseptions are actually different with the truth. I with I would go to the Mars expectation after 20 years, I would be able to try great things on Mars. IF I'm be able to do that, I'm okay. Even I have to miss my parents, friends, and more, I'll be not regreted for this decision. Please select me for the Mar's Expeditation. PS. My height is 161cm, 50kgs

****

MY NAME IS JAMJUREE, and I am from Bangkok Thailand. I'm a healthy 35 yr old single woman of Thai/American nationality and an very sexy and beautiful. I strongly support this mission, one way ticket to Mars. I'm a radio DJ for Virgin Radio Thailand, and I know how to operate a radio and would be very good company. I volunteer to be a part of this mission!

****

HEY, ABOUT THE ONE WAY TRIP to Mars. I'd like to go. I think I would be a great candidate because I could help people cope with the depression, I'm continuously optimistic and happy. Also, because I have nothing to lose. I realize this is late, and your probably already full. But if you can make an acceptance, please do. My name is Bryson Corbett and I'm currently 18, I grew up in a small town in Kansas. I currently live in Anchorage, Alaska in a Transitional Living Program called Rights of Passage. To get in, you have to be homeless and between the ages of 18 and 21. My mom moved me up here in February 2010, then kicked me out shortly after my 18th birthday in June. I lived with some family members I didn't know until November. In September my mom moved to Arizona and currently lives there now. I'm up here alone, I've got no one. It would be exciting and a wonderful experience. I'm aware that I would end up dying on Mars. It doesn't matter. If you need anything else, please contact me.

****

ALTHOUGH YOU NEVER EXPLICITLY invited submissions, I would like to register my interest in the opportunity to participate in a human mission to Mars. I recognise you are already unexpectedly oversubscribed, but if you could add me to the tally, I would be incredible grateful! I am a female, 30 years old, 49kg, healthy, well-educated (1st class MA with honors in Economics & Econometrics) ..and incredibly enthusiastic about all things astronomy and cosmology!

I am in a full-time, well-paid job.. but would leave it all behind me for participation in this programme. Which - to be clear and so you do not misinterpret my assertion, is not a whimsical decision - rather something I consider a very small sacrifice for the opportunity of a lifetime - of a generation, to be among those that are the first of mankind on such an expedition.

****

HEY--I'VE BEEN READY FOR THE MOON or mars for 30 years. 58 year old female with multiple health problems, so I'll never get to go. But, have ALWAYS looked to the stars. I read one time that there are two kinds of people in the world--those who look to the stars and say, "hell, yes"; and those who always look to staying on earth and say, "never". There a very few people who are uncertain about the answer to the question "Would you like to colonize another site?" We're either hard rock certain it's a fantastic, desirable goal, or we consider the people who want to go as nuts. I'd take a one way trip in a heartbeat.

My 28 year old son emailed me this link--glad you at least asked the question. I don't think you'll have a recruiting problem if it ever comes to fruition. I'll yearn for it till the day I die.

Sincerely, J. M. H.

****

I WOULD LIKE TO VOLUNTEER as a crew member for a one-way (suicidal) mission to Mars. As an honorably-discharged veteran in Electronics of the U.S. Naval Submarine Force, I served on a submerged mission aboard a Polaris Class S.S.B.N., and surface craft, and also as a P.A.D.I certified diver, I have experience in some of the same training astronauts must undergo. I am a Red Cross certified First Responder. I have training in many fields of science including Electronics, Chemistry, Biology, Botany, Physics, Psychology, Internal Medicine, and Astronomy. Although I have not achieved an academic degree in any field, I am preparing to return to University to do so. I have been an an enthusiastic spectator and close follower of NASA space travel ever since the Apollo missions, when I was a child. I am a 48 year old Caucasian male in above-average physical and psychological condition. I am a U.S. natural-born citizen currently residing in Greenville, NC. I am well-read, intellectual, conversationalist, an adventurous traveler, and courageous - yet try to remain humble and sincere. I make friends easily, have empathy, compassion, and understanding for others, and I have experience in Biblical religious lifestyles including deep and loving human interactions. I take direction and learn easily, having a recently tested I.Q. of 160 plus. I am a single male, heterosexual, never married, never fathered children, and have no commitments on Earth that would cause me significant distress to leave behind forever. Sign me up for the adventure of a lifetime! Sincerely; J.L.

****

MY NAME IS SIMON L. and I am 27 pushing 28, and I wanted to write in support of a one-way trip to mars for human explorers. I would volunteer for this mission, I have an extensive understanding of computer systems and software design and feel i could not only help develop the software to get us there but also help develop the software needed to keep things running once there. Helping design the software and being one of the colonist has the advantage that i can keep it running once there as well as make adjustments/improvements over time.

I have always been interested in all things science and spend several hours a day reading up on new discoveries and modern theories and feel i have a strong understanding of many topics and would be of assistance in the preparation of the project, especially when it comes to thinking "outside the box." I am very good at thinking quickly and making snap decisions, which would be helpful when dealing with life and death on a daily basis like those who first settle on the red planet.

****

MY NAME IS TOMASZ S. I'm 26 years old computer scientist and software developer with BEng degree (The Bachelor of Engineering in Computer Science).

I also had graduated Secondary Technical School with degree Master of Electronics and I can say that my passion is Create and Space. I have a great wife and three months son right now. I'm well build, strong and healthy, 6'2" tall. My IQ is 138 called as "inspired inventor" so not so low and not too high, also for social behavior. In this case I dont't have problems when I'm alone for a very long time or when I spend time with other peolpe, if there is a posibility I'm very sociable person and always helpfull. I don't really scare of death because this is invariant, no one will live forever, so there is no difference that I will die on Mars on on Earth - I'm not so religious, I'm rather an atheist with Christian roots, but I respect life and religion of other people and I don't meddle with it.

Now I'm building my house with my own hands, also I was build my own solar panels, wind turbine etc. I'm very excited when I do create "something from nothing" and I always want to find a clue and solution.

I can also create and program microcontrollers in low level like assembler or even machine code/opcodes in hex editor - I've a very good practice with virus/antivirus software research, so debugging in low level asm codes it's my one of hobbies. I always want to work in NASA or other space agency, make space explorations, but I think my Polish roots make it impassable.

I think that my person will be the best complacement for the mission and crew. My advantage, fresh mind, software, electronic, electric and building knowledge can be very usefull for mission and adaptation on Mars. I'm ready to go to Mars. Please, sign me up.

Even if you have a lot of volunteers better than me to go to Mars, please sign me up as supporter, planner or problems resolver, I'm good with it.

****

I FULLY SUPPORT A HUMAN MISSION to Mars and would go so far as to even volunteer myself. There isn't a location on Earth that isn't owned by someone and the idea of vast land that has never been stepped on before is inspiring. We need to take advantage of our technology and capabilities and expand into space. Sure, the European settlers could have never sailed out to America, but would anyone take that back today? It is hard to convey to someone, why a human mission to Mars is necessary, but that is because it is such a large idea with endless possibilities and people are not confined to their daily lives and society to deem it necessary. There are those of us who think bigger. Mars is freedom, Mars is the next frontier. I would give up everything to step foot on Mars, but will be satisfied, none-the-less, knowing that someone is making those crucial steps into the future of humanity.

Eric L. Age 20

****

I VOLUNTEER MY LIFE to a Mission on Mars. I believe everyone has a purpose. Most human beings eat, crap, and die. But what if we all fulfilled a significant purpose while we lived. Mankind is doomed to destroy itself. The human race will continue to exist longer if it existed on other planets. My name is Benny Rosario, I'm 40 years old, I'm physically fit, nothing wrong with me, I'm currently in the military, and I love space science.

****

MY NAME IS ROBERT G., and I have dreamt of leaving this planet since I was a child. I have no training whatsoever in any area that would be related to a colonization mission to mars. But what I do have is a burning, almost unbearable, desire to explore. To "get off this rock". I know I would probably never be chosen for a mission like this, given my lack of any useable skillset. But I hope that my boundless enthusiasm and desire to learn and explore everything I can in my short, little life will be enough to merit at least consideration for a colonization mission.

****

MAY NEED SOME KIND of security up there to settle disputes between people and keep the peace. I've been in law enforcement for 21 years. Would love to go and be a part of this great adventure!

****

I AM NOT CLASSICALLY TRAINED in the sciences, however I have a better than average understanding of the science and psychology required for such a mission. I think my life circumstances have prepared me for the isolation and loneliness, and as I understand it the mission would require someone able to function well under such circumstances, though I also note that modern communications would make it less of a lonely experience, though still isolated.

I would be happy to submit myself to tests to determine my aptitude and suitability for such a mission, I understand the need to have someone who can handle the mission parameters of a one-way journey without looking at it like a form of suicide, but also someone who is capable of making that mission a success.

I don't just consider it a good idea, but essential, that we learn as a species to take advantage of space with renewed urgency. I think that if you look at all life, it is easy to see that life's purpose is to spread as far and wide as possible to increase its chance of survival, we are the logical vehicle to make that a reality.

I am not under any illusions that such a mission would be fraught with danger and a high risk of death. I would be a liar and a fool if I said those prospects didn't frighten me, however I still think it is worth the risks, and I am undaunted at the challenges such a mission would entail.

Even if it turns out that I am unsuitable, I wish you and your team the very best of luck at finding suitable candidates, as well as promoting the need to go to Mars (and other places).

****

1. HAVE SUCCESSFULLY OPERATED a hydroponics garden that produced edible food

2. have a green thumb and am an avid gardener

3. know how to grind my own flours and how to bake

4. am a reasonably decent cook

5. can sew and mend clothes

However, given all these skills and abilities does not overcome one problem. I went to your website and read all the details about this proposed trip and concept of populating Mars and the one fact remains: You won't be ready for another 20 years at best. By then I will be 90 years old (if I live that long) so I must disqualify myself on age alone.

However, if any of you want to get serious about starting a Mars Training Camp out in the Nevada desert (or some such place) then I certainly would like to contribute my skills in making that happen.

Then when the time comes, you will have a facility to train potential canidates for this mission.

****

THE ONLY QUALIFYING REASONS I have for going is I feel we need to continue the space program because of the benefits for all humankind, my experiences in the military, leadership abilities, and I am a clergy member of the United Methodist Church, in other words a spiritual advisor, and all that goes with being a pastor. I am not a scientist, nor plan on being one. I do not think that should stop others like myself from going. As for the rigors of the journey, one can be trained for that and I do not believe we have to be that physically fit to go. Give us a job, train us well, give additional instruction when we are enroute and when on the planet. The rocket scientists could stay home. I would like this to be an international effort, rather than just a US effort. I think everyone needs to get in on this opportunity. Perhaps we could explore space together and not in competition with each other.

Would my wife let me go? No! However, that does not make it any less of a desirable destination in the last years of my life. I started my seminary education at 50. I just turned 61. I cannot think of a better place to meet God, again, on our spiritual journeys, than on another planet.

Peace,
Pastor Paul

****

I HAVE SO MANY REASONS why I would want to be able to go to mars on this life mission. I am a single father of my 9 year old boy and I have spoke with him if I was able to be chosen I asked him what he would think? My son looked into my eyes and said "dad you love space and this is a once in a life time chance." I said "well how would you feel if I was chosen?" He then replied "I would be so proud of my dad to be one of the one's chosen to become part of the growth of the human race." As I see it my son at 8 almost 9 years old and from him to be so proud of his father is something any parent would want. Giving one man's life to create the possibility for millions of others is priceless.

If I were chosen I would want to bring one thing and that is my son's picture because yes will I miss him of course and he will miss me. Will I miss everything about earth yes, but the one thing that would keep me going everyday from 1 year to 10 years is that I am not doing this for me, my son, or my family, but for the USA.

I would have to say at this point in mission needs we need people that are able to keep an open mind and see outside the box. We as human complain more about what is wrong with out country but not what we have. We take everything for granted, from food to kindness to even people who have mad bad choices yet we as a nation still see them as bad people. Life is one thing we can never get back so why not give 110% into us, or friends, or family, or kids and the chance to change the world as we know it today.

****

MARS IS THE FUTURE OF HUMANITY, living on Earth may not be possible two or three hundred years from now and that might not be curable, so we have to seriously start thinking about a plan B, let alone the growing needs of all kinds of metals which might not be as rare on Mars as it is on Earth.

As a volunteer I think that this is my chance to do something good for humanity and for my self, I'm a 22 year old guy, I live in Syria, being chosen for that mission I'll make sure that twenty years from now I'll be trained, experienced, and wise enough to make that mission successful, along with all the members of the team.

Many people may consider me a suicidal which is not right because no one knows what happens within twenty years, I may outlive those who said so, but one think I'm sure about is that later on; their children will think of me as a hero, and my friends and my family will be proud.

This task won't be easy, and I am aware that we may never come back, but we have to pave the way for generations to come, we don't fear death in what might be the most important accomplishment in the twenty first century.

Wake up people! It's never too early! If we want to be able to populate and invest the treasures of the red planet within the coming three centuries we have to start working from now.

****

LIFE ON MARS - just think about it. No human being has ever been there. We've been trying to uncover its mysteries for years. There is a possibility of discovering life on a planet besides our Earth. My name is C.H. and I am a sixteen year old male from the United States of America. When I heard of the possibility of humans colonizing of Mars, I was overjoyed. I have always had a great passion for space and all the mystery surrounding it. The idea of humans colonizing Mars is incredible! Putting humans on Mars is not just one small step for mankind, but rather one giant leap for mankind.

Life on Mars would be unique to say the least. Not many people could say they were the first humans to step foot on another planet and form a completely new society. The incredible individuals who should be selected for such a glorious adventure would have to leave everything they cherish here on Earth. Since the proposed journey is only one way, the pioneers must say goodbye to all their family, friends, loved ones, pets, and everything else they hold dear, knowing they will never get to see any of them face to face again. It sounds sad, but really, it isn't. While yes, they are giving up a whole lot to go on a great adventure with such high risks, they will forever go down in history as role models and celebrities until the end of time. People of all generations, young and old, will look up to the brave team who risked their lives to further the science world and our own human world.

Whoever is lucky enough to be selected will have to undergo years of extensive training. Learning to live on a different planet with only a select few other people would be very hard and labor intensive in order to fight off diseases and maintain an appropriate level of oxygen. Although, as having been portrayed in movies before, I wonder if we would have a super intelligent computer to assist the team going to Mars. Being able to interact with a computer similar to the HAL9000 featured in the film "2001: A Space Odyssey" would truly be an incredible experience.

Two decades from now is the proposed trip to Mars. What a special day that will be. I hope that all the necessary details can be completed and that the journey will be suceessful. Just as putting a man on the Moon was a huge landmark for our entire race, so will be setting foot on Mars. Humans will become the first multi-planetarial race - a true accomplishment millions of people only experience in their wildest dreams.

****

I'M SURE YOU ARE INUNDATED with messages from people wishing to volunteer to join any future Mars colony, so I'll keep this brief. From what I've read, it seems that many of the volunteers are single, retired, or otherwise unattached and have a mechanical or other NASA-esque background. That sounds great for a first wave of people to set up, and indeed has a lot of historical precedence, as many colonies were founded by groups of adventurous and often disinherited men who went off and broke ground in a new land. But in order to have a successful and ultimately sustainable colony, at some point it makes sense to also have families.

It is in this capacity that I would like to volunteer myself (J. S.) and my partner (L. J.). We're young, smart, capable of figuring things out, and though we are not close to our families we are very devoted to one another. Between his math/engineering skills and my team-building and knack for making something out of nothing, I think we would be able to make some sort of life for ourselves and anyone else we were to bring into the (new) world. In my humble opinion, in addition to retired mathematicians, any extra-terrestrial colony will ultimately also need people like us to help populate it. Remember the wagon trains out west? I don't think they were full of retired engineers. So if there is any kind of emerging list, would you kindly add us to it?

****

I READ THE ARTICLE about a trip to mars being more than about guts. I find it interesting that, yes, people tend to focus on personalities like MacGyver for such a trip. I agree, you need to basic group of problem solvers, but what about artists? Who is going to be able to express the experience to the world? To help those on earth understand the experience not just the science? Many artists also work well in seclusion...although not all science and arts-based people get along well. You would need at least one person who could bridge that gap in personalities.

My other question is that of the sex of those that go. If a one way trip is probable, then you need to have both sexes for balance, love, companionship, etc. But then what about the possibility of developing families there? Or would you force them all to be sterile? Are we talking just research or also developing a true colony in the future?

I think the hardest part of this type of one-way mission is for the person who outlasts all the others, not so much the not coming home aspect. Humans have survived many different hard situations, but true solidarity I think is where most people have the hardest time. Think of the pioneers crossing the great plains. They were going and probably never going back. They faced loosing everyone they knew. I wouldn't say many of them had that large of a skill set but the will to make it was essential. My suggestion is to study groups like that to figure out the right group of people to send.

Just some thoughts.
jacquelyn.

****

HOW MANY OF YOU ACTUALLY actively seek out information on current affairs? How many of you care about the world you live on? How many of you see things escalating at an exponential rate? How many of you fully understand the almost ridiculous state of fragility we are in? The entire human race can be extinguished in an instant. We've been gathering all the eggs in a single basket for millennia. Does no one understand this? How is that only so few people have the ability to just step back and realize that fact? The fact that if we fuck this place up, we're done. Our puny little civilization(ill come back to that word) will become extinct, snuffed out for the rest of eternity.

Now to combine the points.

This is the beginning of a new Earth. I believe that soon enough the world will be finally thrust into unity by the fear of destroying themselves. Once the people of Earth realize the only way to survive, not just our world but each other, is to leave this planet. We must unite the nations and powers of the world and start building colony ships. A colonization mission to Mars cannot be accomplished by any one nation or even a combination of the most powerful nations. It can only be accomplished by our race as a whole. And if one small mission of only a hundred or so colonists is what it takes to make the world realize this is the solution; then I would give my life to help that mission succeed.

-RJG, 21

****

BY THE END OF THE CURRENT CENTURY, our species will push the very boundry of sustainability on our planet. Now is the time for those among us that still have the pioneering spirit to lead, to explore and expand our species' knowledge of ourselves and the solar system in which we live. I want to be one of the first to step foot on a new world. To break ground on unseen soil, discovering new resources and forging a new way of life on another world. A world without borders, without age-old turf wars and long held prejudices. One that while rugged and inhospitable could be slowly transformed to a new home for future generations in the centuries to come.

If it means that I might die before getting there, then so be it. It is time for our species to think more about the future than the present, more about future needs than present wants or desires. It is a time for sacrifice. if the last image I see is through a visor looking at the sun setting on a Martian landscape then I will die knowing that there is a chance for us and that I was one of the ones who dared, who risked, who succeeded.

****

I AM CURRENTLY A SOPHOMORE in college, but when the time comes I intend to be at the forefront of this mission. I wholeheartedly believe that this proposed mission to Mars is the beginning of the future of our race. My stepfather pointed me in the direction of your article after I had just explained extensively that this is what I desire for the future of our kind. I believe this to be an amazing opportunity just waiting for us to take hold.

It is easy to see that there are a great number of people in support of this mission. In the time that it develops and becomes a reality I am confident that dedicated supporters will aid in advancing technologies that will allow it to be a safer mission. But for the sake of our species this is a risk that I feel we must take. The possibility of not only establishing a colony, but having the space and distance from Earth to conduct research is but one reason to go. The technology that the masses have been taught to fear from science fiction novels, could be developed and worked on without endangering the whole of our species. Exploration on Mars will provide the basis for future expansion, and the isolation will provide focus.

I am excited for the coming years, as this has been my intended future since I first heard of humans exploring space. I wish to travel with my husband so that we may attempt to start a family, possibly a next step in the evolution of our species. Preparation will be key, as this will be the greatest challenge our species has yet to face, but it is one that we must.

Cassandra G. - Age 22

****

Why I Want to Go To Mars
Written by: Brian L.H., male, age 39, U.S.A., Earth

In the course of Human History there have been many great endeavors to advance the cause of the human race. We have come to a pivotal point in our history where we have attained the knowledge, and application of that knowledge, to venture out into our universe. To a planet, that in the terms of the vast endlessness of space, is not only a close neighbor, but is also habitable. Mars is our closet and best option for early colonization outside of the Earth. The reason of course for choosing Mars, rather than the moon, is the presence of water, and a small atmosphere. I believe it is my right and my duty to advance the further cause of our human race. Knowing full well that my life and the life of my collogues may be at perpetual risk. It is a task that is mandatory for our species to not only survive, but to prove that we can and will accomplish colonization there and to the ends of the universe. This is our stepping stone. This is our first attempt. And yes, there will be things that go wrong, people may die, I may die, but it is worth the risk to learn the lessons that we need to. The time to do this is now. We must not delay. Sure we need a few years to get prepared, but we need to take this leap of faith soon.

With the Earth and all its inhabitants in a current state of imbalance. We can only assume with great foreboding that the weather will keep getting worse, the air will become more and more toxic, and the seas will eventually flood our lands. It really doesn't matter how and when it will happen, but the Earth will one day be unable to support life. The colonization of Mars is important now more than ever. We have a chance to do something great for our entire human race. I am volunteering for this mission of my own free will. I have studied abroad, and have lived in three other countries around the world. I believe that I would be good as a Liaison between cultures and various races of colonist. I have an innate ability to keep calm under stress while working with others to accomplish a common goal. I think it would be important for families to join the colonist once the Colony had been formed and established. It is important for humans to live in family units, and would be vital to hierarchy and sustained, long term colonization. We have this wonderful gift called life. If we want our ancestors to carry on our human light, it is up to us to make the right choices, here and now. I hope with all my heart and soul that soon we can come together, a great Family of Humans, joined to achieve what one could not do alone. We must and we will reach out in peace, united, to expand our human light into the Heavens.

****

MARS BY ITSELF ISN'T SPECIAL in any way. Although the view of Olympus Mons and other Martian landscape would be breathtaking, Mars is barren, frigid, and hostile to human life. If colonists were to live on Mars, they'd be scrambling to recycle oxygen and water every day just to cling to life. The colonists would be bombarded with dangerous levels of radiation due to the absence of a planetary magnetic field. The human lifespan would be cut in half. But, colonizing Mars is worth it.

The first colonists, in my opinion, are heroes of mankind and of life itself. I believe their goal will be to spread carbon-based life to many tiny rocks in the galaxy (and beyond). When spread out, there is less of a chance for a natural, cosmological disaster to extinguish life completely. The perseverance of life is echoed throughout our planet's history (as demonstrated during the previous 5 mass extinctions). Before a sixth mass extinction confronts our species, we should be prepared to evade it. Our colonization of Mars is the special, pivotal challenge we need to successfully complete before colonizing other (habitable) planets or moons.

E.D.L, Age 20, United States.

****

WHY MARS?

The idea of humans colonizing Mars is an idea that can inspire and bring hope to a world that suffers from it own lack of original thinking and large aspirations. In the world we live in today, people are so greatly consumed by their daily lives, paying the bills, and making a living, that they never get a chance to dream and think about the bigger picture. The Universe is a beautiful place, vast beyond are comprehension, yet we confine ourselves to a world that is no larger than a spec in comparison to the rest of the Universe.

There are still many places yet to be explored on our own planet, but we live on a world where we are not entirely free to go where we choose and owning land requires currency. We live in a planetary society that is limiting our ability to be truly free individuals. Mars provides us with an entire new world full of unexplored and unclaimed lands that can create a new sense of freedom. New societies and culture will emerge and will give rise to new ideas and ways of thinking that could have a profound effect on Earth as well. It will give human beings a new perspective on life, a much more positive one that enables us to envision the human race as not only surviving, but thriving.

We have the capabilities to take the first step and create the first colony on Mars if the proper funding is provided. There will never be a perfect moment to start funding such an expensive journey, but there will never be time when everyone on Earth is prosperous and the funding is easily available. The colonization of Mars is a long-term investment, one that will provide us with the best possible future and provide us with benefits in the future. People may fail to see the necessity in embarking on such an journey and colonizing the planet, because as a species, we think too often think only in the short-term. We seem to want to solve our short-term problems with short-term solution and ignore the greater good of humanity.

The mission would be completely lacking in short-term benefits, however. It has the potential to inspire and lead to further innovation and greater interest in the sciences. Its direct economic benefits might not be easily found, but that is not to say its wide spread effect would not be beneficial. It could create a golden age of technology and innovation. Scientists and engineers would have realistic applications for the development of new technologies that could benefit future missions and the living conditions on Mars. It could also provide the general public with hope. Realizing that, we as a species, are capable of making such a journey, would give people a reason to believe in ourselves.

Mars is a world that will provide us with a sense of freedom. The idea that humans can have a fresh start are perhaps even create a better way of living could bring about changes on Earth. Whoever takes those first steps will realize something greater when they glance at the night sky from a different perspective. Earth is not permanent and to survive as a race we must expand. While there is no impending threats to end all life in the short-term future, we must realize how fragile we are as a species.

Only with the first mission to Mars can we begin to live these possibilities. Why not let it happen in our lifetime? Why not be a part of the generations of people on Earth who first brought us in to space? What are we working for every day if don't want to take the next step to the advancement of humankind? What do we as humans aspire to become? How can we claim success if we are confined by are own societal limitations? All we need is to wake up and take a step out of the bubble to see the beauty of possibility and these questions can be answered.

****

I LIVE IN THE UNITED KINGDOM aslmost 5 years, but I'm not originally from UK so I'm sorry for my english. I'm 31 years old man, my initials are TK. Few days ago I found the message about Mars human colonisation project. I could't believe something like that is getting real in our century and I can try to be a part of this biggest human project in all history of the Earth. I know this is just one way ticket, but ticket to the something so amazing and unbelievable that this should be a main point of human existence. Not wars and killing people, but discover a new places in the space and get the places ready to life. This will be very hard and dangerous, but what wasn't? Columbus, Amundsen? We have a technology to go above so why don't do that?

I was born at 1979 in socialistic country. Had a usual childhood, nothing bad happened to me, so I was a happy kid. Graduated at 1997 at High school of Engineering, branch of Electromechanic for Automatics management. So very easy. I'm a kind of electrician. I had a few of jobs in my country, but no one made me satisfied. My father was a cop, but was killed on duty 2 days before my 18's birthday. So that completely changed my live. I wanted to be a cop as well, but this got all our family down for long time. I'm divorced, unfortunately, and couldn't see my kids, so that was my point to move myself somewhere far away and try to find "peace" again. It's done. Today I'm a full time graphics designer and my big hobby is buildings. Sounds funny, I know, but it works. Somebody likes a sunny holidays, I like to build houses over my holidays. So mainly I can do everything around the buildings which should be a one of the first tasks on Mars.

The way on the Mars could be the biggest thing in my life I can join, so why don't "sign up"? I know the people can die due to trip. But how many people die for nothing? "Mars people" can die for something big and this is the point, why don't be scared about death in this project. I would like to bring my girlfriend with me, but I'm prepared to hear her NO, as no everybody understand this project. And what about my friends? I think they will be proud of me and this is more that be drinking in pub with them.

And I will be proud of myself if the trip will happen.

****

IX

IMAGINING LIFE ON MARS: A SCIENTIFIC ADVENTURE

- 20 -
Aurora - The First Martian - A Vision of Colonial Life on the Red Planet

Dirk Schulze-Makuch[1], Paul Davies[2], and Joseph Gabriel[3]

[1]School of Earth and Environmental Sciences, Washington State University
[2]Beyond Center, Arizona State University [3]Cosmology.com

Aurora, first daughter of the Red Planet, was excited. Her 12th birthday was coming up, and even the people of Earth were going to celebrate the birthday of the first true Martian. She gazed out into the darkening crimson horizon and her pale blue eyes focused on the little blue star which was Earth. The whole Milky Way was in grandiose display, so twinkling bright with a billion stars winking on and off in the blackness of night. Her mother had told her it was impossible to see the night sky in such clarity on Earth; but here on Mars they were blessed with heavens which sparkled diamond-bright, a heavenly vault brimming with precious jewels. Of course it was different when dust storms were ravaging the Red Planet. Her mother called the dust storms a "vermilion hell."

Aurora couldn't disagree more. Mars was god-like in its forbidding savage majesty. It was no wonder the people of Earth called the Red Planet the "god of war". But if there had been "war" then the people of Earth had won. The prize? Adventure! Her earliest memory was peering out of her little space-suit, picking up rusted red rocks and tossing them into the air. They were like so many giant rubies strewn upon a carmine carpet which was pock-marked with tiny craters; a jagged landscape which jutted out in all directions from steep valleys to rugged mountains which towered into the pink sky.

Her mother was with her that day so long ago, and had violated regulations by taking little Aurora outside the habitat units.

Her mother was beautiful and had always been a bit of a rebel. She was among the first to volunteer for the "one way missions to Mars," but because she was "too independent" and had a propensity for questioning orders she didn't like, the "powers that be" (as her mother like to call them) were loath to chose her. But she was a top notch scientist, a superb athlete, smart and funny, and the media on Earth loved her. Mars had always been her dream. Finally, they had no choice and she was chosen; and lucky for her, for it was the last of the One-Way Missions.

Aurora loved her mother, and wanted to be just like her, a famous scientist. Aurora's mother may have been the last to arrive, but she was the first to get pregnant. The pregnancy was a media sensation back on Earth, and all sorts of self-appointed experts predicted it would be impossible for her to bear a healthy child on Mars. Their reasons were legion: radiation, gravity, the stress, there was no way the pregnancy would be normal. The experts were wrong. Here she was and soon going to be celebrating her 12th birthday.

Yes, she was excited, but also nervous. They were going to send her back to Earth, the Blue Planet. It was not everyday that a Martian had a 12th birthday, and the entire Earth wanted to celebrate with her. The public wanted it, the media couldn't stop talking about it, and the "powers that be" had ordered it.

For Aurora, Earth, that blue twinkling star, had a special attraction, but it was not home. Her home was Mars and she was already adapted to life on the Red

Planet. Whenever she went into the Earth simulation chamber, she felt that confining pressure on her lungs, and her body felt so heavy and weak. She fatigued very quickly under Earth-like conditions. No, she was a Martian and Mars was her planet. She loved the rugged red mountains and crimson desert plains. Despite her bulky pressure suit, she was able to run like a deer. On Mars she was strong. It wouldn't be like this on Earth. But she had no choice, especially now after regular shuttle service had been established to and from Mars.

She was excited but worried. The stories she heard about those who had returned to Earth, were not encouraging. Despite the significant loss of bone density and muscle caused by the six months of weightlessness during transit from Earth to the Red Planet, it was easy for people born on Earth to adapt to Mars because it has only 1/3 the gravity. But the conditions were not the same for those going from Mars to Earth which has almost 3 times the gravity. They were as weak as babies.

And how would she deal with the crowds, with so many people, and the big cities and such? She had been warned: she would be a celebrity, they would never leave her alone. She would suffocate in an ocean of people. There were only 24 colonists on Mars. Earth had 10 billion people!

They might even treat her, the first Martian, as a freak to be studied and probed. She had read the protocols: One month of isolation under observation and quarantine while she adapted to Earth. The scientists on Earth also wanted to make sure she was not carrying Martian microbes that might infect the people of the Blue Planet, with some kind of disease.

It was the same on Mars. Every new human arrivee was thoroughly checked and held in quarantine for two weeks. The lesser gravity on Mars reduced the vigilance of the human immune system. Martian colonists had also adapted to the Red Planet which had few pathogens capable of attacking and sickening humans. If someone from Earth arrived with a cold or the flu, it wouldn't just make everyone sick, but might wipe out the entire colony. It was for the same reason they had never shipped animals to Mars. Aurora would love to have a kitty, but it was impossible. Animals were hothouses of contagion, and everyone was afraid of the possible affects not only on the Martian settlers, but on the fragile ecosystem on Mars. What if microbes from Earth exchanged genes with Martian microbes? Might they produce new strains of a super bug?

Planetary protection issues were a great concern for both Earth and the colonists of Mars. People back on Earth feared that some vicious Martian bug would kill them all. The media and scare-mongers called it: "The Red Death" and wrote horrifying newspaper articles of the pain and suffering, warning: "Better dead than Red."

Aurora's mother had laughed at these scare tactics, but also welcomed them. "If they weren't scared" she warned, "we would be over run with billionaires from the Blue Planet; the rich and famous, tourists by the shipload would be arriving in droves, demanding that we entertain them, and it would be impossible to do any real science."

Her mother was a microbiologist and she had determined long ago that Martian microbes were ideally adapted to the harsh life on Mars. Any Earthly microorganisms that made it outside the habitat units would quickly die once exposed to the brutal Martian conditions, or they would be killed off by Martian life that had evolved for billions of years on Mars and were well adapted to the savage conditions which prevailed.

According to her mother, fear of Martian microbes was also baseless. Microbes from Mars were continually being transferred to Earth, and probably had been since the formation of the solar system and the origin of life. And just as Earth microbes couldn't survive against their Martian counterparts on Mars, Martian microbes were at a decided disadvantage on Earth. People simply feared what they didn't understand.

It was her mother who had first discovered life on Mars; a discovery the experts back on Earth dismissed at first, and ridiculed as contamination. She had found the microbes in the Martian soil transported to the greenhouses. Despite the doubters, she was positive the microbes are indigenous.

Aurora had helped her mother experiment with these Martian microbes in the station lab. Once placed in an Earth-like environment, all that they were good at was to stake out a meek existence in a low nutrient - low organic carbon environment. They were hopelessly outcompeted by Earth life as soon as water and carbon and other nutrients became more plentiful. Martian microbes had little chance for survival on Earth. They were indigenous to the Red Planet, and

now that this had been proven, instead of ridicule, the experts were fanning the flames of fear.

Aurora suited upon, exited the habitat unit, and strolled outside toward the greenhouse dome. True, she could have taken a more indirect path, indoors, but whenever she had the chance, and if there were no weather fronts, her preference was for the outdoors. Mars was so breath-takingly lovely. The early morning and evening skies were a beautiful pale blue -- just like her eyes. Her mother had explained the blue-color was due to less reflective light during these hours. The rest of the day the Martian skies are as red as the soil, due to the abundance of iron particles in the air which captures and reflects the feeble sunlight.

The domed Martian greenhouses had been fully functional for years, and offered ample protection against ultra violet rays and Martian dust storms and weather fronts. Martian water provided more than enough moisture, liberated from frozen water beneath the surface. Hydroponic gardening techniques produced a bounty of beautiful vegetables and fruits. Asparagus, broccoli, beans, and other vegetables do just fine in Martian soil because of its high levels of alkalinity. Soil high in perchlorates also never proved to be a problem; all it took was mixtures of fertilizers and additives and the planting of genetically modified seeds.

The habitat units were boring. If she had to be inside, then she preferred the greenhouse dome. It was her favorite place. All this green! The only place on Mars where there was green.

Aries, who was 3 months younger than her, was inside the greenhouse tending to his chores. As usual he ignored her, so she ignored him. It hadn't always been like this. They had grown up together in the habitat units and had spent countless afternoons playing games, having fun, and exploring outside under the watchful supervision of the adults. They were always together. Even took naps together. They talked about everything and were the best of best friends! And then a few months ago, he started acting different, said she had "cooties" and just like that, he didn't want anything to do with her. She knew why. He told her. "Because you're a girl." He made her so mad! But she said nothing. Instead she looked with admiration at the barley bushes. What a little genetic engineering can do, she thought. She much preferred the barley, wheat, and vegetables to the freeze-dried food from Earth and the snottites. Yes, snottites. Ugh!

The experts on Earth were right about the possibility of gene exchange between the microbes of Earth and Mars. One of the geneticists, Dr. Solstice, after confirming that Martian microbes were in fact indigenous, came up with the idea of growing these microbes in one of the lava tube caves and then tinkering with their genome. Everything he did was against planetary protection protocols, but by mucking around with their genes he was able to create a new strain and began growing them as microbial mats. Then, he harvested the mats which provided a good source of protein, vitamins, and trace minerals.

Aurora hated the taste! "Yuck" she had exclaimed the first time she took a bite. "It tastes like snot!" And thus was born the term, "Snotties" much to the chagrin of Dr. Solstice. Ever since that day, he never smiled at her, and this made her sad. She knew his work was important and wanted to be his friend, and his student.

In order to successfully cultivate vegetables in the greenhouse dome, they had to import an enormous variety of microbes and fungi. Dr. Solstice's experiments with genetic engineering were a life-saver. He had made the colony self-sufficient.

The whole history of the colony was really a story of human ingenuity. The settlers had been faced with many crises, some totally unexpected. "Necessity is the mother of invention" read a plaque on the greenhouse door.

Aurora gazed longingly into the distance. The sky was turning a dark red. Evening was coming and with it, a team of explorers which included her father. Aurora idolized her dad. He was big and tall and handsome. But he never said much, and she never knew what he was thinking. She knew he loved her, and her mother, "but why didn't he say it?" she wondered. When she asked her mother these questions, she always had the same answer: "your father is a thinker, not a talker."

He had been leading a team to the Cydonia region to explore some of the anomalous surface structures and surrounding terrain. Cydonia was a transitional region between the heavily cratered southern highlands and the smooth northern lowlands. Aurora had studied the maps and knew it by heart. She also knew the legends. Back in the 20th century, there had been claims about giant pyramids organized in precise geometric patterns on the outskirts of a titanic face which stared up into the Martian sky. These were illusions captured by the first Viking space craft photographs of Cydonia.

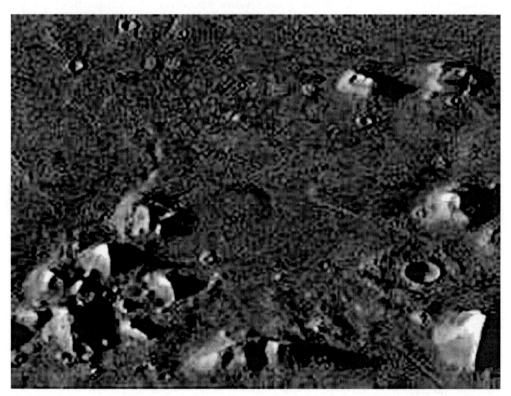

Cydonia

When she questioned her father, he would just smile and say: "Honey, there are no pyramids and there is no face."

But then, why all the secrecy, she wondered. What were they doing out there?

On his return from the last excursion to Cydonia, she peppered him with questions and asked again and again why she couldn't go with him. He just smiled, patted her head, and said: "Ask your mother. She knows all the answers."

It was infuriating. He was treating her like a child. But she wasn't a child. She was going to be 12 and she was going to Earth.

And Earth was coming to Mars, despite the misgivings of her mother and some of the other colonists. With the regular shuttle service now in place, it had been decided, back on Earth, that it was time to grow the colony and expand the base. Wealthy space tourists were claiming it was their right to journey to Mars, and some politicians agreed with them. It seemed there were constant squabbles with the Earth authorities.

Dr. Pauli, her other idol, had complained that "those fools don't have a clue what it means to live on Mars. All they can come up with are unrealistic plans, and then we have to clean up the mess."

Dr. Pauli, a member of the 5-member advisory council, had been elected spokesperson by the adults of the Mars colony. However, the authorities back on Earth preferred to call him "Governor." Governor or not, Dr. Pauli had a lot of difficulty making sure the concerns of the Mars colony were heard back on Earth. Aurora's mother was a member of the advisory council and Aurora had overheard her discussing with her father, some of the heated arguments they had with the "powers that be" back on Earth.

One of the craziest ideas was to send garbage to Mars. Yes, Earth was drowning in garbage, and some politician had come up with the brilliant idea of sending it all to Mars. Dr. Pauli grumbled about this for weeks. "First they fear creating Martian monsters out of microbes, and now they want to send garbage contaminated with every sort of contagion to this planet!"

Beautiful, desolate, majestic Mars, covered with Earth garbage? Aurora shuddered at the idea.

Fortunately, the "garbage-to-Mars" initiative was dropped due to the expense involved.

She felt a touch at her elbow. "Would you like to come with us to Skeleton Bay?" her mother asked.

"Before dad comes home?" she asked in surprise.

"Tomorrow morning" her mother replied. "On the hopper. Tomorrow will be a special day!"

Aurora smiled with excited delight. "Oh yes, Mom", she replied. "yes, yes yes." That was one of Aurora's favorite things to do. Exploring the planet and flying over these grandiose crimson canyons. But going on an exploratory excursion to take samples at Skeleton Bay was a special treat.

The next morning was bright and gay. The sky glimmered in a red-pinkish hue as the sun rose over the eastern mountains. Aurora was up before everybody and ready to go: Skeleton Bay!

The exploration team included four people in total. Aurora and her mother met the others at the gas hopper and together they went through the supply and equipment checklist. She was thrilled that she was allowed to tag along. Although her father treated her like a child, her mother and the rest of the colony were treating her more and more like an adult and were allowing her to accompany them on a variety of scientific excursions. Her mother was right. Today was a special day.

Aurora had learned everything she could about the field equipment. She wanted to be an important member of the team. Some day she hoped they would let her fly a hopper. But when she asked for lessons, everyone just laughed.

The gas hoppers required an experienced pilot to fly them. Winds were unpredictable on Mars and could send a hopper and an inexperienced pilot crashing to the Martian surface. Martian dust also had a nasty habit of getting into mechanical components and clogging them up. If there was sand and wind the outcome could be death for all onboard.

The gas hoppers were the preferred mode of transportation for traveling long distances in the shortest amount of time. The first model had quite a few bugs, but they were able to finally work them all out. Necessity had again gave birth to invention. The technicians were understandably proud that they were able to optimally adjust the pump to an ideal setting for drawing in carbon dioxide from the atmosphere, storing it in a liquid form, and then sending it through a preheated pellet bed turning it into hot rocket exhaust to produce lots of thrust. Once this had been mastered, the hoppers stopped hoping.

Aurora giggled. She remembered what her mother told her when she first asked. "They were called hoppers" her mother told her, "because the first flying vehicles were actually just hopping over the ground, several kilometers per jump." That was quite a funny thought. Hop Hop Hopping! The hopper they were flying in today was very sophisticated and included large solar panels and a ground rover in the belly of the hopper.

At 0600 Mars Standard Time they lifted off toward Valles Marineris.

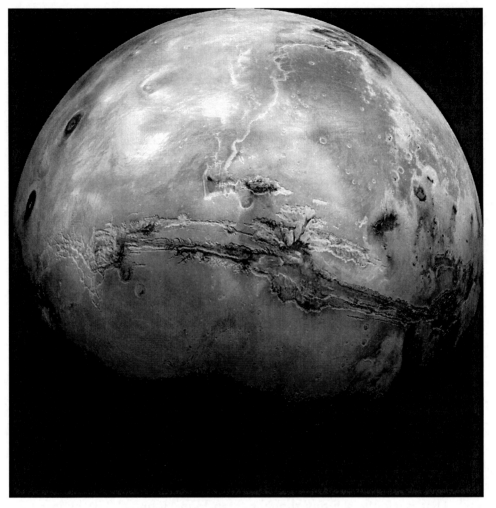

Valles Marineris

Her mother explained they were going to take samples in the bottom of one of the side canyons. Aurora was glued to one of the observation ports to soak in as much of the desolate glory as she could. The Martian landscape was amazingly beautiful. Craters, canyons, and lonely mesas shimmered in the bluish-pink light of the rising sun. Everywhere she looked there were shades of scarlet, red and rust. Aurora liked to imagine she could see the ruins of castles and the remains of ancient temples; but it was just an illusion. So far, no one had discovered any evidence of intelligent past life on Mars; or had they? She wondered again about what her father was doing in Cydonia. Why all the secrecy?

Why was Mars a vast crimson desert?

These were questions she had asked of her mother. Why were there no trees, no plants, no animals? Why had complex life evolved on Earth but not Mars?

Even she could see that Mars once had mighty rivers and rushing streams. There was undeniable evidence of ancient shorelines of great oceans. What became of them?

Gazing down from the hopper, as they flew over the lower parts of Valles Marineris, Aurora could see yet more evidence that Mars had once been a very wet planet. Down below was a vast canyon system with steep cliffs which could have only been carved by oceans of water. It was easy to imagine the huge catastrophic outbursts of water that had so long ago cut through them, carving these canyons. But since there had been water, why was there no evidence of intelligent life?

Valles Marineris

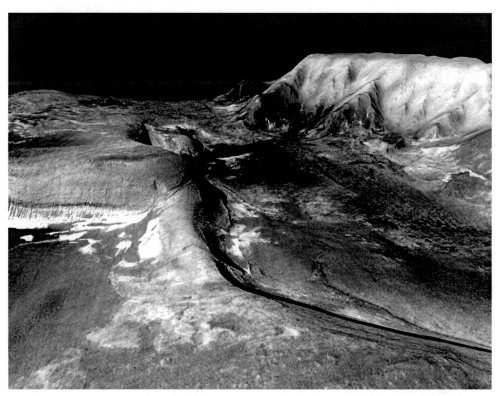

"Why?" she asked her mother, again and again.

Her mother just smiled and stroked her head. "Tiger got to hunt, bird got to fly, Aurora asks, why why why?"

If there had been intelligent life on Mars, it died out a long time ago. Now all that was left were vast red deserts.

Aurora drifted off into her thoughts. She loved the stark scarlet beauty of Mars. But what had it been like when there were oceans, she wondered? Aurora knew that many serious scientists back on Earth, believed Mars may have once harbored complex life, and that the Red Planet could be restored to its once and former living majesty. The answer was "terraforming." Mars would be transformed and given an atmosphere to make it like Earth.

The possibility of terraforming Mars was gaining support with some wealthy and influential members of society back on Earth; particularly those that desired an exciting off-world experience. The entire Mars council was against it.

Any major change in the Martian environment and widespread introduction of foreign flora and fauna outside of the station habitat would likely mean extinction for the endemic life already existing on Mars, and any complex life forms which may dwell deep beneath the soil or in areas of Mars not yet explored. Aside from the moral issues, who could possible bear these immense costs connected with a global terra-forming project, especially when the governments on Earth had already enough problems to keep their enormous population fed and happy? Besides, such a project would need at least centuries to come to fruition, and even then Mars would never truly be earthlike. No, that just didn't make any sense.

Aurora agreed. Mars is just fine the way it is, she thought.

As the hopper approached the deep canyons, a movement down below caught Aurora's eye. Yes, there it was, something, near the edge of cliffs, burrowing up from out of the red soil. She pressed herself against the observation port and caught her breath. It looked like it was alive.

It was twisting, turning, a blur of movement, growing larger and reaching up in a monstrous cloud of spinning red dust; they were heading right for it!

"Mom", Aurora said nervously.

And then it had them, and the hopper shook, sputtered, and fell, as if a great hand had grabbed hold and was dragging them down intending to crush them against the blood red surface of the cliffs.

And then, just as suddenly, the hopper leveled off and ascended into the crimson skies.

"Dust devil," her mother said, squeezing her daughter's hand reassuringly. "A big one."

"Winds are always bad near canyons" the pilot muttered. "But that was a bad one."

"Fortunately we have a great pilot," her mother replied cheerfully.

"Sure," he answered. "Lets go back and do it again!"

Everybody laughed.

Aurora took a deep breath and relaxed.

They approached the landing site, a plateau on top of one of the side canyons, and the gas hopper set down with a jolt. The crew got out, opened the cargo bay and charged up the rover's power sequencer. In nearly no time the rover was ready. Her mother took the driver's seat and drove it out of the belly of the gas hopper. Meanwhile, the exploration team put on their environmental gear and unloaded their equipment.

"Make sure that you have all your gear for the base camp setup with you, particularly the oxygen tanks, hook-ups, and ropes," her mother instructed the team as the crew boarded the rover. All four members of the exploration team held onto the railing while the rover made its way through the rough terrain.

After they reached the cliff, they jumped off the rover and attached thin nylon ropes to rocks at the cliff-edges for the downward climb. The terrain became even more difficult to traverse, with boulders and large rocks chaotically jumbled at the cliff side near the bottom of the canyon.

One after the other they made their way down and then through the red rocky terrain. As they stumbled along the path, they noticed wet patches of fine sedimentary material from which small gas bubbles gurgled up. Aurora pointed to one of them and her mother answered via the radio link: "That's methane gas, which is produced by Martian life. But we already have samples for analysis in the lab, so we'll just move on."

Finally, they reached their destination for setting up the base camp. They unloaded their gear from the rover and quickly erected the self-contained tent. It had the shape of an oversized soda can and could accommodate four, a power generator, and an oxygen generator. The crew pumped it up with oxygenated air, so they would not have to wear their suits and helmets inside. That's what the adults called "camping on Mars". Of course, they very rarely spent the night in the tent, only in emergency situations. Once the base camp was set-up, they moved further ahead toward their final destination: Skeleton Bay!

Skeleton Bay was easily identifiable from a distance. The whitish crust consisting of gypsum and other salts sparkled in the sun. The bright whitish color was in stark contrast to the rust-red Martian soil and the dark cliffs surrounding it. The site, a few square kilometers in extension, was reminiscent of White Sands National Monument near the New Mexico Space Harbor in Upham, her mother explained. The similarity was particularly apparent with Lake Lucero, a salt playa within White Sands, which was occasionally flooded but dry most of the year.

Not only could one make out crescent shaped dunes ahead, but gypsum crystals peaked out of the ground everywhere, and gave the whole place a mystical appearance. Some of the gypsum crystals grew to large sizes and formed intricate patterns, occasionally reminiscent of skeletons from large animals, like the rib cages of whales. "This is why they call this place Skeleton Bay" her mother said over the radio.

Aurora was excited. She was sure she would make a big discovery today. Maybe even find the skeletons of ancient animals that long ago swam beneath the oceans of Mars! Maybe if they dug down deep enough, they would even discover something still alive! But she kept these thoughts to herself.

The exploration team walked further toward the center of the dry salt lake. Their goal was to find clues as to why the playa was located here, in an outflow channel of Valles Marineris. She heard several hypotheses put forward by the team members. One was that the playa was the remnant of an ancient lake. Aurora's mother didn't agree and said this was hard to believe unless the side arm of the canyon was dammed at some point. More credible was that upwelling groundwater was intersecting the valley floor, evaporating, and leaving the salt crust behind. It would be very exciting if that were still the case today. However, although they took several samples and dug deep beneath the surface, they did not discover any evidence of recent water. After sampling the soil to a depth of about two meters, all they found were bone-dry sediments. "If the ground had once been inundated with water," said her mother, "it was surely a long time ago." Nevertheless, they bagged some samples for further analysis and Aurora took several pictures of odd formations with her digital recording device. Then as she gazed through the view finder she saw the yellow sparkle.

She picked it up and dusted it off with her gloved hand. It looked so soft, but was hard and had a yellow sheen, "like gold" she thought. It reminded her of yellow cookie dough. "Mom, what's this?" she asked, holding it out between her gloved fingers. She handed it to her mother who turned it this way and that. "I don't know what it is sweetie" her mother said softly. "Show me where you found it." Aurora pointed, her mother nodded, and then she bagged it and said nothing more about it.

Finally, they marched back to the base camp. For reasons Aurora couldn't grasp, everyone seemed in a rush to get everything packed. Aurora looked around, her eyes searching the ground and the scarlet skies for signs of impending danger. The sun was moving closer to the horizon and Aurora knew how quickly the temperatures would drop, but they had plenty of heating canisters to keep warm, so why the rush? But her mother said they needed to get back to the hopper and then back to the base as soon as possible. "Why?" Aurora asked. "Because it is going to get extremely cold," her mother replied.

Hurry. Hurry. Hurry. And in no time at all they were in the hopper and had made it home in time for dinner.

Her father was obviously relieved to see them. "There is a dust storm moving in," he exclaimed, giving her mother a serious look. "And for you to keep her out, on this day of all days!" he admonished. Her mother just smiled and kissed him on the cheek. Aurora could see they were giving each other those "looks"; and sure enough, they went off together, her mother leading her father by the hand.

"Yuck", Aurora said out loud.

Dr. Solstice who was just walking by turned and gave her a serious look. Then he smiled and said, "no snotties today, kid."

Aurora looked at him questioningly.

"Don't you know what day it is?" He asked.

"No," she said suspiciously.

"It's greenhouse day," he said. "It's a very special day!"

The "greenhouse days" were enormously important for the morale of the crew, and were held once every two weeks, or more often if the supply would allow it.

"Yay!" she said aloud. "Greenhouse day." This meant fruits and fresh vegetables from the station greenhouse. Tomatoes, cucumbers, carrots, turnips, green beans, radishes, and corn, together with dehydrated beef and chicken; they were going to feast tonight!

An hour later, much of the crew had gathered around the great table and everybody was talking excitedly about Cydonia, Skeleton Bay, the political situation on Earth, and telling jokes and reminiscing about the first days on Mars. Aurora loved these social get togethers, and listening to the old stories, even if she didn't always understand the jokes.

And then her father got up, left the room, and came back with a bottle of wine, which he set upon the table. "Wow" she thought. "I wonder what that's for?" Then she noticed that someone had switched on the cameras. Everything was being recorded. Why? Again, she wondered what her father might have discovered at Cydonia. But he said nothing about it.

Aurora swallowed her last bite and washed it down with a glass of cold water. Feeling tired, but satisfied and surrounded by friendship, camaraderie and love, she leaned back in her chair and closed her eyes... but then quickly opened them and sat upright when she became aware of the sudden silence. Everyone was looking at her!

Her father stood up, holding a glass of wine outstretched in his hand. Then, everybody stood up, each with a glass of wine, their eyes on her.

"What?" she said, feeling confused. "What's the matter?"

"Nothings the matter, sweetheart," said her mother. "Don't you know what day it is?"

"Greenhouse day?" Aurora replied feeling uncertain.

Everybody laughed.

"No," her father answered. "It's first Martian day. A very special day!"

And then out came the presents and everybody began to sing: "Happy birthday to you...."

To Be Continued...

Lightning Source UK Ltd.
Milton Keynes UK
UKOW011215140312

188971UK00001B/10/P